POLYNOMIAL METHODS AND INCIDENCE THEORY

The past decade has seen numerous major mathematical breakthroughs for topics such as the finite field Kakeya conjecture, the cap set conjecture, Erdös's distinct distances problem, the joints problem, as well as others, thanks to the introduction of new polynomial methods. There has also been significant progress on a variety of problems from additive combinatorics, discrete geometry, and more. This book gives a detailed yet accessible introduction to these new polynomial methods and their applications, with a focus on incidence theory.

Based on the author's own teaching experience, the text requires a minimal background, allowing graduate and advanced undergraduate students to get to grips with an active and exciting research front. The techniques are presented gradually and in detail, with many examples, warm-up proofs, and exercises included. An appendix provides a quick reminder of basic results and ideas.

Adam Sheffer is Mathematics Professor at the City University of New York (CUNY)'s Baruch College and the CUNY Graduate Center. Previously, he was a postdoctoral researcher at the California Institute of Technology. Sheffer's research work is focused on polynomial methods, discrete geometry, and additive combinatorics.

CAMBRIDGE STUDIES IN ADVANCED MATHEMATICS

All the titles listed below can be obtained from good booksellers or from Cambridge University Press.
For a complete series listing, visit www.cambridge.org/mathematics.

Polynomial Methods and Incidence Theory

ADAM SHEFFER

Baruch College and The Graduate Center, City University of New York

CAMBRIDGE
UNIVERSITY PRESS

CAMBRIDGE
UNIVERSITY PRESS

University Printing House, Cambridge CB2 8BS, United Kingdom

One Liberty Plaza, 20th Floor, New York, NY 10006, USA

477 Williamstown Road, Port Melbourne, VIC 3207, Australia

314–321, 3rd Floor, Plot 3, Splendor Forum, Jasola District Centre,
New Delhi – 110025, India

103 Penang Road, #05–06/07, Visioncrest Commercial, Singapore 238467

Cambridge University Press is part of the University of Cambridge.

It furthers the University's mission by disseminating knowledge in the pursuit of
education, learning, and research at the highest international levels of excellence.

www.cambridge.org
Information on this title: www.cambridge.org/9781108832496
DOI: 10.1017/9781108959988

First published 2022

A catalogue record for this publication is available from the British Library.

ISBN 978-1-108-83249-6 Hardback

To Liora, Daniel, and Amanda.

Contents

Introduction

Algebra is the offer made by the devil to the mathematician. The devil says: I will give you this powerful machine, it will answer any question you like. All you need to do is give me your soul: give up geometry and you will have this marvellous machine.

Michael Atiyah (2005).

In his famous essay on how to write mathematics, Paul Halmos (1970) states, "Just as there are two ways for a sequence not to have a limit (no cluster points or too many), there are two ways for a piece of writing not to have a subject (no ideas or too many)." The book that you are now starting has two main subjects, which is hopefully a reasonable amount. These two subjects, *the polynomial method* and *incidence theory*, are tied together and difficult to separate.

Geometric incidences are a family of problems that have existed in discrete geometry for many decades. Starting around 2009, these problems have been experiencing a renaissance. New and interesting connections between incidences and other parts of mathematics are constantly being exposed. Incidences already have a variety of applications in harmonic analysis, theoretical computer science, model theory, number theory, and more. At the same time, significant progress is being made on long-standing open incidence problems. The study of geometric incidences is currently an active and exciting research field. One purpose of this book is to survey this field, the recent developments in it, and its connections to other fields.

What are incidences? Consider a set of points $\mathcal{P}$ and a set of lines $\mathcal{L}$ in the plane $\mathbb{R}^2$. An *incidence* is a pair $(p, \ell) \in \mathcal{P} \times \mathcal{L}$ such that the point p is on the line ℓ. For example, see Figure 1. One fundamental incidence result states that n points and n lines in $\mathbb{R}^2$ form at most $2.5n^{4/3}$ incidences. While the exponent $4/3$ cannot be improved, it is possible that the coefficient 2.5 could be replaced with a slightly smaller one.

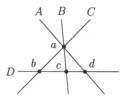

Figure 1 A configuration of four points, four lines, and nine incidences. For example, the point a forms an incidence with each of the lines A, B, and C.

In other incidence problems, we replace the lines with circles, parabolas, or other types of curves. Additional variants include incidences with higher-dimensional objects in $\mathbb{R}^d$, incidences with semi-algebraic sets, incidences with complex objects in $\mathbb{C}^d$, in spaces over finite fields, o-minimal structures, and more. In most of these cases, finding the maximum possible number of incidences remains an open problem.

An incidence result of a different flavor states that there exists a positive constant $c \in \mathbb{R}$ that satisfies the following. For every sufficiently large n, every set of n points in $\mathbb{R}^2$ satisfies at least one of the following statements:

- There exists a line that is incident to at least cn of the points.
- There exist at least cn^2 lines that are incident to at least two of the points.

Sylvester (1868) studied incidence problems back in the 1860s. The earliest incidence problem that we are aware of appears in a book of riddles (Jackson, 1821). This book contains 10 problems of the form that is presented in Figure 2. In modern English, the problem in Figure 2 asks for the following: Place points in the plane, such that the number of lines that contain exactly three points is at least the number of points.

> **2. Fain would I plant a grove in rows,**
> But how must I its form compose
> With three trees in each row ;
> To have as many rows as trees;
> Now tell me, artists, if you please;
> 'Tis all I want to know.

Figure 2 A riddle from the 1821 book *Rational Amusement for Winter Evenings, Or, A Collection of Above 200 Curious and Interesting Puzzles and Paradoxes Relating to Arithmetic, Geometry, Geography.*

Most of the recent progress in incidence theory is due to new algebraic techniques. One may describe the philosophy behind these techniques as

Collections of objects that exhibit extremal behavior often have hidden algebraic structure. This algebraic structure can be exploited to gain a better understanding of the original problem.

For example, in a point-line configuration with many incidences, we might expect the points to form a lattice structure. Intuitively, we expose the algebraic structure by defining polynomials according to the problem, and then studying properties of these polynomials. In an incidence problem, we might study a polynomial that vanishes on all the points. This approach is called *the polynomial method*. In this book, we explore a wide variety of such polynomial proofs. We use these techniques to study incidence bounds, the finite field Kakeya problem, the cap set problem, distinct distances problems, the joints problem, and more.

Polynomial methods have existed for several decades. One well-known polynomial method is Alon's Combinatorial Nullstellensatz, as described in Alon (1999). As long ago as 1970, Rédei introduced an elegant polynomial proof. This book is focused on the new wave of polynomial methods that started to appear around 2009. These methods are quite different from the preceding ones.

This book aims to be an accessible introduction to the new polynomial methods and to incidence theory. For that reason, the book includes many examples, warm-up proofs, figures, and intuitive ways of thinking about tricky ideas. Many techniques are presented gradually and in detail. Readers who wish to dig deeper into a particular topic can find references in the relevant chapter.

Incidence theory and the polynomial methods are still developing. There are many interesting open problems, and, in some sense, the foundations are not completely established yet. For that reason, most of the chapters of this book end with an open problems section. These sections focus mostly on long-standing difficult problems. Their goal is to illustrate the current research fronts and the main difficulties that researchers are currently facing.

Several sections are defined as *optional*. Some sections, such as Section 7.3, are optional because they consist of standard technical proofs that may not provide any new insights. Other sections require familiarity with a topic that is orthogonal to the topics of this book. For example, the optional Section 9.3 requires basic familiarity with differential topology, which does not appear anywhere else in the book.

Two other good sources for polynomial methods in discrete geometry are the book *Polynomial Methods in Combinatorics* (Guth, 2016) and the survey

"Incidence theorems and their applications" (Dvir, 2012). While these sources and the current book study similar topics, the overlap between them is smaller than one might expect.

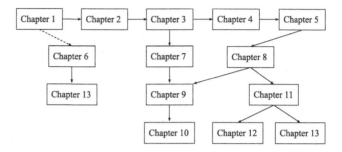

Figure 3 Chapter dependencies. The dashed edge marks a dependency that is recommended but not necessary.

How to Read This Book

Throughout this book, we rely heavily on asymptotic notation such as $x = O(y)$. The appendix contains an introduction to asymptotic notation, together with exercises. This appendix also briefly surveys basic graph theory notation and the Cauchy–Schwarz inequality.

There are many ways to read this book, depending on the goal of the reader. One way is to start from the beginning and read the chapters consecutively. The beginning of the book contains more introductory material. The end of the book contains mostly optional advanced topics. Figure 3 illustrates the chapter dependencies. Some reading options are:

- **A brief introduction to discrete geometry.** For an introduction to problems and techniques from classical discrete geometry, read Chapter 1. This chapter does not involve polynomial methods.

- **An introduction to polynomial partitioning.** To learn how to prove incidence results by using polynomial methods, read Chapters 1–3. Chapter 2 is a minimal introduction to algebraic curves in the real plane. Chapter 3 consists of the basics of the *polynomial partitioning* technique, and how to use this technique to prove incidence bounds.

- **A variety of polynomial methods in combinatorics.** To see a variety of polynomial methods in combinatorics, read Chapters 1–6. In addition to the polynomial partitioning technique, Chapters 5 and 6 contain several other polynomial breakthroughs. Chapter 4 introduces basic concepts from real algebraic geometry, and can be quickly skimmed by a reader who does not intend to read beyond Chapter 6. Chapter 5 contains the polynomial proof of the joints theorem. Chapter 6 contains polynomial proofs for problems in finite fields, such as the finite field Kakeya problem and the cap set problem.

- **The distinct distances theorem.** To understand the distinct distances theorem of Guth and Katz, read Chapters 1–5 and 7–10. Chapter 7 reduces the distinct distances problem to an incidence problem in $\mathbb{R}^3$. Chapter 8 introduces the *constant-degree polynomial-partitioning* technique and uses it to prove incidence bounds in the complex plane. Chapter 9 extends this technique and uses it to prove the distinct distances theorem. Chapter 10 studies a few variants of the distinct distances problem.

- **Incidences and polynomial methods over finite fields.** To study incidences and polynomial methods over finite fields, read Chapters 6 and 13. You might wish to first read Chapter 1, but this is not necessary. Chapter 13 studies point-line incidences over finite fields.

- **Incidences in $\mathbb{R}^d$.** To understand advanced incidence techniques in $\mathbb{R}^d$, read Chapters 1–5, 8, 11, 12, and 14. Chapter 11 studies more advanced techniques for deriving incidence bounds in $\mathbb{R}^d$. Chapter 12 consists of applications for such incidence bounds. Chapter 14 introduces more advanced tools for studying incidences and related problems. In particular, this final chapter studies properties of ruled surfaces.

Acknowledgments

This book would not have been written without Micha Sharir and Joshua Zahl. Micha Sharir was my guide to the world of discrete geometry. Joshua Zahl was my guide to the world of real algebraic geometry. I am also indebted to Frank de Zeeuw for carefully reading and commenting on earlier versions of this book, and to Nets Katz.

I am grateful for having so many people who helped improve parts of this book. These include Moaaz AlQady, Boris Aronov, Abdul Basit, Alan Chang, Zachary Chase, Ana Chavez Caliz, Alex Cohen, Daniel Di Benedetto, Jordan Ellenberg, Esther Ezra, Evan Fink, Davey Fitzpatrick, Nora Frankl, Marina Iliopoulou, Alex Iosevich, Jongchon Kim, Bob Krueger, Brett Leroux, Shachar Lovett, Ben Lund, Michael Manta, Guy Moshkovitz, Brendan Murphy, Jason O'Neill, Yumeng Ou, Jonathan Passant, Cosmin Pohoata, Piotr Pokora, Anurag Sahay, Steven Senger, Olivine Silier, Shakhar Smorodinsky, Noam Solomon,

Samuel Speas, Samuel Spiro, Sophie Stevens, Jonathan Tidor, Bartosz Walczak, Audie Warren, Chengfei Xie, and Ruixiang Zhang.

While working on this book, I have been partially supported by the NSF award DMS–1802059 (Polynomial Methods in Discrete Geometry).

I also thank the nonmathematicians who helped make this book happen: Avner Itzhaki, Amanda Schneier, and Sofia Tolmach.

1

Incidences and Classical Discrete Geometry

My most striking contribution to geometry is, no doubt, my problem on the number of distinct distances. This can be found in many of my papers on combinatorial and geometric problems.

Paul Erdős, in a survey of his favorite contributions to mathematics, compiled for the celebration of his 80th birthday (Erdős, 1993).

1.1 Introduction to Incidences

We begin our study of geometric incidences by surveying the field and deriving a few first bounds. In this chapter we only discuss classical discrete geometry from before the discovery of the new polynomial methods. This makes the current chapter rather different from the rest of the book (outrageously, it even includes some graph theory). We also learn basic tricks that are used throughout the book, such as double counting, applying the Cauchy–Schwarz inequality, and dyadic decomposition. These techniques are presented in full detail in this chapter, while some details are omitted in the following chapters.

Consider a set $\mathcal{P}$ of points and a set $\mathcal{L}$ of lines, both in $\mathbb{R}^2$. An *incidence* is a pair $(p, \ell) \in \mathcal{P} \times \mathcal{L}$ such that the point p is contained in the line ℓ. We denote the number of incidences in $\mathcal{P} \times \mathcal{L}$ as $I(\mathcal{P}, \mathcal{L})$. For example, Figure 1 (in the Introduction) depicts a configuration with nine incidences. For any m and n, Erdős constructed a set $\mathcal{P}$ of m points and a set $\mathcal{L}$ of n lines with $\Theta(m^{2/3}n^{2/3} + m + n)$ incidences. Erdős (1985) conjectured that no point-line configuration has an asymptotically larger number of incidences. This conjecture was proved by Szemerédi and Trotter in 1983.

Theorem 1.1 (The Szemerédi–Trotter theorem) *Let $\mathcal{P}$ be a set of m points and let $\mathcal{L}$ be a set of n lines, both in $\mathbb{R}^2$. Then $I(\mathcal{P}, \mathcal{L}) = O(m^{2/3}n^{2/3} + m + n)$.*

1

The original proof of the Szemeredi–Trotter theorem is rather involved. In this chapter we present a later elegant proof by Székely (1997). A more general algebraic proof is presented in Chapter 3.

Finding the maximum number of point-line incidences in $\mathbb{R}^2$ is one of the simplest incidence problems. It is also one of very few incidence problems that are solved asymptotically. Other problems involve incidences with circles or other types of curves, incidences with varieties in $\mathbb{R}^d$, with semi-algebraic objects in $\mathbb{R}^d$, in complex spaces $\mathbb{C}^d$, in spaces over finite fields, and much more. In each of these problems, we wish to find the maximum number of incidences between a set of points and a set of geometric objects. If you ever need to snub a discrete geometer, try pointing out how they can barely solve any of these problems after decades of work.

One reason for studying incidence problems is that they are natural combinatorial problems. Throughout this chapter, we start to see additional reasons for studying incidence problems, including:

- *Incidence problems are not purely combinatorial, but also require an understanding of the underlying geometry.* One example of this appears in Section 1.5, where we introduce the unit distances problem. This problem involves studying properties that distinguish the Euclidean metric from almost all other distance metrics.

- *Incidence results are also useful for problems that may not seem related to geometry.* In Section 1.8, we use incidences to study the sum-product problem. This problem started as a number-theoretic problem that does not involve any geometry.

1.2 First Proofs

We now develop some initial intuition about incidences. We begin by deriving our first bound for an incidence problem. This is a weak bound, but it is still useful in some cases.

Lemma 1.2 *Let $\mathcal{P}$ be a set of m points and let $\mathcal{L}$ be a set of n lines, both in $\mathbb{R}^2$. Then $I(\mathcal{P}, \mathcal{L}) = O(m\sqrt{n} + n)$ and $I(\mathcal{P}, \mathcal{L}) = O(n\sqrt{m} + m)$.*

Why do we say that Lemma 1.2 is weaker than Theorem 1.1? For some intuition, consider the case where $m = n$. In this case, Theorem 1.1 leads to the bound $O(n^{4/3})$, while Lemma 1.2 only gives $O(n^{3/2})$.

Proof of Lemma 1.2 We only derive $I(\mathcal{P}, \mathcal{L}) = O(m\sqrt{n} + n)$. The other bound is obtained in a symmetric manner. Consider the set of triples

$$T = \left\{ (a, b, \ell) \in \mathcal{P}^2 \times \mathcal{L} \ : \ a \text{ and } b \text{ are both incident to } \ell \right\}.$$

Note that T also contains triples (a, b, ℓ) where $a = b$.

Let m_j be the number of points of $\mathcal{P}$ that are incident to the jth line of $\mathcal{L}$. Then the number of triples of T that include the jth line of $\mathcal{L}$ is m_j^2. This implies that $|T| = \sum_{j=1}^{n} m_j^2$. Also, note that $I(\mathcal{P}, \mathcal{L}) = \sum_{j=1}^{n} m_j$. We apply the Cauchy–Schwarz inequality (Theorem A.1). We present this first application of the inequality in full detail. Throughout the rest of the book, we skip the intermediary steps. For $1 \le j \le n$, we set $a_j = m_j$ and $b_j = 1$. The Cauchy–Schwarz inequality implies that

$$\sum_{j=1}^{n} m_j \le \left(\sum_{j=1}^{n} m_j^2 \right)^{1/2} \left(\sum_{j=1}^{n} 1 \right)^{1/2} = \left(\sum_{j=1}^{n} m_j^2 \right)^{1/2} \cdot n^{1/2}.$$

Squaring both sides and rearranging leads to

$$|T| = \sum_{j=1}^{n} m_j^2 \ge \frac{\left(\sum_{j=1}^{n} m_j \right)^2}{n} = \frac{I(\mathcal{P}, \mathcal{L})^2}{n}. \tag{1.1}$$

The number of triples $(a, b, \ell) \in T$ with $a = b$ is $I(\mathcal{P}, \mathcal{L})$. The number of triples $(a, b, \ell) \in T$ with $a \ne b$ is at most $\binom{m}{2}$, since each pair of distinct a, $b \in \mathcal{P}$ is contained in at most one line of $\mathcal{L}$. Thus, $|T| \le \binom{m}{2} + I(\mathcal{P}, \mathcal{L})$. Combining this with Equation (1.1) gives

$$\frac{I(\mathcal{P}, \mathcal{L})^2}{n} \le \binom{m}{2} + I(\mathcal{P}, \mathcal{L}). \tag{1.2}$$

When $\binom{m}{2} \ge I(\mathcal{P}, \mathcal{L})$, rearranging Equation (1.2) leads to $I(\mathcal{P}, \mathcal{L}) = O(mn^{1/2})$. Otherwise, rearranging Equation (1.2) leads to $I(\mathcal{P}, \mathcal{L}) = O(n)$. $\square$

To prove Lemma 1.2, we used a common combinatorial method called *double counting*. In this method, we bound some quantity X in two different ways and then compare the two bounds. This leads to new information that does not involve X. In the proof of Lemma 1.2, we derived upper and lower bounds for the size of T. By comparing these two bounds, we obtained a bound for the number of incidences. Double counting is ubiquitous in this book.

In the proof of Lemma 1.2, we did not use any geometry beyond observing that two points are contained in one line. This implies that the proof still holds after removing all the other geometric properties of the problem. That is, when replacing the lines with abstract sets of points, such that every two sets have at most one common element. For example, instead of the lines in Figure 1 (in the Introduction), we can consider the sets

$$A = \{a, d\}, \quad B = \{a, c\}, \quad C = \{a, d\}, \quad D = \{b, c, d\}.$$

In this abstract setting, the bounds of Lemma 1.2 are asymptotically tight. There exist n subsets of m elements with the above property and $\Theta(mn^{1/2})$ incidences (or $\Theta(nm^{1/2})$). Thus, to derive a stronger upper bound for point-line incidences, we must rely on additional geometric properties of lines.

We now consider an asymptotically tight lower bound for Theorem 1.1. Instead of Erdős's original construction, we present a simpler construction due to Elekes (2001).

Claim 1.3 *For every m and n there exist a set $\mathcal{P}$ of m points and a set $\mathcal{L}$ of n lines, both in $\mathbb{R}^2$, such that $I(\mathcal{P}, \mathcal{L}) = \Theta(m^{2/3}n^{2/3} + m + n)$.*

Proof The term m dominates the bound $\Theta(m^{2/3}n^{2/3}+m+n)$ when $m = \Omega(n^2)$. In this case we can simply take m points on a single line to obtain m incidences. Similarly, the term n dominates the bound when $n = \Omega(m^2)$. In this case we take n lines that pass through a single point to obtain n incidences. It remains to construct a configuration with $\Theta(m^{2/3}n^{2/3})$ incidences when $m = O(n^2)$ and $n = O(m^2)$.

Let $r = (m^2/4n)^{1/3}$ and $s = (2n^2/m)^{1/3}$ (for simplicity, instead of taking the ceiling function of s and r, we assume that these are integers). We set

$$\mathcal{P} = \{ (i, j) : 1 \le i \le r \quad \text{and} \quad 1 \le j \le 2rs \},$$

and

$$\mathcal{L} = \{ y = ax + b : 1 \le a \le s \quad \text{and} \quad 1 \le b \le rs \}.$$

Note that $\mathcal{P}$ is a rectangular section of the integer lattice. The slopes and y-intercepts of the lines of $\mathcal{L}$ also form such a lattice. Figure 1.1 depicts an example configuration rotated by $90°$. We also have that

$$|\mathcal{P}| = 2r^2s = 2 \cdot \frac{m^{4/3}}{(4n)^{2/3}} \cdot \frac{(2n^2)^{1/3}}{m^{1/3}} = m,$$

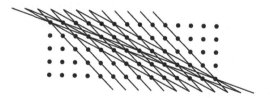

Figure 1.1 Elekes's construction, rotated by $90°$.

and

$$|\mathcal{L}| = rs^2 = \frac{m^{2/3}}{(4n)^{1/3}} \cdot \frac{(2n^2)^{2/3}}{m^{2/3}} = n.$$

Consider a line $\ell \in \mathcal{L}$ that is defined by the equation $y = ax + b$. For any $x \in \{1, \ldots, r\}$, there exists $y \in \{1, \ldots, 2rs\}$ such that the point (x, y) is incident to ℓ. That is, every line of $\mathcal{L}$ is incident to exactly r points of $\mathcal{P}$, which in turn implies that

$$I(\mathcal{P}, \mathcal{L}) = r \cdot |\mathcal{L}| = \frac{m^{2/3}}{(4n)^{1/3}} \cdot n = 2^{-2/3} m^{2/3} n^{2/3}. \qquad \square$$

1.3 The Crossing Lemma

One elegant proof of Theorem 1.1 is based on the *crossing lemma*. We study this proof in Section 1.4. Here, we first go over some required preliminaries. For a brief review of graph theory notation, see Section A.2.

The *crossing number* of a graph $G = (V, E)$, denoted cr(G), is the smallest integer k such that we can draw G in the plane with k edge crossings. Figure 1.2(a) depicts a drawing of K_5 with a single crossing. Since K_5 cannot be drawn without crossings, we have that cr(K_5) = 1. Intuitively, we expect a graph with a lot more edges than vertices to have a large crossing number. Given a graph $G = (V, E)$, we are interested in a lower bound for cr(G) with respect to $|V|$ and $|E|$.

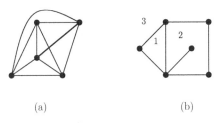

(a)   (b)

Figure 1.2 (a) A drawing of K_5 with a single crossing. (b) A graph with two bounded faces and one unbounded face.

A graph G is *planar* if cr(G) = 0. We consider a connected planar graph $G = (V, E)$ with v vertices and e edges. More specifically, we consider a drawing of G in the plane with no crossings. The *faces* of this drawing are the maximal two-dimensional connected regions that are bounded by the edges. This includes one outer, infinitely large region. For an example, see Figure 1.2(b). Denote by f the number of faces in the drawing of G. Then *Euler's formula* states that

$$v + f = e + 2. \tag{1.3}$$

For planar graphs that are not connected, we instead have that $v + f > e + 2$.

Every edge of G is either on the boundary of two faces or has both of its sides on the boundary of the same face. Moreover, the boundary of every face of G consists of at least three edges. Thus, we have $2e \geq 3f$. Plugging this into Equation (1.3) yields

$$e \leq v + f - 2 \leq v + \frac{2e}{3} - 2.$$

That is, for any planar graph $G = (V, E)$, we have that

$$|E| \leq 3|V| - 6. \tag{1.4}$$

The above leads to our first lower bound on $\text{cr}(G)$.

Lemma 1.4 *For any graph $G = (V, E)$, we have $\text{cr}(G) \geq |E| - 3|V| + 6$.*

Proof Consider a drawing of G in the plane that minimizes the number of crossings. Let $E' \subset E$ be a maximum subset of the edges such that no two edges of E' intersect in the drawing. By Equation (1.4), we have that $|E'| \leq 3|V| - 6$. Since every edge of $E \backslash E'$ intersects at least one edge of E', and since $|E \backslash E'| \geq |E| - 3|V| + 6$, there are at least $|E| - 3|V| + 6$ crossings in the drawing. □

Since K_5 has 5 vertices and 10 edges, Lemma 1.4 gives the correct value $\text{cr}(K_5) = 1$. However, in general the bound of this lemma is rather weak. For example, it is known that $\text{cr}(K_n) = \Theta(n^4)$, while Lemma 1.4 only implies that $\text{cr}(K_n) = \Omega(n^2)$. We can amplify the lower bound of Lemma 1.4 by combining it with a probabilistic argument. The following lemma was originally derived in Ajtai et al. (1982); Leighton (1983), with different proofs.

Lemma 1.5 (The crossing lemma) *Let $G = (V, E)$ be a graph with $|E| \geq 4|V|$. Then $\text{cr}(G) = \Omega(|E|^3 / |V|^2)$.*

Proof Consider a drawing of G with $\text{cr}(G)$ crossings. Set $p = \frac{4|V|}{|E|}$. The assumption of the lemma implies that $0 < p \leq 1$. We remove every vertex of V from the drawing with probability $1 - p$ (together with the edges adjacent to the vertex). Let $G' = (V', E')$ denote the resulting subgraph. Let c' denote the number of crossings in the drawing of G that have both of their edges in E'.

To avoid confusion with the edge set E, we denote expectation of a random variable as $\mathbb{E}[\cdot]$. Since every vertex remains with probability p, we have that $\mathbb{E}[|V'|] = p|V|$. Since every edge remains if and only if its two endpoints remain, we have that $\mathbb{E}[|E'|] = p^2|E|$. Finally, since each crossing remains if and only if the two corresponding edges remain, we have that $\mathbb{E}[c'] = p^4 \text{cr}(G)$. By linearity of expectation,

$$\mathbb{E}[c' - |E'| + 3|V'|] = p^4 \mathrm{cr}(G) - p^2|E| + 3p|V|$$

$$= \frac{4^4|V|^4}{|E|^4}\mathrm{cr}(G) - \frac{4^2|V|^2}{|E|^2} \cdot |E| + 3 \cdot \frac{4|V|}{|E|} \cdot |V|$$

$$= \frac{4^4|V|^4}{|E|^4}\mathrm{cr}(G) - \frac{4|V|^2}{|E|}.$$

Since this is the expected value, there exists a subgraph $G^* = (V^*, E^*)$ with c^* crossings remaining from the drawing of G, such that

$$c^* - |E^*| + 3|V^*| \le \frac{4^4|V|^4}{|E|^4}\mathrm{cr}(G) - \frac{4|V|^2}{|E|}. \tag{1.5}$$

By Lemma 1.4, we have $c^* \ge |E^*| - 3|V^*| + 6$. Combining this with Inequality (1.5) implies

$$0 < 6 \le c^* - |E^*| + 3|V^*| \le \frac{4^4|V|^4}{|E|^4}\mathrm{cr}(G) - \frac{4|V|^2}{|E|}.$$

That is, $\frac{4|V|^2}{|E|} < \frac{4^4|V|^4}{|E|^4}\mathrm{cr}(G)$. Tidying up this inequality leads to the required bound. □

Lemma 1.5 implies the asymptotically tight bound $\mathrm{cr}(K_n) = \Omega(n^4)$.

1.4 Szemerédi–Trotter via the Crossing Lemma

We are now ready to prove Theorem 1.1. We first restate this theorem.

Theorem 1.1 *Let $\mathcal{P}$ be a set of m points and let $\mathcal{L}$ be a set of n lines, both in $\mathbb{R}^2$. Then $I(\mathcal{P}, \mathcal{L}) = O(m^{2/3}n^{2/3} + m + n)$.*

Proof We write $\mathcal{L} = \{\ell_1, \ldots, \ell_n\}$ and denote by m_j the number of points of $\mathcal{P}$ that are on ℓ_j. Notice that $I(\mathcal{P}, \mathcal{L}) = \sum_{j=1}^n m_j$. We may remove any line ℓ_j that satisfies $m_j = 0$, since this would not change the number of incidences.

We build a graph $G = (V, E)$ as follows. Every vertex of V corresponds to a point of $\mathcal{P}$. For $v, u \in V$, we add (v, u) to E if v and u correspond to consecutive points along a line of $\mathcal{L}$. For an example, see Figure 1.3. A line ℓ_j contributes exactly $m_j - 1$ edges of E. Thus, we have $|V| = m$ and $|E| = \sum_{j=1}^n (m_j - 1) = I(\mathcal{P}, \mathcal{L}) - n$.

If $|E| < 4|V|$ then $I(\mathcal{P}, \mathcal{L}) = O(m + n)$, which completes the proof. We may thus assume that $|E| \ge 4|V|$. Then, Lemma 1.5 leads to

$$\mathrm{cr}(G) = \Omega\left(\frac{(I(\mathcal{P}, \mathcal{L}) - n)^3}{m^2}\right). \tag{1.6}$$

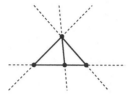

Figure 1.3 (Solid segment) The edges of the graph. (Dashed segment) The portions of the lines ℓ_j that do not form graph edges.

We draw G according to the point-line configuration: Every vertex is at the corresponding point and every edge is the corresponding line segment. Every crossing in this drawing is an intersection of two lines of $\mathcal{L}$. Since every two lines intersect at most once, we have that $\mathrm{cr}(G) \le \binom{n}{2} = O(n^2)$. Combining this with Equation (1.6) implies that

$$\frac{(I(\mathcal{P}, \mathcal{L}) - n)^3}{m^2} = O(n^2).$$

Rearranging this equation gives $I(\mathcal{P}, \mathcal{L}) = O(m^{2/3}n^{2/3} + n)$. ☐

The proof of Theorem 1.1 is another example of the double counting method. We counted $\mathrm{cr}(G)$ in two different ways. By combining the two resulting bounds, we obtained a bound on the number of incidences.

In the proof of Theorem 1.1, we used the geometric property that two lines intersect at most once. This is similar to the observation that any two points are contained in one line,[1] which was used in the proof of Lemma 1.2. In the proof of Theorem 1.1 we used a second geometric property when stating that the line ℓ_j corresponds to exactly $m_j - 1$ edges of E. This statement relies on the observation that a line consists of a single connected component and does not intersect itself. When replacing the lines with other curves that satisfy the same geometric properties, the proof of Theorem 1.1 remains valid.

1.5 The Unit Distances Problem

The *unit distances problem* is one of the main open problems in discrete geometry. While it is extremely difficult to solve this problem, it easy to state:

In a set of n points in the plane, what is the maximum possible number of pairs of points at distance 1 from each other?

---

[1] These two geometric properties are equivalent when studying point-line incidences, due to point-line duality. We discuss this concept in Section 1.10.

We denote this maximum number of pairs as $u(n)$. By taking a set of n points equally spaced on a line, we immediately obtain that $u(n) \geq n-1$. Erdős (1946) introduced the problem, while also deriving the bounds $u(n) = O(n^{3/2})$ and $u(n) = \Omega(n^{1+c/\log\log n})$, for some constant c. While many mathematicians have studied this problem, the lower bound for $u(n)$ has not been improved since 1946 and the upper bound was last improved in 1984. That was when Spencer et al. (1984) derived the bound $u(n) = O(n^{4/3})$.

Consider a set $\mathcal{P} \subset \mathbb{R}^2$ of n points such that the number of unit distances between pairs of points of $\mathcal{P}$ is $u(n)$. We draw a unit circle (a circle of radius one) around each point of $\mathcal{P}$, and denote the set of these n circles as C. Every two points $p, q \in \mathcal{P}$ that determine a unit distance correspond to two incidences in $\mathcal{P} \times C$: The circle around p is incident to q and vice versa. See Figure 1.4 for an example. Thus, to bound $u(n)$ it suffices to bound the maximum number of incidences between n points and n unit circles (it is not difficult to show that this maximum number of incidences is asymptotically equivalent to $u(n)$).

Figure 1.4 Every two points that are at a unit distance correspond to two point-circle incidences.

Theorem 1.6 *Let $\mathcal{P}$ be a set of n points and let C be a set of n unit circles, both in $\mathbb{R}^2$. Then $I(\mathcal{P}, C) = O(n^{4/3})$.*

Theorem 1.6 immediately implies the current best bound $u(n) = O(n^{4/3})$.

Proof of Theorem 1.6 We imitate the proof of Theorem 1.1. Let $C = \{c_1, \ldots, c_n\}$ and let m_j denote the number of points of $\mathcal{P}$ on c_j. Note that $I(\mathcal{P}, C) = \sum_{j=1}^{n} m_j$. We may remove any circle c_j that satisfies $m_j < 3$, since this reduces the number of incidences by at most $2n$.

We build a graph $G = (V, E)$ as follows. Every vertex of V corresponds to a point of $\mathcal{P}$. For $v, u \in V$, the edge (v, u) is in E if v and u are consecutive points along at least one circle of C. A circle c_j corresponds to exactly m_j edges of E, and every edge originates from at most two unit circles. Note that $|V| = n$ and $|E| \geq (\sum_{j=1}^{n} m_j)/2 = I(\mathcal{P}, C)/2$.

If $|E| < 4|V|$ then $I(\mathcal{P}, C) = O(n)$, which completes the proof. We may thus assume that $|E| \geq 4|V|$. By Lemma 1.5, we have that

$$\mathrm{cr}(G) = \Omega\left(\frac{I(\mathcal{P}, C)^3}{n^2}\right). \tag{1.7}$$

We draw G according to the point-circle configuration: Every vertex is at the corresponding point and every edge is one of the corresponding circle arcs. Every crossing in this drawing is the intersection of two circles of C. Since every two circles intersect at most twice, we have that $cr(G) \le 2\binom{n}{2} = O(n^2)$. Combining this with Equation (1.7) implies that

$$\frac{I(\mathcal{P}, C)^3}{n^2} = O(n^2).$$

Rearranging this equation leads to $I(\mathcal{P}, C) = O(n^{4/3})$.   □

Erdős offered \$250 for proving the following conjecture (Erdős, 1985).

Conjecture 1.7 (Erdős, 1985) $u(n) = O(n^{1+\varepsilon})$ *for any* $\varepsilon > 0$.

This is an example of how little we currently know about incidences. While the problem of point-line incidences in $\mathbb{R}^2$ has been settled for decades, the case of unit circles remains wide open. Hardly any other incidence problems have been solved.

The answer to the unit distances problem significantly depends on the metric:

- For Euclidean distance, this is a long-standing difficult problem.
- For some metrics, there exist sets of n points that span $\Theta(n^2)$ unit distances. See Exercise 1.3.
- Valtr (2005) discovered a well-behaved metric for which $u(n) = \Theta(n^{4/3})$.
- Matoušek (2011) showed that, for most metrics,[2] $u(n) = O(n \log n \log \log n)$.

The bound that is conjectured for the Euclidean distance is different from all other bounds stated above. One may thus say that the unit distances problem is about studying properties of the underlying geometry. A proof of Conjecture 1.7 is likely to require properties that are unique for the Euclidean metric.

1.6   The Distinct Distances Problem

The *distinct distances* problem is a close relative of the unit distances problem. Both problems were introduced in the same 1946 paper of Erdős. For a set $\mathcal{P} \subset \mathbb{R}^2$, let $\Delta(\mathcal{P})$ denote the set of distances spanned by pairs of points of $\mathcal{P}$. Every distance appears in $\Delta(\mathcal{P})$ at most once, no matter how many pairs of points span it. This is why we refer to $\Delta(\mathcal{P})$ as the *set of distinct distances of* $\mathcal{P}$. See Figure 1.5 for an example. The distinct distances problem asks for $\min_{|\mathcal{P}|=n} |D(\mathcal{P})|$. In other words:

---

[2] The exact meaning of "most metrics" is beyond the scope of this chapter.

What is the minimum number of distinct distances that can be determined by a set of n points in $\mathbb{R}^2$.

We denote this quantity as $d(n)$.

Figure 1.5 When $\mathcal{P}$ consists of these five points, $D(\mathcal{P}) = \left\{\sqrt{2}, 2, \sqrt{8}, \sqrt{10}\right\}$.

Note that a set of n points that are equally spaced on a line determines $n - 1$ distinct distances. This implies that $d(n) \leq n - 1$. An asymptotically better bound appeared in Erdős's original paper. Erdős considered the set

$$\mathcal{P} = \left\{(a, b) \in \mathbb{Z}^2 : 0 \leq a, b < \sqrt{n}\right\}.$$

The number of distinct distances determined by this set is an immediate corollary of the following result from number theory.

Theorem 1.8 (Berndt and Rankin, 1995; Landau, 1909) *The number of integers smaller than n that can be written as $a^2 + b^2$ with $a, b \in \mathbb{Z}$ is* $\Theta(n/\sqrt{\log n})$.

Every distance in $\mathcal{P}$ is the square root of a sum of two squares between 0 and $n - 1$. By Theorem 1.8, there are $\Theta(n/\sqrt{\log n})$ such numbers. Thus, $D(\mathcal{P}) = \Theta(n/\sqrt{\log n})$.

Theorem 1.9 (Erdős, 1946) $d(n) = O(n/\sqrt{\log n})$.

Erdős conjectured that $d(n) = \Theta(n/\sqrt{\log n})$, but deriving lower bounds for $d(n)$ turned out to be significantly more difficult. We now derive Erdős's original lower bound for $d(n)$ (using a different proof).

Claim 1.10 $d(n) = \Omega(n^{1/2})$.

Proof Consider a set $\mathcal{P}$ of n points in $\mathbb{R}^2$ and two points $v, u \in \mathcal{P}$. Let d_v denote the number of distinct distances between v and $\mathcal{P}\backslash\{v\}$.[3] The points of $\mathcal{P}\backslash\{v\}$ are contained in d_v circles that are centered at v. We denote this set of circles as C_v. We define d_u and C_u symmetrically. Each of the $n - 2$ points of $\mathcal{P}\backslash\{v, u\}$ is contained in the intersection of a circle of C_v with a circle of C_u (see Figure 1.6 for an example). Since the number of such intersections is at most $2|C_v||C_u| = 2d_v d_u$, we have that $2d_v d_u \geq n - 2$. This in turn implies that $\max\{d_v, d_u\} = \Omega(n^{1/2})$. $\qquad\square$

---

[3] We use the symbol $\backslash$ to denote set subtraction.

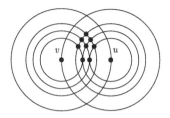

Figure 1.6 The points of $\mathcal{P}\backslash\{v, u\}$ are contained in the intersections of C_v and C_u.

We can derive a better lower bound for $d(n)$ by using incidences. This bound was originally proved in Moser (1952), but we study a different proof.

Claim 1.11 $d(n) = \Omega(n^{2/3})$.

Proof Consider a set $\mathcal{P}$ of n points in $\mathbb{R}^2$. Set $d = d(\mathcal{P})$ and denote the distances spanned by $\mathcal{P}$ as $D = \{\delta_1, \ldots, \delta_d\}$. Let C denote the set of $n \cdot d$ circles with a center in $\mathcal{P}$ and a radius in D. The claim is proved by double counting $I(\mathcal{P}, C)$. For every point $v \in \mathcal{P}$, the points of $\mathcal{P}\backslash\{v\}$ are contained in the d circles of C centered at v. Thus, $I(\mathcal{P}, C) = n(n-1)$.

Let C_j denote the subset of circles of C with radius δ_j. By Theorem 1.6, we have $I(\mathcal{P}, C_j) = O(n^{4/3})$ (Theorem 1.6 holds for any set of circles with the same radii). This leads to

$$I(\mathcal{P}, C) = \sum_{j=1}^{d} I(\mathcal{P}, C_j) = O(dn^{4/3}).$$

Combining our two bounds for $I(\mathcal{P}, C)$ gives that $n^2 = O(dn^{4/3})$. Rearranging this equation completes the proof. □

We can prove Claim 1.11 in a simpler way. Each of the $\binom{n}{2}$ pairs of points determines a distance. By Theorem 1.6, every distance occurs $O(n^{4/3})$ times. To have distances for $\Theta(n^2)$ pairs of points, there must be $\Omega(n^{2/3})$ distinct distances. We first present the longer proof, since it demonstrates a common way of using incidences.

Both proofs of Claim 1.11 show that the distinct distances problem can be reduced to the unit distances problem. The bound $u(n) = O(n^{1+c/\log\log n})$ would immediately lead to an almost tight lower bound for $d(n)$.

Guth and Katz (2015) proved the bound $d(n) = \Omega(n/\log n)$, almost settling Erdős's conjecture. The proof of Guth and Katz is a deep result that combines tools from multiple mathematical subfields. One of the peaks of this book is a proof of this result (see Chapters 7 and 9).

Although the distinct distances problem is solved (up to a gap of $\sqrt{\log n}$), interesting variants of it are still wide open. These include problems asked by Erdős decades ago. A couple of examples:

- For $d \geq 3$, what is the minimum number of distinct distances spanned by n points in $\mathbb{R}^d$? Erdős constructed a set of n points that determines $\Theta(n^{2/d})$ distinct distances. He conjectured that no set determines an asymptotically smaller number. The problem remains wide open even in $\mathbb{R}^3$. See also Section 9.5.
- Characterize the sets of n points in $\mathbb{R}^2$ that determine $O(n/\sqrt{\log n})$ distinct distances. The past decades produced many conjectures for this problem, but hardly any results. See also Section 10.2.

For a list of many other related open problems, see Sheffer (2014).

1.7 A Problem about Unit Area Triangles

We briefly consider one of the many generalizations of the unit distances problem:

What is the maximum number of unit area triangles that have their vertices in a set of n points in $\mathbb{R}^2$?

When do two points p and q in $\mathbb{R}^2$ form a unit area triangle with a third point r? A key observation is that r must be on one of the two lines that are parallel to the segment pq and at a distance of $2/|pq|$ from this segment ($|pq|$ is the distance between the points p and q). For an example, see Figure 1.7(a). By taking two parallel lines at a distance of 2 from each other, and placing $n/2$ points at unit intervals on each, we obtain $\Theta(n^2)$ unit area triangles (see Figure 1.7(b)). Erdős and Purdy (1971) showed that a $\sqrt{\log n} \times (n/\sqrt{\log n})$ section of the integer lattice determines $\Omega(n^2 \log \log n)$ triangles of the same area. By applying a uniform scaling of the plane, we obtain $\Omega(n^2 \log \log n)$ unit area triangles.

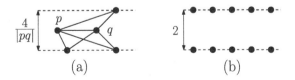

(a) (b)

Figure 1.7 (a) The points that form a unit area triangle with p and q are on two parallel lines. (b) A configuration with $\Theta(n^2)$ unit area triangles.

Claim 1.12 (Pach and Sharir, 1992)   *Every set of n points in* $\mathbb{R}^2$ *determines* $O(n^{7/3})$ *unit triangles.*

Proof Consider a set $\mathcal{P}$ of n points and a point $p \in \mathcal{P}$. We first bound the number of unit area triangles that are determined by p and two other points of $\mathcal{P}$. For every $q \in \mathcal{P}\backslash\{p\}$, let $\ell_{p,q}$ and $\ell'_{p,q}$ be the lines parallel to the segment pq and at a distance of $2/|pq|$ from this segment. We set $\mathcal{L}_p = \{\ell_{p,q}, \ell'_{p,q} : q \in \mathcal{P}\}$. Note that a line of $\mathcal{L}_p$ can originate from two distinct points of $\mathcal{P}\backslash\{p\}$, but not from more than two. When a line originates from two such points q, we include only one copy of the line in $\mathcal{L}_p$. This implies that $n - 1 \leq |\mathcal{L}_p| \leq 2n - 2$.

If the points $p, q, r \in \mathcal{P}$ form a unit area triangle, then r is incident to $\ell_{p,q}$ or $\ell'_{p,q}$. Similarly, q is incident to $\ell_{p,r}$ or $\ell'_{p,r}$. In the other direction, if a point r is incident to $\ell_{p,q}$ or $\ell_{p,q'}$ then p, q, r are the vertices of a unit area triangle. Thus, the number of unit area triangles that involve p is at least $I(\mathcal{P}, \mathcal{L}_p)/2$. Theorem 1.1 implies that $I(\mathcal{P}, \mathcal{L}_p) = O(n^{4/3})$. Therefore, the total number of unit area triangles spanned by $\mathcal{P}$ is at most

$$\sum_{p \in \mathcal{P}} I(\mathcal{P}, \mathcal{L}_p) = \sum_{p \in \mathcal{P}} O(n^{4/3}) = O(n^{7/3}). \qquad \square$$

Recently, Raz and Sharir (2017) improved the bound of Claim 1.12 to $O(n^{20/9})$. Their proof involves incidences with two-dimensional surfaces in $\mathbb{R}^4$.

1.8 The Sum-Product Problem

Incidence results are sometimes useful when studying problems that at first do not seem related to geometry. In this section, we meet our first example of such a problem. Given a set A of n real numbers, we consider the sets

$$A + A = \{a + b : a, b \in A\}, \quad \text{and} \quad AA = \{ab : a, b \in A\}.$$

When $A = \{1, 2, 3, \ldots, n\}$, we have that $A + A = \{2, 3, 4, \ldots, 2n\}$. More generally, $|A + A| = 2n - 1$ if and only if A is an arithmetic progression. There are sets A that are not arithmetic progressions but still satisfy $|A + A| = \Theta(n)$. Similarly, to obtain $|AA| = \Theta(n)$, we can take A to be a geometric progression. Erdős and Szemerédi (1983) made the following conjecture.

Conjecture 1.13   *For any $\varepsilon > 0$, there exists n_0 such that any set A of $n > n_0$ integers satisfies*

$$\max\{|A + A|, |AA|\} = \Omega(n^{2-\varepsilon}).$$

Intuitively, Conjecture 1.13 predicts that no set can simultaneously have a small set of sums and a small set of products. This problem received the name *the sum-product problem.* It started as a number-theoretic problem, dealing with sums and products of integers. Elekes (1997) introduced a geometric approach to the problem, also extending it to sets of real numbers. This was a major breakthrough that influenced most of the future works on the problem. We now study Elekes's result.

Theorem 1.14 *Let A be a set of n real numbers. Then*

$$\max\{|A + A|, |AA|\} = \Omega(n^{5/4}).$$

Proof Consider the point set

$$\mathcal{P} = \{(c, d) \ : \ c \in A + A \quad \text{and} \quad d \in AA\}.$$

Note that $|\mathcal{P}| = |A + A| \cdot |AA|$. We also consider the set of lines

$$\mathcal{L} = \{y = a(x - a') \ : \ a, a' \in A\}.$$

(By $y = a(x - a')$ we refer to the line that is defined by this expression.) Note that $|\mathcal{L}| = n^2$. The point set $\mathcal{P}$ is a Cartesian product. The slopes and y-intercepts of the lines of $\mathcal{L}$ also form a Cartesian product. See Figure 1.8 for an example.

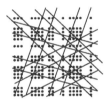

Figure 1.8 The configuration in Elekes's sum-product proof consists of Cartesian products of points and lines.

The proof is based on double counting $I(\mathcal{P}, \mathcal{L})$. A line defined by the equation $y = a(x - a')$ (with $a, a' \in A$) contains the point $(a' + b, ab) \in \mathcal{P}$ for every $b \in A$. Since each of the n values of b leads to a different point, every line of $\mathcal{L}$ is incident to at least n points of $\mathcal{P}$. Therefore, we have that

$$I(\mathcal{P}, \mathcal{L}) \geq |\mathcal{L}| \cdot n = n^3.$$

On the other hand, Theorem 1.1 gives

$$I(\mathcal{P}, \mathcal{L}) = O\left(|\mathcal{P}|^{2/3}|\mathcal{L}|^{2/3} + |\mathcal{P}| + |\mathcal{L}|\right)$$

$$= O\left(|A + A|^{2/3}|AA|^{2/3}n^{4/3} + |A + A| \cdot |AA| + n^2\right).$$

By combining the two bounds for $I(\mathcal{P}, \mathcal{L})$, we get that

$$n^3 = O\left(|A + A|^{2/3}|AA|^{2/3}n^{4/3} + |A + A| \cdot |AA| + n^2\right),$$

or

$$|A + A| \cdot |AA| = \Omega(n^{5/2}). \tag{1.8}$$

If both $|A + A|$ and $|AA|$ are asymptotically smaller than $n^{5/4}$, then $|A + A| \cdot |AA|$ is asymptotically smaller than $n^{5/2}$. This contradicts Equation (1.8), so $\max\{|A + A|, |AA|\} = \Omega(n^{5/4})$.   □

The bound of Theorem 1.14 has been improved many times, always by using geometric arguments. Most notably, Solymosi (2009) derived the bound $\Omega\left(n^{4/3} / \log^{1/3} n\right)$.

1.9 Rich Points

We now study an alternative way of formulating incidence bounds. Consider a set $\mathcal{L}$ of lines in $\mathbb{R}^2$ and an integer $r \geq 2$. We say that a point $p \in \mathbb{R}^2$ is *r-rich with respect to* $\mathcal{L}$ if at least r lines of $\mathcal{L}$ are incident to p. When the set of lines is clear from the context, we simply say that p is r-rich. Note that a 4-rich point is also 3-rich and also 2-rich. For a set of n lines that intersect at the origin there is a single 2-rich point: the origin. Let $M_{\geq r}(n)$ denote the maximum number of r-rich points that a set of n lines in $\mathbb{R}^2$ can have. For example, Figure 1.9 demonstrates that $M_{\geq 3}(n) = \Omega(n^2)$. Rich points have an important role in advanced proofs in Chapter 9.

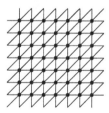

Figure 1.9 A construction with $n/4$ vertical lines, $n/4$ horizontal lines, $n/2$ diagonal lines, and $\Theta(n^2)$ points that are 3-rich.

Rich points lead us to an equivalent formulation of the Szemerédi–Trotter theorem (Theorem 1.1). By "equivalent," we mean that each formulation can be easily derived from the other by using only basic arguments.

Lemma 1.15 *Theorem 1.1 is equivalent to the claim: For every $r \geq 2$, we have that*

$$M_{\geq r}(n) = O\left(\frac{n^2}{r^3} + \frac{n}{r}\right). \tag{1.9}$$

Proof We first prove that Theorem 1.1 implies Equation (1.9) for every $r \geq 2$. Since there are $O(n^2)$ intersection points in a set of n lines, we have that $M_{\geq r}(n) = O(n^2)$ for every $r \geq 2$. This completes the proof when r is constant. We may thus assume that r is sufficiently larger than the constant in the $O(\cdot)$-notation of the bound of Theorem 1.1.

Consider a set $\mathcal{L}$ of n lines in $\mathbb{R}^2$ and fix a sufficiently large r as described above. Let $\mathcal{P}_r$ be the set of r-rich points and set $m_r = |\mathcal{P}_r|$. By the definition of rich points, we have that $I(\mathcal{P}_r, \mathcal{L}) \geq m_r r$. On the other hand, Theorem 1.1 implies $I(\mathcal{P}_r, \mathcal{L}) = O(m_r^{2/3} n^{2/3} + n + m_r)$. Combining these two bounds leads to

$$m_r r = O\left(m_r^{2/3} n^{2/3} + n + m_r\right).$$

Since r is sufficiently larger than the constant in the $O(\cdot)$-notation, the right-hand side cannot be dominated by m_r. We obtain that $m_r r = O(m_r^{2/3} n^{2/3} + n)$. This implies that $m_r = O\left(\frac{n^2}{r^3} + \frac{n}{r}\right)$, as required.

We now assume that $M_{\geq r}(n) = O(n^2/r^3 + n/r)$ holds for every $r \geq 2$, and rely on this to prove Theorem 1.1. Consider a set $\mathcal{P}$ of m points and a set $\mathcal{L}$ of n lines. If $m = \Omega(n^2)$, then Lemma 1.2 implies that $I(\mathcal{P}, \mathcal{L}) = O(n\sqrt{m} + m) = O(m)$. Thus, we may assume that $m = O(n^2)$.

Let $\hat{m}_j$ denote the number of points of $\mathcal{P}$ that are incident to at least 2^j lines of $\mathcal{L}$ and to fewer than 2^{j+1} such lines. Let $\mathcal{P}_+$ be the set of points of $\mathcal{P}$ that $(\sqrt{n} + 1)$-rich, and set $k = \lceil \log(n^{2/3}/m^{1/3}) \rceil$. Since $m = O(n^2)$, we get that $k \geq 1$. Thus,

$$I(\mathcal{P}, \mathcal{L}) \leq \sum_{j \geq 0}^{(\log n)/2} \hat{m}_j 2^{j+1} + I(\mathcal{P}_+, \mathcal{L}) = \sum_{j=0}^{k} \hat{m}_j 2^{j+1} + \sum_{j=k+1}^{(\log n)/2} \hat{m}_j 2^{j+1} + I(\mathcal{P}_+, \mathcal{L}).$$
(1.10)

If $k \geq \frac{1}{2} \log n$, then we ignore the second sum in the right-hand side of Inequality (1.10) and have the first sum stop at $j = \frac{1}{2} \log n$. Since $\hat{m}_j \leq m$ holds for every j, we have that

$$\sum_{j=0}^{k} \hat{m}_j 2^{j+1} \leq \sum_{j=0}^{k} m 2^{j+1} = O\left(m^{2/3} n^{2/3}\right).$$
(1.11)

When $r = 1$, we cannot apply the upper bound for $M_{\geq r}(n)$. Instead, we use the trivial observation that at most m points of $\mathcal{P}$ are 1-rich. The assumption $m = O(n^2)$ implies that $m = O(m^{2/3} n^{2/3})$.

When $j \leq \sqrt{n}$, the assumed upper bound for $M_{\geq r}(n)$ leads to $\hat{m}_j = O(n^2/2^{3j})$. We get that

$$\sum_{j=k+1}^{(\log n)/2} \hat{m}_j 2^{j+1} = \sum_{j=k+1}^{(\log n)/2} O\left(\frac{n^2}{2^{2j}}\right) = O\left(m^{2/3} n^{2/3}\right). \qquad (1.12)$$

It remains to bound $I(\mathcal{P}_+, \mathcal{L})$. By the assumed upper bound for $M_{\geq r}(n)$, we have $|\mathcal{P}_+| = O(\sqrt{n})$. By plugging this into the bound of Lemma 1.2, we obtain $I(\mathcal{P}_+, \mathcal{L}) = O(n)$. Combining this with Equations (1.10)–(1.12) leads to the bound of Theorem 1.1.   □

For the argument in Equation (1.12) to work, we partition the points of $\mathcal{P}$ according to their richness. Each part contains points with approximately the same richness (up to a factor of 2). When dealing with higher richness, the value of 2^j becames larger. This large value is balanced with the observation that there is a smaller number of rich points. This technique is called *dyadic decomposition*. We use it throughout the book.

1.10 Point-Line Duality

We complete this chapter with a useful tool. Consider a point $p = (p_x, p_y) \in \mathbb{R}^2$ and a line ℓ that is defined by $y = sx - t$ (with $s, t \in \mathbb{R}$). The *dual* p^* of p is the line defined by $y = p_x x - p_y$. The *dual* ℓ^* of ℓ is the point (s, t). Note that p is incident to ℓ if and only if ℓ^* is incident to p^*. Indeed, both incidences are equivalent to $p_y = p_x s - t$. This idea is referred to as *point-line duality*.

Let $\mathcal{P}$ be a set of points and let $\mathcal{L}$ be a set of lines, both in $\mathbb{R}^2$. Set $\mathcal{P}^* = \{p^* : p \in \mathcal{P}\}$ and $\mathcal{L}^* = \{\ell^* : \ell \in \mathcal{L}\}$. By the above, the number of incidences is preserved: $I(\mathcal{P}, \mathcal{L}) = I(\mathcal{L}^*, \mathcal{P}^*)$. Other useful information is preserved by the duality. For example, the points p, q, r are contained in a common line if and only if the lines p^*, q^*, r^* intersect at a common point. A point p is 100-rich with respect to $\mathcal{L}$ if and only if the line p^* contains at least 100 points of $\mathcal{L}^*$.

Consider a point set $\mathcal{P} \subset \mathbb{R}^2$ and an integer r. A line $\ell \subset \mathbb{R}^2$ is *r-rich with respect to* $\mathcal{P}$ if ℓ is incident to at least r points of $\mathcal{P}$. Let $\overline{M}_{\geq r}(n)$ denote the maximum number of r-rich lines that a set of n points in $\mathbb{R}^2$ can have. This leads to yet another equivalent formulation of the Szemerédi–Trotter theorem.

Lemma 1.16   *Theorem 1.1 is equivalent to the claim: For every $r \geq 2$, we have that*

$$\overline{M}_{\geq r}(n) = O\left(\frac{n^2}{r^3} + \frac{n}{r}\right).$$

Proof After applying point-line duality, Lemma 1.16 becomes Lemma 1.15.

□

When trying to prove a result involving points or lines in the plane, it is often helpful to try point-line duality. The dual problem is equivalent to the original one, but the new perspective might make some aspects of the problem easier to observe.

1.11 Exercises

Some exercises ask for "asymptotically tight bounds" or "matching lower and upper bounds." In such cases, you are asked to provide both a proof and a construction. For example, consider how we obtained an asymptotically tight bound for incidences between m points and n lines in $\mathbb{R}^2$. Claim 1.3 provides a construction with $\Theta(m^{2/3}n^{2/3} + m + n)$ incidences, and Theorem 1.1 proves that the number of incidences is always $O(m^{2/3}n^{2/3} + m + n)$. Combining the proof and the construction implies that the maximum number of incidences is $\Theta(m^{2/3}n^{2/3} + m + n)$. In this case, the construction led to a lower bound for the maximum number of incidences and the proof led to a matching upper bound. In some problems, constructions lead to upper bounds and proofs to lower bounds.

A set of points is said to be *collinear* if there exists a line that contains all the points.

Exercise 1.1 Construct a set $\mathcal{P}$ of m points and a set Γ of n parabolas that are defined by equations of the form $y = ax^2 + bx + c$, such that $I(\mathcal{P}, \Gamma) = \Theta(m^{1/2}n^{5/6})$. (Hint: Adapt the proof of Claim 1.3.)

Figure 1.10 The graph created by connecting consecutive and antipodal vertices.

Exercise 1.2 Let $n \geq 6$ be an even integer. Let $G = (V, E)$ be the graph obtained by taking n vertices equally spaced on a circle, and connecting every two vertices that are consecutive along the circle or antipodal. Figure 1.10 depicts the case of $n = 8$. (We only use the circle to easily describe G. As an abstract graph, G can also be drawn in other ways.)

(a) Prove that $\text{cr}(G) \geq 1$.

(b) Prove that $\text{cr}(G) \leq 1$.

Exercise 1.3   With the ℓ_1 metric (or *Manhattan distance*), the distance between the points (x_1, y_1) and (x_2, y_2) is $|x_1 - x_2| + |y_1 - y_2|$. Show that in this case there exists a set of n points that spans $\Theta(n^2)$ unit distances.

Exercise 1.4   Define a *unit chain* as a triple of points $(a, b, c) \in \mathbb{R}^2$ such that $|ab| = |bc| = 1$ and $a \neq c$ (recall that $|ab|$ is the distance between a and b). Derive asymptotically matching upper and lower bounds for the maximum number of unit chains that a set of n points in $\mathbb{R}^2$ can span. While the unit distances problem is an extremely difficult open problem, this variant is not.

Exercise 1.5   After reading Sections 1.4 and 1.5, one might assume that the same technique could be used to derive incidence bounds with most types of curves. This is not the case.

(a) Explain why the proof of Theorem 1.6 does not hold for arbitrary circles (as opposed to the case where all circles have the same radii).

(b) Show that the the proof of Theorem 1.6 still holds when replacing the circles with hyperbolas defined by $(x - a)^2 - (y - b)^2 = 1$ (each hyperbola has different values for $a, b \in \mathbb{R}$). How can we address the issue from part (a)?

Exercise 1.6   Let $\mathcal{P}$ be a set of n points in $\mathbb{R}^2$. Prove that the following holds for at least $n - 1$ points $p \in \mathcal{P}$: There are $\Omega(n^{1/2})$ distinct distances between p and the points of $\mathcal{P} \backslash \{p\}$.

Exercise 1.7   We saw that the point set $\{1, 2, 3, \ldots, \sqrt{n}\}^2 \subset \mathbb{R}^2$ spans $\Theta(n/\sqrt{\log n})$ distinct distances. Prove that the point set $\{1, 2, 3, \ldots, n^{1/3}\} \times \{1, 2, 3, \ldots, n^{2/3}\}$ spans $\Omega(n)$ distinct distances. (Hint: Consider only pairs of points with a large difference between their y-coordinates.)

Exercise 1.8   Find an asymptotically tight bound for the distinct distances problem when using the ℓ_1 metric (as defined in Exercise 1.3). That is, prove matching lower and upper bounds.

Exercise 1.9   Prove that every set $\mathcal{P}$ of n points in $\mathbb{R}^3$ spans $\Omega(n^{1/3})$ distinct distances. (Hint: Adapt the proof of Claim 1.10. Use the following property: Every three spheres with centers that are not collinear intersect in at most two points.)

Exercise 1.10   Instead of the Euclidean distance, we consider the metric where the distance between (x_1, y_1) and (x_2, y_2) is $\sqrt{4(x_1 - x_2)^2 + (y_1 - y_2)^2}$. Use a transformation of $\mathbb{R}^2$ to show that the distinct distances problem with this metric is equivalent to the distinct distances problem with the Euclidean metric.

(0,0)

Figure 1.11 Three triangles from the same congruency class.

Exercise 1.11 Given a set $\mathcal{P}$ of n points in $\mathbb{R}^2\backslash\{(0,0)\}$, we consider the set of triangles whose vertices are the origin and two points from $\mathcal{P}$ (including degenerate triangles where the three points are collinear). We partition these $\binom{n}{2}$ triangles into congruency classes. That is, all triangles in the same class are congruent. See Figure 1.11. We are interested in the minimum number of congruency classes that $\mathcal{P}$ can span. Prove that this number is $O(n)$ and $\Omega(n/\log n)$. You may use the Guth–Katz distinct distances bound.

Exercise 1.12 Let $\mathcal{P}$ be a set of n points in $\mathbb{R}^2$, such that the origin is not in $\mathcal{P}$. We consider right-angled triangles whose vertices are the origin and two points from $\mathcal{P}$. In addition, the origin should not be the vertex that is next to the right angle.
(a) Prove that the maximum number of such triangles is $O(n^{4/3})$.
(b) Show that it is possible to have $\Theta(n^{4/3})$ such triangles. (Hint: Start with a point-line configuration.)

Exercise 1.13 Let A be a set of n real numbers.

(a) We define

$$A + AA = \{a + bc \ : \ a,b,c \in A\}.$$

Prove that $|A + AA| = \Omega(n^{3/2})$. (Hint: Consider the point set $A \times (A + AA)$.)

(b) We define

$$A(A + A) = \{a \cdot (b + c) \ : \ a,b,c \in A\}.$$

Prove that $|A(A + A)| = \Omega(n^{3/2})$.

(c) Find a set B of n real numbers that satisfies $|B + BB| = O(n^2)$. Find a set C of n real numbers that satisfies $|C(C + C)| = O(n^2)$.

Exercise 1.14 Let $\mathcal{P}$ be a set of m points and let $\mathcal{L}$ be a set of n lines, both in $\mathbb{R}^2$. The lines of $\mathcal{L}$ are not necessarily distinct, but the same line may appear in $\mathcal{L}$ at most k times. Prove that $I(\mathcal{P},\mathcal{L}) = O(m^{2/3}n^{2/3}k^{1/3} + mk + n)$. (Hint: Use dyadic decomposition on the number of times a line repeats.)

Exercise 1.15   Change the proof of Theorem 1.14, as follows. Instead of applying Theorem 1.1, find a way to apply Lemma 1.16.

Exercise 1.16   Let A be a set of m real numbers. Prove that the number of collinear triples of points in the lattice $A \times A \subset \mathbb{R}^2$ is $O(m^4 \log m)$. (Hint: Use rich lines and dyadic decomposition.)

Exercise 1.17   In the proof of Lemma 1.2, we showed that $I(\mathcal{P}, \mathcal{L}) = O(m\sqrt{n}+n)$. Find a one-line argument for proving that $I(\mathcal{P}, \mathcal{L}) = O(n\sqrt{m}+m)$.

Exercise 1.18 (Beck, 1983)   Let $\mathcal{P}$ be a set of m points in $\mathbb{R}^2$. Let $\mathcal{L}$ denote the set of 2-rich lines with respect to $\mathcal{P}$. Prove that either there exists a line containing $\Omega(m)$ points of $\mathcal{P}$ or $|\mathcal{L}| = \Omega(m^2)$.

To prove the claim, use dyadic composition: Set $\mathcal{L}_j = \{\ell \in \mathcal{L} : 2^j \le |\mathcal{P} \cap \ell| < 2^{j+1}\}$. There are $\Theta(m^2)$ pairs of points of $\mathcal{P}$, and each pair is contained in one line of $\mathcal{L}$. Prove that there exists a sufficiently large constant c, such that at most $m^2/100$ pairs are on the lines of $\bigcup_{j=\log c}^{\log(m/c)} \mathcal{L}_j$. Then consider the remaining pairs.

Exercise 1.19   The following theorem is from Hablicsek and Scherr (2016).

> **Theorem.**   *Consider a set $\mathcal{P} \subset \mathbb{R}^3$ of m points and an integer $r \ge 2$. If the number of r-rich lines is $\Omega(m^2/r^4 + m/r)$ then there exists a plane containing $\Omega(m/r^2)$ points of $\mathcal{P}$.*

Rely on this theorem to prove the following corollary, while also finding what the question marks should be replaced with.

> **Corollary.**   *Let $\mathcal{L}$ be a set of n lines and let $\mathcal{P}$ be a set of m points, both in $\mathbb{R}^3$, such that every plane contains $O(???)$ points of $\mathcal{P}$. Then $I(\mathcal{P}, \mathcal{L}) = O\left(m^{1/2}n^{3/4} + m\log m + n\right)$.*

Hint: Use dyadic decomposition. Separately handle lines that are very rich and lines that are not so rich (as in the proof of Lemma 1.15).

Exercise 1.20 (Farber et al., 2014)   A matrix is said to be *totally positive* if all its minors are positive. Let M be an $n \times 2$ totally positive matrix. Prove that the number of 2×2 minors of M that are equal to 1 is $O(n^{4/3})$.

1.12 Open Problems

We mentioned several main open problems throughout the chapter: the unit distances problem, multiple distinct distances problems, the maximum number of unit area triangles, and the sum-product problem. We now study another difficult open problem: a structural Szemerédi–Trotter result.

In extremal combinatorics problems, we first work towards obtaining an extremal tight bound: finding the maximum number of edges in a graph that does not contain a specific subgraph H, finding the minimum possible size of $A + A$, and so on. Once the extremal bound is found, the problem changes to characterizing the configurations that achieve this bound. This second step is often called *the structural problem*, since we are looking for the structure of the optimal configurations. For example, we may wish to characterize all graphs that do not contain the subgraph H and have the maximum possible number of edges. In additive combinatorics, one main structural problem is to characterize all sets A that satisfy $|A + A| = O(|A|)$.

For discrete geometry problems such as incidence and distance problems, the structural problems tend to be unusually difficult. Hardly any of the structural problems have been solved. Moreover, hardly anything nontrivial is known for many of these structural problems. One celebrated exception is a structural result of Green and Tao (2013) for the ordinary lines problem.

So far, we have focused mainly on point-line incidences in $\mathbb{R}^2$, one of the few incidence problems for which a tight bound is known. While this tight bound has been known for several decades, the structural problem remains wide open. In this problem, we wish to characterize the configurations of m points and n lines that have $\Theta(m^{2/3}n^{2/3})$ incidences.

In Claim 1.3, we saw Elekes's construction, which also has $\Theta(m^{2/3}n^{2/3})$ incidences. Both Elekes's construction and Erdős's earlier construction consider a point set that is a rectangular section of the integer lattice $\mathbb{Z}^2$. We can obtain somewhat different point sets by applying projective transformations to these constructions. Using point-line duality we get a point set that is not in a section of the integer lattice, but then the slopes and y-intercepts of the lines form a lattice. We can also add additional random points that are not incident to any lines.

Conjecture 1.17 *Consider sufficiently large positive integers m and n that satisfy $m = O(n^2)$ and $m = \Omega(\sqrt{n})$. Let $\mathcal{P}$ be a set of m points and $\mathcal{L}$ be a set of n lines, both in $\mathbb{R}^2$, such that $I(\mathcal{P}, \mathcal{L}) = \Theta(m^{2/3}n^{2/3})$. Then there exists a subset $\mathcal{P}' \subset \mathcal{P}$ such that $|\mathcal{P}'| = \Theta(m)$ and $\mathcal{P}'$ is contained in a section of the integer lattice of size $\Theta(m)$, possibly after applying a projective transformation or point-line duality.*

In Conjecture 1.17, we ask $\mathcal{P}'$ to be contained in a section of the integer lattice. There exist configurations with $\Theta(m^{2/3}n^{2/3})$ incidences where the largest lattice contained in $\mathcal{P}$ is of size $O(\log m) \times O(\log m)$. Thus, we may not ask for $\mathcal{P}'$ to contain a large section of a lattice.

The following is the only significant structural result that we currently know. It was obtained by Solymosi (2006).

Theorem 1.18   *For every constant integer k, the following holds for every sufficiently large n. Let $\mathcal{P}$ be a set of n points and let $\mathcal{L}$ be a set of n lines, both in $\mathbb{R}^2$, such that $I(\mathcal{P}, \mathcal{L}) = \Theta(n^{4/3})$. Then there exists a set of k of the points, no three on a line, such that there is a line passing through each of the $\binom{k}{2}$ point pairs.*

2

Basic Real Algebraic Geometry in $\mathbb{R}^2$

Everyone knows what a curve is, until he has studied enough mathematics to become confused through the countless number of possible exceptions.

Attributed to Felix Klein (Boyer, 1949).

After seeing some basics of incidence theory, we wish to discuss how polynomial methods are used to study incidences. For that, we first need a basic introduction to algebraic geometry over the reals. In this chapter we focus mainly on $\mathbb{R}^2$, postponing the treatment of $\mathbb{R}^d$ to Chapter 4. This allows us to discuss several planar results in Chapter 3, before dealing with more involved algebraic geometry.

2.1 Varieties

Algebraic geometry can be thought of as the study of geometries that arise from algebra (more specifically, from polynomials). The *polynomial ring* $\mathbb{R}[x_1, \ldots, x_d]$ is the set of polynomials in the variables $x_1, \ldots, x_d$ and with coefficients in $\mathbb{R}$. Given a (possibly infinite) set of polynomials $f_1, \ldots, f_k \in \mathbb{R}[x_1, \ldots, x_d]$, the *affine variety* $\mathbf{V}(f_1, \ldots, f_k)$ is defined as

$$\mathbf{V}(f_1, \ldots, f_k) = \left\{ (a_1, \ldots, a_d) \in \mathbb{R}^d \ : \ f_j(a_1, \ldots, a_d) = 0 \text{ for all } 1 \leq j \leq k \right\}.$$

The adjective "affine" distinguishes these varieties from projective varieties. At this point we only consider affine varieties, and for brevity we refer to those simply as varieties.[1] For example, some varieties in $\mathbb{R}^3$ are a torus, the union of a circle and a line, and a set of 1,000 points.

The following is a special case of *Hilbert's basis theorem* (for example, see Cox et al., 2013, Section 2.5).

Theorem 2.1 *Every variety can be described by a finite set of polynomials.*

---

[1] Some authors call these objects *algebraic sets*, while using the term "varieties" to denote the objects that we call irreducible varieties.

25

Theorem 2.1 is valid in every field. When working over the reals, we can make a stronger claim.

Corollary 2.2 *Every variety in* $\mathbb{R}^d$ *can be described by a single polynomial.*

Proof Consider a variety $U \subset \mathbb{R}^d$. By Theorem 2.1, there exist $f_1, \ldots, f_k \in \mathbb{R}[x_1, \ldots, x_d]$ such that $U = \mathbf{V}(f_1, \ldots, f_k)$. We set $f = f_1^2 + f_2^2 + \cdots + f_k^2$. For any point $p \in \mathbb{R}^d$ we have that $f(p) = 0$ if and only if $f_1(p) = \cdots = f_k(p) = 0$. Thus, $U = \mathbf{V}(f)$. □

Let us consider some basic properties of varieties.

Claim 2.3 *Let* $U, W \subset \mathbb{R}^d$ *be two varieties. Let* $\tau \colon \mathbb{R}^d \to \mathbb{R}^d$ *be an invertible linear map (such as a translation, rotation, reflection, or stretching). Then*
(a) $U \cap W$ *is a variety.*
(b) $U \cup W$ *is a variety.*
(c) $\tau(U)$ *is a variety.*

Proof Since U and W are varieties, there exist $f_1, \ldots, f_k, g_1, \ldots, g_m \in \mathbb{R}[x_1, \ldots, x_d]$ such that $U = \mathbf{V}(f_1, \ldots, f_k)$ and $W = \mathbf{V}(g_1, \ldots, g_m)$.[2] For (a), we note that

$$U \cap W = \mathbf{V}(f_1, \ldots, f_k, g_1, \ldots, g_m).$$

For (b), we note that $U \cup W$ is the variety defined by the set of polynomials

$$\bigcup_{\substack{1 \le i \le k \\ 1 \le j \le m}} \{f_i \cdot g_j\}.$$

For (c), we write the inverse of τ as $\psi \colon \mathbb{R}^d \to \mathbb{R}^d$. Note that ψ consists of d linear polynomials. We have that

$$\tau(U) = \mathbf{V}(f_1 \circ \psi, \ldots, f_d \circ \psi). \qquad \square$$

At this point it is helpful to ask what subsets of $\mathbb{R}^d$ are not varieties.

Claim 2.4 *The set* $X = \{(x, x) : x \in \mathbb{R}, \ x \ne 1\} \subset \mathbb{R}^2$ *is not a variety.*

Proof Note that X is the line defined by $y = x$, with the point $(1, 1)$ missing. Assume for contradiction that there exist $f_1, \ldots, f_k \in \mathbb{R}[x, y]$ such that $X = \mathbf{V}(f_1, \ldots, f_k)$. For every $1 \le j \le k$, we set $g_j(t) = f_j(t, t)$ and note that $g_j \in \mathbb{R}[t]$. Recall that 0 is the only univariate polynomial that evaluates to zero for infinitely many values. Since $g_j(t)$ vanishes on every $t \ne 1$, we have that $g_j(t) = 0$. This in turn implies that $f_j(1, 1) = 0$. Since this holds for every

---

[2] By Corollary 2.2, it suffices to use a single polynomial for each variety. We present this slightly less elegant proof since it holds over every field.

$1 \le j \le k$, we get that $(1, 1) \in X$. Since this is a contradiction to $(1, 1) \notin X$, we conclude that X is not a variety. $\square$

By imitating the proof of Claim 2.4, we can create many other sets that are not varieties: line segments, half circles, a sphere in $\mathbb{R}^3$ with one circle removed from it, and so on. See also Exercise 2.1.

A set $U' \subset \mathbb{R}^d$ is a *subvariety* of a variety U if $U' \subseteq U$ and U' is a variety. A set U' is a *proper subvariety* of U if U' is a nonempty subvariety of U' and $U' \ne U$. A variety U is *reducible* if there exist two proper subvarieties $U', U'' \subset U$ such that $U = U' \cup U''$. Otherwise, U is *irreducible*. For example, the union of the two axes $\mathbf{V}(xy) \subset \mathbb{R}^2$ is reducible since $\mathbf{V}(xy) = \mathbf{V}(x) \bigcup \mathbf{V}(y)$. A single line is irreducible since it is not the union of two proper subvarieties.

Every variety $U \subset \mathbb{R}^d$ can be decomposed into distinct irreducible subvarieties $U_1, U_2, \ldots, U_k$ such that $U = \bigcup_{j=1}^{k} U_j$. After removing every U_j that is a proper subvariety of another $U_{j'}$, we obtain a unique decomposition of U. The subvarieties of this decomposition are the *irreducible components of U* (or *components*, for brevity).

2.2 Curves in $\mathbb{R}^2$

In Chapter 4, we study in detail degrees, dimensions, singular points, and other properties of varieties in $\mathbb{R}^d$. Here, we only consider the case of $\mathbb{R}^2$, where these concepts are significantly simpler.

We say that an irreducible variety in $\mathbb{R}^2$ is a *curve* if it is not a single point, the empty set, or all of $\mathbb{R}^2$ (note that a set of several points is reducible). A reducible variety in $\mathbb{R}^2$ is a curve if each of its components is a curve. This definition corresponds to what we would intuitively call a polynomial curve.

An irreducible variety in $\mathbb{R}^2$ is of *dimension* one if it is a curve. The empty variety is of dimension -1, a single point is of dimension zero, and the entire plane $\mathbb{R}^2$ is of dimension two. The dimension of a reducible variety $U \subset \mathbb{R}^2$ is the maximum dimension of a component of U.

Degrees and intersections: We say that the *degree* of a curve $\gamma \subset \mathbb{R}^2$ is the minimum integer k such that there exists a polynomial $f \in \mathbb{R}[x, y]$ of degree k with $\mathbf{V}(f) = \gamma$.

We now present a result that is constantly applied throughout the book. We refer to it as *Bézout's theorem*, although it is only a special case of that theorem. For a proof, see Gibson (1998, Section 14.4).

Theorem 2.5 (Bézout's theorem) *Let f and g be two polynomials in $\mathbb{R}[x, y]$ of degrees k_f and k_g, respectively. If f and g do not have common factors, then $\mathbf{V}(f) \cap \mathbf{V}(g)$ consists of at most $k_f \cdot k_g$ points.*

For our purposes, we can rephrase Theorem 2.5: Consider two curves of degrees k_1 and k_2. If the curves have a common component, then their intersection is infinite. Otherwise, the intersection of the two curves consists of at most $k_1 \cdot k_2$ points. As simple examples, note that two lines indeed intersect in at most one point and that two ellipses (which are of degree 2) intersect in at most four points.

Irreducible components: Let us prove a few basic properties of irreducible components of curves. While these properties may seem obvious, we will see other properties of varieties that seem obvious but are false.

Lemma 2.6

(a) Let $\gamma \subset \mathbb{R}^2$ *be an irreducible curve of degree* d. *Then there exists an irreducible polynomial* $f \in \mathbb{R}[x, y]$ *of degree* d *that satisfies the following. For every* $g \in \mathbb{R}[x, y]$ *with* $\gamma \subseteq \mathbf{V}(g)$ *there exists* $h \in \mathbb{R}[x, y]$ *such that* $g = f \cdot h$.
(b) A curve of degree d' *contains at most* d' *irreducible components.*

Proof (a) By definition, there exists a degree d polynomial that defines γ. We arbitrarily set $f \in \mathbb{R}[x, y]$ to be one such polynomial. Assume for contradiction that there exist nonconstant $f_1, f_2 \in \mathbb{R}[x, y]$ such that $f = f_1 \cdot f_2$. Note that $\deg f_1 < d$ and $\deg f_2 < d$. We have that $\mathbf{V}(f_1) \subsetneq \mathbf{V}(f)$ and $\mathbf{V}(f_2) \subsetneq \mathbf{V}(f)$, since otherwise $\deg \gamma < d$. For the same reason, we have that $\mathbf{V}(f_1) \neq \emptyset$ and $\mathbf{V}(f_2) \neq \emptyset$. Since $\gamma = \mathbf{V}(f) = \mathbf{V}(f_1) \cup \mathbf{V}(f_2)$, we conclude that γ is reducible. This contradiction implies that f is irreducible.

Consider $g \in \mathbb{R}[x, y]$ with $\gamma \subseteq \mathbf{V}(g)$. Since both $\mathbf{V}(f)$ and $\mathbf{V}(g)$ have the component γ, Bézout's theorem (Theorem 2.5) implies that f and g have a common factor. Since f is irreducible, we obtain that f is a component of g. In other words, there exists $h \in \mathbb{R}[x, y]$ such that $g = f \cdot h$.

(b) Let $U \subset \mathbb{R}^2$ be a curve of degree d', and let $\gamma \subset U$ be an irreducible curve. Let $f \in \mathbb{R}[x, y]$ be as in part (a) of this lemma, with respect to γ. Consider a degree d' polynomial $f' \in \mathbb{R}[x, y]$ that satisfies $\mathbf{V}(f') = U$. By part (a) of this lemma, f is a factor of f'. Since f' has at most d' irreducible factors, we conclude that U consists of at most d' components. $\square$

Lemma 2.6(b) is stated only for curves since it is false for arbitrary varieties in $\mathbb{R}^2$. For an integer $n > 2$, consider the polynomial

$$f(x, y) = \prod_{j=1}^{n} (x - j)^2 + \prod_{k=1}^{n} (y - k)^2. \qquad (2.1)$$

Note that $\deg f = 2n$. Because of the squares, $f(x, y) = 0$ if and only if both products equal to zero. This implies that $\mathbf{V}(f) = \{1, 2, \ldots, n\}^2$. Thus, $\mathbf{V}(f)$ is a variety of dimension zero and with n^2 components. Unlike Lemma 2.6(b), the number of components is significantly larger than the degree of f.

Singular points: Consider a curve $\gamma \subset \mathbb{R}^2$ and a point $p \in \gamma$. Intuitively (and with exceptions), p is a *singular point* of γ if one of the following holds:

- The tangent line to γ at p is not well defined. For example, see Figure 2.1.

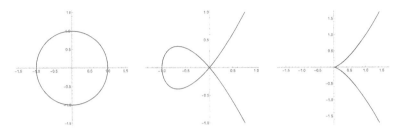

Figure 2.1 Every point of the circle has a well-defined tangent line. In the other two curves the tangent is not well defined at the origin. The curve in the middle is $\mathbf{V}(y^2 - x^3 - x^2)$, and the curve on the right is $\mathbf{V}(x^3 - y^2)$.

- The point p is contained in more than one irreducible component of γ. For example, consider the union of two circles that intersect at a point p and have the same tangent line at p. While one might say that the tangent line is well defined at p, it is a singular point of γ.
- The point p is an *isolated point* of γ. That is, there exists a disk[3] centered at p that contains no other point of γ.

Some readers may complain that the third bullet seems redundant. We defined a curve as not having components that are a single point. Doesn't this mean that a curve cannot contain isolated points? Unfortunately not. For example, consider the cubic curve $\mathbf{V}(y^2 - x^3 + x^2)$, depicted in Figure 2.2. This curve is an irreducible variety that includes an isolated point at the origin. After removing

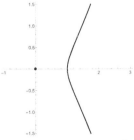

Figure 2.2 The irreducible curve $\mathbf{V}(y^2 - x^3 + x^2)$.

---

[3] A *disk* is the set of points that satisfy $(x - a)^2 + (y - b)^2 \leq r^2$, for some $a, b, r \in \mathbb{R}$.

this point, the set is no longer a variety. By the third bullet, the origin is a singular point of the curve $\mathbf{V}(y^2 - x^3 + x^2)$.

We now provide a rigorous definition of a singular point of a curve $\gamma \subset \mathbb{R}^2$. Given a polynomial $f \in \mathbb{R}[x, y]$, the *gradient* of f is

$$\nabla f = \left(\frac{\partial f}{\partial x}, \frac{\partial f}{\partial y}\right).$$

Consider a minimum-degree polynomial $f \in \mathbb{R}[x, y]$ such that $\mathbf{V}(f) = \gamma$. Then $p \in \gamma$ is a singular point of γ if and only if $\nabla f(p) = (0, 0)$. We denote the set of singular points of γ as γ_{sing}. A point of γ that is not singular is a *regular* point of γ.

To see why f must be of a minimum degree, let $\gamma \subset \mathbb{R}^2$ be the x-axis. When writing $\gamma = \mathbf{V}(f)$ with $f(x, y) = y$, we have that $\nabla f = (0, 1)$. Since this gradient is never zero, the x-axis has no singular points. However, when writing $\gamma = \mathbf{V}(g)$ with $g(x, y) = y^2$, we have that $\nabla g = (0, 2y)$. This gradient would imply that the origin is a singular point, contradicting the above. Note that using the minimum degree polynomial $f(x, y) = y$ led to a result that fits the above intuition.

Our first intuitive definition of a singular point is not accurate. The three cases in that definition always lead to singular points, but there are singular points that do not fit any of these cases. As an example, consider the polynomial $f = y^3 + 2x^2y - x^4$. The curve $\gamma = \mathbf{V}(f)$ is depicted in Figure 2.3. Note that γ is an irreducible variety that does not intersect itself and has a well-defined tangent line at every point. However, $\nabla f = (4xy - 4x^3, 3y^2 + 2x^2)$, which implies that $\nabla f(0, 0) = (0, 0)$. Thus, the origin is a singular point of γ. (For a discussion of this phenomenon, see for example Bochnak et al., 2013, Section 3.3.)

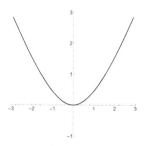

Figure 2.3 The variety $\mathbf{V}(y^3 + 2x^2y - x^4) \subset \mathbb{R}^2$.

A polynomial $f \in \mathbb{R}[x_1, \ldots, x_d]$ is *square-free* if the factorization of f into irreducible factors does not contain any factor more than once. Let $f \in \mathbb{R}[x_1, \ldots, x_d]$ be a square-free polynomial, and let g be an irreducible factor of

f that depends on x_j. Then $\frac{\partial f}{\partial x_j}$ is not divisible by g. Indeed, write $f = g \cdot h$ for some $h \in \mathbb{R}[x_1, \ldots, x_d]$ that does not have g as a factor. We have that

$$\frac{\partial f}{\partial x_j} = \frac{\partial g}{\partial x_j} \cdot h + g \cdot \frac{\partial h}{\partial x_j}.$$

Since the second summand is divisible by g but the first summand is not, this expression does not have g as a factor.

We rely on square-free polynomials to prove that a curve cannot have too many singular points.

Theorem 2.7 *Let $\gamma \subset \mathbb{R}^2$ be an irreducible curve of degree k. Then γ_{sing} is a set of at most $k(k-1)$ points.*

Proof Consider a minimum-degree polynomial $f \in \mathbb{R}[x, y]$ such that $\mathbf{V}(f) = \gamma$. Since removing repeated factors from f does not affect $\mathbf{V}(f)$, the polynomial f is square-free. Since γ is irreducible, f is also irreducible. Without loss of generality, we assume that f contains the variable x.

By the definition of a singular point, both $f_x = \frac{\partial f}{\partial x}$ and f vanish on every singular point of γ. Since f is square-free, it has no common components with f_x. By Bézout's theorem (Theorem 2.5), $\mathbf{V}(f) \cap \mathbf{V}(f_x)$ consists of at most $k(k-1)$ points. Thus, γ has at most $k(k-1)$ singular points. $\square$

Connected components: Let $U \subset \mathbb{R}^2$ be a variety. A *connected component* of U is a maximal subset $C \subset \gamma$ that satisfies the following: For any $p, q \in C$, there is a path between p and q that consists only of points from C. Make sure not to confuse connected components with irreducible components. A hyperbola has one irreducible component and two connected components. The union of the axes $\mathbf{V}(xy)$ has two irreducible components and one connected component.

A connected component of C is *bounded* if there exists a disk that contains C. A connected component that is not bounded is *unbounded*. Both hyperbolas and $\mathbf{V}(xy)$ are unbounded.

Theorem 2.8 (Harnack's curve theorem) *Let $f \in \mathbb{R}[x, y]$ be a polynomial of degree k. Then the number of connected components of $\mathbf{V}(f)$ is $O(k^2)$.*

The exact bound of Harnack's theorem is $1 + \binom{k-1}{2}$. The following proof leads to a slightly worse bound.

Proof A rotation of $\mathbb{R}^2$ around the origin does not change the number of connected components of $\mathbf{V}(f)$ and does not increase the degree of f. We may thus assume that $\mathbf{V}(f)$ does not contain any vertical lines. This implies that every factor of f includes x. We may also assume that f is square-free, since removing repeated factors does not change $\mathbf{V}(f)$.

Let C be a bounded connected component of $\mathbf{V}(f)$, and let p be a point of C with a maximal x value. Set $f_y = \frac{\partial f}{\partial y}$, and note that $f_y(p) = 0$. Indeed, if $f_y(p) \neq 0$ then $\mathbf{V}(f)$ has a well-defined non-vertical tangent line at p. This would contradict p having a maximal x-coordinate. A symmetric argument holds for a point of C with a minimal x-coordinate. Since f is square-free, it has no common components with f_y. Bézout's theorem (Theorem 2.5) implies that $\mathbf{V}(f) \cap \mathbf{V}(f_y)$ consists of at most $k(k-1)$ points. Every bounded connected component contains at least two of those points: a point with a minimal x-coordinate and a point with a maximal x-coordinate. We conclude that the number of bounded connected components of $\mathbf{V}(f)$ is at most $k(k-1)/2$.

For $c \in \mathbb{R}$, consider the four lines $\mathbf{V}(x-c), \mathbf{V}(x+c), \mathbf{V}(y-c)$, and $\mathbf{V}(y+c)$. By taking c to be sufficiently large, we may assume that no bounded component of $\mathbf{V}(f)$ intersects these lines. We may also assume that every unbounded component intersects at least one of the lines. By slightly changing the value of c, we may assume that every point on these four lines is contained in at most one connected component of $\mathbf{V}(f)$. By Bézout's theorem, $\mathbf{V}(f)$ intersects a line in at most k points. Since every unbounded component of $\mathbf{V}(f)$ contains at least one such intersection point, there are at most $4k$ unbounded connected components. $\qquad\square$

2.3 An Application: Pascal's Theorem

Several mathematical results from the nineteenth century could be considered as precursors of the current polynomial methods. We now consider one such result as another application of Bézout's theorem. The result was originally proved by Pascal in 1640. The polynomial proof was discovered in 1847 by Plücker. For points $a, b \in \mathbb{R}^2$, we denote the line incident to a and b as $\overline{ab}$.

Theorem 2.9 (Pascal's theorem) *Let* $\gamma \subset \mathbb{R}^2$ *be an ellipse, parabola, or hyperbola. Consider six distinct points* $a, b, c, d, e, f \in \gamma$. *Then the three points* $\overline{ab} \cap \overline{de}$, $\overline{cd} \cap \overline{af}$, *and* $\overline{ef} \cap \overline{bc}$ *are collinear.*

Figure 2.4 is an example of Pascal's theorem.

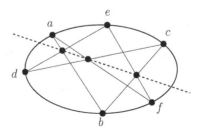

Figure 2.4 An example of Pascal's theorem.

Proof Note that γ is an irreducible curve of degree 2. Let $h \in \mathbb{R}[x, y]$ be a degree 2 polynomial that satisfies $\gamma = \mathbf{V}(h)$. We define two more curves:

$$L_1 = \overline{ab} \cup \overline{cd} \cup \overline{ef},$$
$$L_2 = \overline{de} \cup \overline{af} \cup \overline{bc}.$$

Both L_1 and L_2 are curves of degree 3, since each can be defined as the product of three linear equations. Let $g_1, g_2 \in \mathbb{R}[x, y]$ be degree 3 polynomials that satisfy $L_1 = \mathbf{V}(g_1)$ and $L_2 = \mathbf{V}(g_2)$. Let p be an arbitrary point of γ that is not one of a, b, c, d, e, f. There exists $\alpha \in \mathbb{R}$ such that the polynomial $g = g_1 + \alpha g_2$ vanishes on p. Indeed, one can think of $g(p)$ as a linear polynomial in α and set α to be the unique root of this polynomial. We note that g is of degree 3 and vanishes on a, b, c, d, e, f, p.

By Bézout's theorem, either g and h have a common factor or

$$|\mathbf{V}(g) \cap \mathbf{V}(h)| \le \deg g \cdot \deg h = 3 \cdot 2 = 6.$$

Since both g and h vanish on the seven points a, b, c, d, e, f, p, we get that the two polynomials have a common factor. Since h is irreducible, h is a factor of g. This implies that $\mathbf{V}(g)$ is the union of γ and a line ℓ. Since γ cannot contain the three intersection points $\overline{ab} \cap \overline{de}$, $\overline{cd} \cap \overline{af}$, and $\overline{ef} \cap \overline{bc}$, these points are contained in ℓ. That is, the three intersection points are collinear. $\qquad \square$

2.4 Exercises

Exercise 2.1 For each of the following sets, prove that it is not a variety:
(a) The sine wave $\{(x, \sin x) : x \in \mathbb{R}\}$ (hint: consider lines that intersect this set).
(b) The disk $\left\{(a, b) \in \mathbb{R}^2 : \sqrt{a^2 + b^2} \le 1\right\}$.
(c) A point set in $\mathbb{R}^2$ whose cardinality is countably infinite.

Exercise 2.2 Prove Bézout's theorem (Theorem 2.5) when $k_g = 1$. (Hint: Work inside the line $\mathbf{V}(g)$.)

Exercise 2.3 Consider a polynomial $f \in \mathbb{R}[x, y]$ of degree d. Let U be a variety obtained by removing one irreducible component from $\mathbf{V}(f)$. We now explore a claim that may seem obvious but is false: The variety U' is defined by a polynomial of degree at most d.

Let $f \in \mathbb{R}[x, y]$ be as defined in Equation (2.1). Recall that $\deg f = 2n$ and that $\mathbf{V}(f)$ consists of n^2 points. Let U be the variety obtained by removing one point from $\mathbf{V}(f)$. Find a minimal degree polynomial $g \in \mathbb{R}[x, y]$ that satisfies $U = \mathbf{V}(g)$.

Exercise 2.4   Let $\gamma \subset \mathbb{R}^2$ be a curve of degree k. Prove that the number of singular points of γ is $O(k^2)$ (Theorem 2.7 applies only to the case where γ is irreducible).

Exercise 2.5   Prove or provide a counterexample:
(a) If $U, V \subset \mathbb{R}^2$ are varieties then the Cartesian product $U \times V \subset \mathbb{R}^4$ is a variety.
(b) If $U \subset \mathbb{R}^3$ is a variety then the projection of U onto the xy-plane is a variety (that is, the set $\{(x, y) \in \mathbb{R}^2 : (x, y, z) \in U$ for some $z \in \mathbb{R}\}$).
(c) The set of regular points of every curve is a variety.

Exercise 2.6   Let $\mathcal{P}$ be a set of n points in $\mathbb{R}^2$ and let $f \in \mathbb{R}[x, y]$ satisfy $\mathbf{V}(f) = \mathcal{P}$. Prove that $\deg f = \Omega(n^{1/2})$.

3

Polynomial Partitioning

In this chapter we meet our first polynomial method. In particular, we study the polynomial partitioning theorem and use this theorem to obtain an incidence result for curves in $\mathbb{R}^2$.

3.1 The Polynomial Partitioning Theorem

Consider a set $\mathcal{P}$ of m points in $\mathbb{R}^d$ and a real number $r > 1$. A polynomial $f \in \mathbb{R}[x_1, \ldots, x_d]$ is an *r-partitioning polynomial* for $\mathcal{P}$ if every connected component of $\mathbb{R}^d \backslash \mathbf{V}(f)$ contains at most m/r^d points of $\mathcal{P}$.[1] There is no restriction on the number of points of $\mathcal{P}$ that lie in $\mathbf{V}(f)$. Figure 3.1 depicts a 2-partitioning polynomial for a set of 12 points in $\mathbb{R}^2$. Each cell contains at most $12/2^2 = 3$ points.

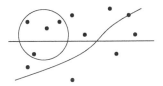

Figure 3.1 A 2-partitioning polynomial for a set of 12 points in $\mathbb{R}^2$.

The following result is due to Guth and Katz (2015).

Theorem 3.1 (Polynomial partitioning) *Let $\mathcal{P}$ be a set of m points in $\mathbb{R}^d$ and let $1 < r \leq m^{1/d}$. Then there exists an r-partitioning polynomial $f \in \mathbb{R}[x_1, \ldots, x_d]$ of degree $O(r)$.*

---

[1] Currently, there is no standard definition for an r-partitioning polynomial. Some authors ask each cell to contain at most m/r points, and others use the notation $1/r$-partitioning polynomial. We chose the definition that we believe is the easiest to work with.

In some cases, no cells of the polynomial partition contain points of $\mathcal{P}$. That is, all the points are contained in the partition $\mathbf{V}(f)$. For example, consider a set $\mathcal{P}$ of m points on the x-axis in $\mathbb{R}^2$. Let f be an r-partitioning polynomial for $\mathcal{P}$, with r asymptotically smaller than $m^{1/2}$. If $\mathbf{V}(f)$ does not contain the x-axis, then Bézout's theorem (Theorem 2.5) implies that $\mathbf{V}(f)$ intersects the x-axis in $O(r)$ points. See Figure 3.2. This in turn implies that the cells of the partition contain $m - O(r) = \Theta(m)$ points of $\mathcal{P}$. When traveling along the x-axis, we enter a new cell only after visiting one of the $O(r)$ intersection points. Thus, $\Theta(n)$ points of $\mathcal{P}$ are contained in $O(r)$ cells. This contradicts each cell containing at most m/r^2 points of $\mathcal{P}$. We conclude that $\mathbf{V}(f)$ contains the x-axis and all of $\mathcal{P}$.

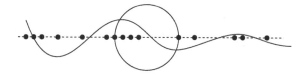

Figure 3.2 (Dashed line) The x-axis. (Solid line) The polynomial partition.

To estimate the number of cells in a polynomial partition, we rely on the following result of Warren (1968).

Theorem 3.2 (Warren) *For a polynomial $f \in \mathbb{R}[x_1, \ldots, x_d]$ of degree k, the number of connected components of $\mathbb{R}^d \backslash \mathbf{V}(f)$ is $O(k^d)$.*

By Theorem 3.2, a polynomial of degree $O(r)$ leads to $O(r^d)$ cells. Since each cell contains at most m/r^d points of $\mathcal{P}$, we need $\Omega(r^d)$ cells for a set of m points. Thus, the bound of Theorem 3.1 is asymptotically tight. (In this argument we ignore the case where most points are contained in the zero set of the partitioning polynomial. The probability of this case happening for a random point set is zero.)

To study incidences with curves in $\mathbb{R}^2$, we require the special case of Theorem 3.1 where $d = 2$. In Section 3.3 we prove this theorem for arbitrary d, since the proof easily extends to this more general case. In particular, the proof does not require any properties of varieties in $\mathbb{R}^d$ that are presented in the following chapters.

3.2 Incidences with Curves in $\mathbb{R}^2$

Consider a point set $\mathcal{P}$ and a set of curves Γ, both in $\mathbb{R}^2$. The *incidence graph* of $\mathcal{P} \times \Gamma$ is a bipartite graph $G = (V_1 \cup V_2, E)$. The vertices of V_1 correspond to the points of $\mathcal{P}$ and the vertices of V_2 correspond to the curves of Γ. An edge $(v_j, u_k) \in V_1 \times V_2$ is in E if the point that corresponds to v_j is incident to the curve that corresponds to u_k. We can think of E as the set of incidences in $\mathcal{P} \times \Gamma$. An example is depicted in Figure 3.3.

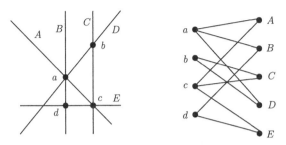

Figure 3.3 (Left) A point-line configuration and (Right) its incidence graph.

Recall that $K_{s,t}$ is a complete bipartite graph with s vertices on one side and t vertices on the other. Our goal in this section is to prove the following theorem (variants of this result originally appeared in Clarkson et al., 1990; Pach and Sharir, 1998).

Theorem 3.3 *Let* $\mathcal{P}$ *be a set of m points and let* Γ *be a set of n irreducible curves of degree at most k, both in* $\mathbb{R}^2$. *If the incidence graph of* $\mathcal{P} \times \Gamma$ *contains no copy of* $K_{s,t}$, *then*

$$I(\mathcal{P}, \Gamma) = O_{s,t,k}\left(m^{\frac{s}{2s-1}} n^{\frac{2s-2}{2s-1}} + m + n\right).$$

In Theorem 3.3, curves are defined as in Section 2.2. We also implicitly assume that no two points are identical and that no two curves are identical. We may remove from Theorem 3.3 the irreducibility and distinctness assumptions. We may also allow the curves to contain components that are single points. We include these restrictions only to simplify our analysis (see also Exercise 3.7).

To demonstrate the strength of Theorem 3.3, we consider several common types of curves:

- **The case where Γ is a set of lines.** Since two points are contained in exactly one line, the incidence graph contains no $K_{2,2}$. Applying Theorem 3.3 with $s = 2$ leads to the bound $O(m^{2/3}n^{2/3} + m + n)$. That is, Theorem 3.3 generalizes the Szemerédi–Trotter theorem (Theorem 1.1).
- **The case where Γ is a set of unit circles.** Two points may have two unit circles passing through both, but not three unit circles. See Figure 3.4(a). Thus, the incidence graph contains no $K_{2,3}$. Since $s = 2$, we again obtain the bound $O(m^{2/3}n^{2/3} + m + n)$. That is, Theorem 3.3 generalizes the current best bound for the unit distances problem (Theorem 1.6).

- **The case where Γ is a set of arbitrary circles.** In this case, the incidence graph may contain $K_{2,t}$ for arbitrarily large t. See Figure 3.4(b). However, the incidence graph does not contain $K_{3,2}$. Applying Theorem 3.3 with $s = 3$ leads to the bound $I(\mathcal{P}, \Gamma) = O(m^{3/5}n^{4/5} + m + n)$.

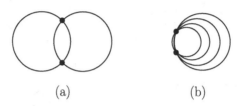

(a)   (b)

Figure 3.4 (a) At most, two unit circles contain the same two points. (b) Infinitely many arbitrary circles can contain the same two points.

In the above examples, did we get the same bound for no $K_{2,2}$ and for no $K_{2,3}$? Asymptotically, yes. However, the constant hidden by the $O(\cdot)$-notation increases with t. We know that Theorem 3.3 is not tight in some cases, such as when Γ is a set of arbitrary circles. For more about the current best bounds and common conjectures, see Section 3.6.

Before proving Theorem 3.3, we study a purely combinatorial proof that leads to a weaker incidence bound. This bound can be seen as a special case of the Kővari–Sós–Túran theorem from extremal graph theory (for example, see Matoušek, 2013, Section 4.5).

Lemma 3.4 *Let $\mathcal{P}$ be a set of m points and let Γ be a set of n curves, both in $\mathbb{R}^2$. If the incidence graph of $\mathcal{P} \times \Gamma$ contains no $K_{s,t}$, then*

$$I(\mathcal{P}, \Gamma) = O_{s,t}\left(mn^{1-\frac{1}{s}} + n\right).$$

Proof The current proof is a simple generalization of the proof of Lemma 1.2. Consider the set of $(s + 1)$-tuples, where we have denoted the curves of Γ as $\gamma_1, \ldots, \gamma_n$.

$$T = \{(a_1, \ldots, a_s, \gamma) \in \mathcal{P}^s \times \Gamma \; : \; a_1, \ldots, a_s \in \gamma\}.$$

We prove the lemma by double counting $|T|$. There are $\binom{m}{s}$ subsets of s points of $\mathcal{P}$. Since the incidence graph contains no $K_{s,t}$, every such subset is contained in at most $t - 1$ curves of Γ. This implies that

$$|T| \leq \binom{m}{s}(t - 1) = O_{s,t}(m^s). \tag{3.1}$$

For each $\gamma_j \in \Gamma$, we set $d_j = |\mathcal{P} \cap \gamma_j|$. Note that $I(\mathcal{P}, \Gamma) = \sum_{j=1}^{n} d_j$ and

$$|T| = \sum_{j=1}^{n} \binom{d_j}{s} = \Omega_s \left(\sum_{j=1}^{n} (d_j - s)^s \right). \tag{3.2}$$

By applying Hölder's inequality (Theorem A.3) with $a_j = d_j - s$, $b_j = 1$, and $p = s$, we get that

$$\sum_{j=1}^{n} (d_j - s) \leq \left(\sum_{j=1}^{n} (d_j - s)^s \right)^{1/s} \left(\sum_{j=1}^{n} 1 \right)^{(s-1)/s} = \left(\sum_{j=1}^{n} (d_j - s)^s \right)^{1/s} n^{(s-1)/s}.$$

Combining this with Equation (3.2) and with $I(\mathcal{P}, \Gamma) = \sum_{j=1}^{n} d_j$ leads to

$$|T| = \Omega \left(\sum_{j=1}^{n} (d_j - s)^s \right) = \Omega \left(\frac{\left(\sum_{j=1}^{n} (d_j - s) \right)^s}{n^{s-1}} \right) = \Omega \left(\frac{(I(\mathcal{P}, \Gamma) - sn)^s}{n^{s-1}} \right). \tag{3.3}$$

By combining Equations (3.1) and (3.3), we obtain

$$\frac{(I(\mathcal{P}, \Gamma) - sn)^s}{n^{s-1}} = O_{s,t} \left(m^s \right).$$

Rearranging leads to $I(\mathcal{P}, \Gamma) = O_{s,t} \left(mn^{(s-1)/s} + n \right)$.   □

In discrete geometry, some proofs first derive a weak bound and then amplify it. When studying crossing numbers, we first derived a weak bound in Lemma 1.4. We then amplified this weak bound in Lemma 1.5, by combining the bound with a probabilistic argument. To prove Theorem 3.3, we amplify the weak incidence bound of Lemma 3.4 by combining it with polynomial partitioning. We partition $\mathbb{R}^2$ into cells by using Theorem 3.1 and then apply the weak bound separately in each cell. That is, we go over the cells one by one, applying Lemma 3.4 only with the points of the current cell and with the curves that intersect this cell.

Why would applying the bound of Lemma 3.4 separately in each cell lead to a stronger bound? Intuitively, one may think of Lemma 3.4 as stating that an average point of $\mathcal{P}$ contributes $O\left(n^{(s-1)/s} \right)$ incidences (where n is the number of curves). For a point $p \in \mathcal{P}$, let n_p be the number of curves that intersect the cell containing p. When applying Lemma 3.4 separately in each cell, intuitively, a point p contributes $O\left(n_p^{(s-1)/s} \right)$ incidences. Since a curve cannot intersect many cells, we expect $\sum_p n_p^{(s-1)/s}$ to be significantly smaller than $mn^{(s-1)/s}$.

Proof of Theorem 3.3 By Theorem 3.1, there exists an r-partitioning polynomial $f \in \mathbb{R}[x, y]$ for $\mathcal{P}$ of degree $O(r)$. The value of r is determined below.

We may assume that f is a minimum-degree polynomial that defines $\mathbf{V}(f)$. In particular, this means that f is square-free.

Let c be the number of cells in the partition (that is, the number of connected components of $\mathbb{R}^2 \setminus \mathbf{V}(f)$). We set $\mathcal{P}_0 = \mathbf{V}(f) \cap \mathcal{P}$. In other words, $\mathcal{P}_0$ is the set of points of $\mathcal{P}$ that are not in any cell but rather on the partition. Let Γ_0 be the set of curves of Γ that are fully contained in $\mathbf{V}(f)$. For $1 \leq j \leq c$, let $\mathcal{P}_j$ denote the set of points that are contained in the jth cell and let Γ_j denote the set of curves of Γ that intersect the jth cell. Note that

$$I(\mathcal{P}, \Gamma) = I(\mathcal{P}_0, \Gamma_0) + I(\mathcal{P}_0, \Gamma \setminus \Gamma_0) + \sum_{j=1}^{c} I(\mathcal{P}_j, \Gamma_j).$$

We separately derive an upper bound for each of these three expressions.

Theorem 3.2 implies that $c = O(r^2)$. We set $m_j = |\mathcal{P}_j|$ and $n_j = |\Gamma_j|$. By definition, $m_j \leq m/r^2$ for every j. By applying Lemma 3.4 separately in each cell, we have that

$$\sum_{j=1}^{c} I(\mathcal{P}_j, \Gamma_j) = O_{s,t}\left(\sum_{j=1}^{c} \left(m_j n_j^{\frac{s-1}{s}} + n_j \right) \right) = O_{s,t}\left(\frac{m}{r^2} \sum_{j=1}^{c} n_j^{\frac{s-1}{s}} + \sum_{j=1}^{c} n_j \right).$$

$$(3.4)$$

We claim that every curve $\gamma \in \Gamma$ intersects $O_k(r)$ cells of the partition. When traveling along a connected component of γ, to enter a new cell of the partition we must first intersect $\mathbf{V}(f)$. By Bézout's theorem, the number of intersection points between a curve $\gamma \in \Gamma$ and $\mathbf{V}(f)$ is $O_k(r)$. By Harnack's curve theorem (Theorem 2.8), γ has $O_k(1)$ connected components. We wish to rely on these two bounds to claim that γ intersects $O_k(r)$ cells. However, the curve γ may split into several cells in an intersection between γ and $\mathbf{V}(f)$. See Figure 3.5. Consider an intersection point $p \in \gamma \cap \mathbf{V}(f)$ and let C_p be a circle centered at p and of a sufficiently small radius. By Bézout's theorem, γ and C_p intersect in at most $2k$ points. Thus, γ splits into at most $2k$ cells in an intersection point. This completes our claim that γ intersects $O_k(r)$ cells of the partition.

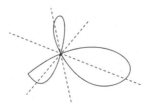

Figure 3.5 (Dashed lines) The partition. (Solid line) The curve splits into six cells at the intersection point.

Since any curve $\gamma \in \Gamma$ intersects $O_k(r)$ cells of the partition, we have that $\sum_{j=1}^{c} n_j = O_k(nr)$. Hölder's inequality with $p = s/(s-1)$ and $q = s$ implies that

$$\sum_{j=1}^{c} n_j^{\frac{s-1}{s}} \leq \left(\sum_{j=1}^{c} n_j \right)^{\frac{s-1}{s}} \left(\sum_{j=1}^{c} 1 \right)^{\frac{1}{s}} = O_k \left((nr)^{\frac{s-1}{s}} (r^2)^{\frac{1}{s}} \right) = O_k \left(n^{\frac{s-1}{s}} r^{\frac{s+1}{s}} \right).$$

Combining this with Equation (3.4) leads to

$$\sum_{j=1}^{c} I(\mathcal{P}_j, \Gamma_j) = O_{s,t,k} \left(\frac{m}{r^2} \sum_{j=1}^{c} n_j^{\frac{s-1}{s}} + \sum_{j=1}^{c} n_j \right) = O_{s,t,k} \left(\frac{mn^{\frac{s-1}{s}}}{r^{\frac{s-1}{s}}} + nr \right). \quad (3.5)$$

Next, consider a curve $\gamma \in \Gamma \backslash \Gamma_0$. Since the number of intersection points between γ and $\mathbf{V}(f)$ is $O_k(r)$, we get that

$$I(\mathcal{P}_0, \Gamma \backslash \Gamma_0) = O_k(nr). \quad (3.6)$$

It remains to derive an upper bound for $I(\mathcal{P}_0, \Gamma_0)$. Note that points that are zero-dimensional components of $\mathbf{V}(f)$ do not contribute to $I(\mathcal{P}_0, \Gamma_0)$. By Lemma 2.6(a), every one-dimensional component of $\mathbf{V}(f)$ corresponds to a factor of f. Let $f_1 \in \mathbb{R}[x, y]$ be the product of the factors of f that correspond to one-dimensional components. By definition, $\deg f_1 \leq \deg f = O(r)$. Since the curves of Γ are irreducible and distinct, each component of $\mathbf{V}(f)$ corresponds to at most one curve of Γ. Recall that a point that is contained in more than one component of $\mathbf{V}(f)$ is a singular point of $\mathbf{V}(f)$. Thus, every regular point of $\mathbf{V}(f)$ is incident to at most one curve of Γ_0. That is, there are $O(m)$ incidences between curves of Γ_0 and points of $\mathcal{P}_0$ that are regular points of $\mathbf{V}(f)$.

We rotate $\mathbb{R}^2$ so that no line in $\mathbf{V}(f_1)$ is parallel to one of the axes. This does not change the incidences in $\mathbf{V}(f_1)$ and does not increase the degree of f_1. It is impossible for both first partial derivatives of f to be identically zero. Without loss of generality, we assume that $f_x = \frac{\partial f}{\partial x}$ is not identically zero. By definition, f_x vanishes on every singular point of $\mathbf{V}(f)$. Since f is square-free, it has no common components with f_x. Consider $\gamma \in \Gamma_0$ and note that γ and $\mathbf{V}(f_x)$ also have no common components. By Bézout's theorem, $\gamma \cap \mathbf{V}(f_x)$ consists of $O_k(r)$ points. That is, γ is incident to $O_k(r)$ singular points of $\mathbf{V}(f)$. By summing this number of singular points over every $\gamma \in \Gamma_0$ and adding the bound from the preceding paragraph, we get that

$$I(\mathcal{P}_0, \Gamma_0) = O_k(nr + m). \quad (3.7)$$

By combining Equations (3.5)–(3.7), we obtain

$$I(\mathcal{P}, \Gamma) = O_{s,t,k} \left(\frac{mn^{\frac{s-1}{s}}}{r^{\frac{s-1}{s}}} + nr + m \right). \quad (3.8)$$

It remains to find the value of r that minimizes Equation (3.8). Since the first term in this bound is decreasing in r while the second is increasing in r, the optimal bound is obtained when these two terms are equivalent. Setting $mn^{\frac{s-1}{s}}/r^{\frac{s-1}{s}} = nr$ gives $r = \Theta\left(m^{\frac{s}{2s-1}}/n^{\frac{1}{2s-1}}\right)$. Plugging this value into Equation (3.8) leads to the bound in the statement of the theorem.

One minor issue: When $m = O(n^{1/s})$ we have that $m^{\frac{s}{2s-1}}/n^{\frac{1}{2s-1}} < 1$, which may lead to an invalid value of $r < 1$. Fortunately, it is easy to handle this special case. Plugging $m = O(n^{1/s})$ into the bound of Lemma 3.4 implies $I(\mathcal{P}, \Gamma) = O_{s,t}(n)$.   □

The proof of Theorem 3.3 is a simple example of using a partitioning polynomial. We briefly repeat the main steps of this proof technique:

(a) First, we obtain a weak incidence bound by using a combinatorial argument (see Lemma 3.4).
(b) We partition the space into cells by using a partitioning polynomial.
(c) We apply the weak incidence bound separately in each cell of the partition.
(d) Finally, we bound the number of incidences on the partition itself.

3.3 Proving the Polynomial Partitioning Theorem

In this section we prove the polynomial partitioning theorem. We first recall the statement of this theorem.

Theorem 3.1 *Let $\mathcal{P}$ be a set of m points in $\mathbb{R}^d$ and $1 < r \le m^{1/d}$. Then there exists an r-partitioning polynomial $f \in \mathbb{R}[x_1, \ldots, x_d]$ of degree $O(r)$.*

Intuitively, we build the partition in steps. We first partition $\mathcal{P}$ into two disjoint sets $\mathcal{P}_1, \mathcal{P}_2 \subset \mathcal{P}$. Some points of $\mathcal{P}$ may not be in $\mathcal{P}_1 \cup \mathcal{P}_2$, but rather on the partition. We then partition each of $\mathcal{P}_1$ and $\mathcal{P}_2$ into two, obtaining four sets. We then partition each of those four sets, and so on. An example of this process is depicted in Figure 3.6.

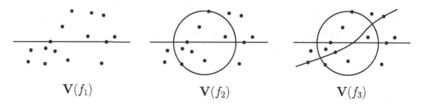

$\mathbf{V}(f_1)$ $\mathbf{V}(f_2)$ $\mathbf{V}(f_3)$

Figure 3.6 Repeatedly partitioning a set of 15 points in the plane. At the end of this process, every cell contains at most one point.

A *hyperplane* in $\mathbb{R}^d$ is a variety that is defined by one linear polynomial (and thus looks like a copy of $\mathbb{R}^{d-1}$). A hyperplane $h \subset \mathbb{R}^d$ *bisects* a finite point set $\mathcal{P} \subset \mathbb{R}^d$ if each of the two open halfspaces bounded by h contains at most $|\mathcal{P}|/2$ points of $\mathcal{P}$. The bisecting hyperplane may contain any number of points of $\mathcal{P}$. For example, the leftmost part of Figure 3.6 depicts a line that bisects a point set. The following is a discrete version of the *ham sandwich theorem* (for example, see Lo et al., 1994).

Theorem 3.5 *Every d finite sets $\mathcal{P}_1, \ldots, \mathcal{P}_d \subset \mathbb{R}^d$ can be simultaneously bisected by a hyperplane.*

A planar example of Theorem 3.5 is depicted in Figure 3.7. To iteratively partition $\mathcal{P}$, we can apply Theorem 3.5. However, after about $\log_2 d$ steps we obtain more than d sets, and can no longer apply the theorem. It is not difficult to find $d + 1$ sets in $\mathbb{R}^d$ that cannot be simultaneously bisected by a hyperplane. To overcome this issue, we use a discrete version of the *polynomial ham sandwich theorem*.

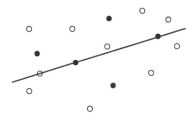

Figure 3.7 The line simultaneously bisects the set of black points and the set of white points.

A polynomial $f \in \mathbb{R}[x_1, \ldots, x_d]$ *bisects* a finite point set $\mathcal{P} \subset \mathbb{R}^d$ if at most $|\mathcal{P}|/2$ points $p \in \mathcal{P}$ satisfy $f(p) > 0$ and at most $|\mathcal{P}|/2$ points satisfy $f(p) < 0$. The variety $\mathbf{V}(f)$ may contain any number of points of $\mathcal{P}$.

Theorem 3.6 (Stone et al., 1942) *Consider finite sets $\mathcal{P}_1, \ldots, \mathcal{P}_t \subset \mathbb{R}^d$ and an integer D such that $\binom{D+d}{d} - 1 \geq t$. Then there exists a nonzero polynomial $f \in \mathbb{R}[x_1, \ldots, x_d]$ of degree at most D that simultaneously bisects all the sets $\mathcal{P}_j$.*

Proof A polynomial of degree at most D in $\mathbb{R}[x_1, \ldots, x_d]$ has at most $\binom{D+d}{d}$ monomials. Indeed, a brief combinatorial proof is depicted in Figure 3.8. We set

$$U_D = \Big\{ (u_1, \ldots, u_d) \in \mathbb{Z}^d \ : \ 1 \leq u_1 + \cdots + u_d \leq D \ \text{ and }$$

$$u_j \geq 0 \ \text{ for every } \ 1 \leq j \leq d \Big\}.$$

Intuitively, U_D is the set of exponents of nonconstant monomials of degree at most D in $x_1, \ldots, x_d$. Set $m = |U_D|$ and note that $m = \binom{D+d}{d} - 1$. The *Veronese map* $v_D : \mathbb{R}^d \to \mathbb{R}^m$ is defined as

$$v_D(x_1, \ldots, x_d) = (x_1^{u_1} x_2^{u_2} \cdots x_d^{u_d})_{u \in U_D}.$$

Figure 3.8 Every monomial of degree at most D in $x_1, \ldots, x_d$ corresponds to a unique choice of d blocks out of a total of $D + d$. The number of unchosen blocks between the jth and $(j + 1)$th chosen blocks is the power of x_{j+1}.

Every coordinate in $\mathbb{R}^m$ corresponds to a nonconstant monomial of degree at most D in $x_1, \ldots, x_d$. The map $v_D(\cdot)$ takes a point $p \in \mathbb{R}^d$ to the values of these m monomials at p. For example, the Veronese map $v_2 : \mathbb{R}^2 \to \mathbb{R}^5$ is

$$v_2(x, y) = (x^2, xy, y^2, x, y).$$

For every $1 \le j \le t$, we set $\mathcal{P}'_j = v_D(\mathcal{P}_j)$. That is, every $\mathcal{P}'_j$ is a finite set in $\mathbb{R}^m$. By the assumption on D, we have that $m \ge t$. Thus, by Theorem 3.5 there exists a hyperplane $\Pi \subset \mathbb{R}^m$ that simultaneously bisects all the sets $\mathcal{P}'_j$. We denote the coordinates of $\mathbb{R}^m$ as y_u, for each $u \in U$. The hyperplane Π is defined by a linear polynomial $h_0 + \sum_{u \in U} y_u h_u$, for some constants $h_u \in \mathbb{R}$.

Returning to $\mathbb{R}^d$, we consider the polynomial

$$f(x_1, \ldots, x_d) = h_0 + \sum_{u \in U} h_u x_1^{u_1} x_2^{u_2} \cdots x_d^{u_d}.$$

For every point $a \in \mathbb{R}^d$, we have that $f(a) = h_0 + (h_u)_{u \in U} \cdot v_D(a)$ (where '$\cdot$' is a dot product). This implies that $f(a) > 0$ if and only if $v_D(a)$ is on the positive side of Π. Symmetrically, $f(a) < 0$ if and only if $v_D(a)$ is on the negative side of Π. Since Π bisects every $\mathcal{P}'_j$, the polynomial f bisects every $\mathcal{P}_j$. Observing that $\deg f \le D$ completes the proof. $\square$

Guth and Katz proved the polynomial partitioning theorem by using the polynomial ham sandwich theorem.

Proof of Theorem 3.1 We recall that the condition of Theorem 3.6 is $\binom{D+d}{d} - 1 \ge t$. We set $c_d = (2(d!))^{1/d}$ and note that $\binom{c_d t^{1/d} + d}{d} - 1 > t$. This implies the following corollary of Theorem 3.6: For any finite $\mathcal{P}_1, \ldots, \mathcal{P}_t \subset \mathbb{R}^d$, there exists a nonzero $f \in \mathbb{R}[x_1, \ldots, x_d]$ of degree at most $c_d t^{1/d}$ that simultaneously bisects all the sets $\mathcal{P}_j$.

We show that there exists a sequence of polynomials $f_1, f_2, \ldots$ that satisfy the following:

(i) The degree of f_j is smaller than $c_d 2^{(j+1)/d}/(2^{1/d}-1)$.

(ii) Every connected component of $\mathbb{R}^d \backslash \mathbf{V}(f_j)$ contains at most $m/2^j$ points of $\mathcal{P}$.

An example is depicted in Figure 3.6. To complete the proof of the theorem, we then set $f = f_s$ with the minimum integer s that satisfies $2^s \geq r^d$. By property (i), we have that $\deg f = O(r)$. By property (ii), we have that every cell of $\mathbb{R}^d \backslash \mathbf{V}(f)$ contains fewer than m/r^d points of $\mathcal{P}$.

We prove the existence of f_j by induction on j. For the base case, the existence of f_1 is immediate from Theorem 3.5. We move to the induction step, assuming that f_j exists and proving that f_{j+1} also exists. We have that $\deg f_j < c_d 2^{(j+1)/d}/(2^{1/d}-1)$ and that every connected component of $\mathbb{R}^d \backslash \mathbf{V}(f_j)$ contains at most $m/2^j$ points of $\mathcal{P}$. Let t be the number of connected components of $\mathbb{R}^d \backslash \mathbf{V}(f_j)$ that contain more than $m/2^{j+1}$ points of $\mathcal{P}$. We denote the subsets of $\mathcal{P}$ in these connected components as $\mathcal{P}_1, \ldots, \mathcal{P}_t$. Since $|\mathcal{P}| = m$, we get that $t < 2^{j+1}$. By the first paragraph of this proof, there exists a nonzero $g_j \in \mathbb{R}[x_1, \ldots, x_d]$ of degree at most $c_d 2^{(j+1)/d}$ that simultaneously bisects all the sets $\mathcal{P}_j$.

We set $f_{j+1} = f_j \cdot g_j$. By definition, every connected component of $\mathbb{R}^d \backslash \mathbf{V}(f_{j+1})$ contains at most $m/2^{j+1}$ points of $\mathcal{P}$. We also have that

$$\deg f_{j+1} = \deg f_j + \deg g_j < \frac{c_d 2^{(j+1)/d}}{2^{1/d}-1} + c_d 2^{(j+1)/d}$$

$$= c_d 2^{(j+1)/d} \cdot \left(\frac{1}{2^{1/d}-1} + 1 \right)$$

$$= \frac{c_d 2^{(j+2)/d}}{2^{1/d}-1}.$$

This completes the induction step, and thus the proof of the theorem. $\square$

3.4 Curves Containing Lattice Points

We conclude this chapter with a simple application of Theorem 3.3. Let $\mathcal{G}$ be a $\sqrt{n} \times \sqrt{n}$ section of the integer lattice in $\mathbb{R}^2$. It is not difficult to show that an irreducible curve of degree k is incident to $O_k(\sqrt{n})$ points of $\mathcal{G}$ (see Exercise 3.14). This bound is asymptotically tight, since some lines contain $\sqrt{n}$ points of $\mathcal{G}$. We now show that curves that are not lines are incident to an asymptotically smaller number of points of $\mathcal{G}$. The following is similar to ideas of Iosevich (2008).

Claim 3.7 *Let $\mathcal{G}$ be a $\sqrt{n} \times \sqrt{n}$ section of the integer lattice in $\mathbb{R}^2$. Let $\gamma \subset \mathbb{R}^2$ be an irreducible curve of degree $k \geq 2$. Then γ contains $O_k\left(n^{k^2/(2k^2+1)}\right)$ points of $\mathcal{G}$.*

Proof Set $x = |\mathcal{G} \cap \gamma|$. Fix a point $p \in \mathcal{G} \cap \gamma$ and let $\mathbb{T}$ be the set of translations of $\mathbb{R}^2$ that take p to another point of $\mathcal{G}$. When also counting the identity as a translation, we have that $|\mathbb{T}| = n$. We apply each of the translations of $\mathbb{T}$ on γ to obtain a set Γ of n copies of γ. An example is depicted in Figure 3.9(a,b).

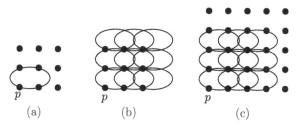

Figure 3.9 (a) A curve γ containing lattice points. (b) Applying the translations of $\mathbb{T}$ on γ. (c) Applying the translations of $\mathbb{T}$ on $\mathcal{G}$.

Some translated copies of γ might contain fewer than x points of $\mathcal{G}$. To fix this, we apply each translation of $\mathbb{T}$ on the points of $\mathcal{G}$. See Figure 3.9(c). We keep one copy of each resulting point, even when the same point is obtained from multiple translations. This leads to a lattice $\mathcal{G}'$ with $2n - 1$ rows and $2n - 1$ columns. That is, $|\mathcal{G}'| < 4n$. To complete the proof, we double count $I(\mathcal{G}', \Gamma)$.

After including the translated points, each of the n copies of γ contains at least x points of $\mathcal{G}$. This implies that

$$I(\mathcal{G}', \Gamma) = \Omega(nx).$$

Two translated copies of an irreducible curve that is not a line cannot have a common component. Bézout's theorem implies that any two curves of Γ have at most k^2 points in common. Thus, the incidences graph of $\mathcal{G}' \times \Gamma$ does not contain a copy of $K_{k^2+1,2}$. Theorem 3.3 with $s = k^2 + 1$ leads to

$$I(\mathcal{G}', \Gamma) = O_k\left(n^{(3k^2+1)/(2k^2+1)}\right).$$

Combining our two bounds for $I(\mathcal{G}', \Gamma)$ implies the bound of the claim. $\square$

In Exercise 3.15, we improve the bound of Claim 3.7 to $O_k(n^{1/3})$. Bombieri and Pila (1989) proved the stronger bound $O_k(n^{1/(2k)})$ by using number-theoretic methods. It would be interesting to obtain bounds stronger than $O_k(n^{1/3})$ by using combinatorial methods.

3.5 Exercises

Exercise 3.1   Show that Theorem 3.1 is no longer true when also asking $V(f)$ not to contain a given point. Specifically, construct a set $\mathcal{P}$ of n points in $\mathbb{R}^2$, a point $q \notin \mathcal{P}$, and an integer r that satisfy the following: Every r-partitioning polynomial of $\mathcal{P}$ of degree $O(r)$ also contains q.

Exercise 3.2   Show that Theorem 3.1 is no longer true when asking f to be irreducible. That is, find a set $\mathcal{P}$ of n points in $\mathbb{R}^2$ and an integer r that satisfy the following: Every r-partitioning polynomial of $\mathcal{P}$ of degree $O(r)$ is reducible.

Exercise 3.3   Let $\mathcal{P}$ be a set of n points in $\mathbb{R}^2$. Prove that $\mathcal{P}$ spans $\Omega(n^{3/4})$ distinct distances. (Hint: Recall the proof of Claim 1.11 and use Theorem 3.3.)

Exercise 3.4   Let $\mathcal{P}$ be a set of n points in $\mathbb{R}^2$. Prove that the number of isosceles triangles whose three vertices are in $\mathcal{P}$ is $O(n^{7/3})$. It is possible to solve the problem by using point-line incidences. However, you are asked to use only point-circle incidences.

Exercise 3.5   Let $\mathcal{P}$ be a set of m points and let C be a set of n circles, both in $\mathbb{R}^2$. Theorem 3.3 implies that $I(\mathcal{P}, \Gamma) = O(m^{3/5}n^{4/5} + m + n)$.
(a) Derive a stronger bound for the special case where the centers of all the circles are on the x-axis.
(b) Derive a stronger bound for the special case where no line contains more than 1,000 circle centers.
(Hint: Use Theorem 3.3 without changing its proof.)

Exercise 3.6   For $m \leq n$, let $\mathcal{P}$ be a set of m points on the x-axis and let $\mathcal{P}'$ be an arbitrary set of n points. Let $D(\mathcal{P}, \mathcal{P}')$ denote the number of distinct distances spanned by pairs of points from $\mathcal{P} \times \mathcal{P}'$. That is, we ignore pairs of points from the same set. Prove that $D(\mathcal{P}, \mathcal{P}') = \Omega(\sqrt{mn})$. (Hint: Recall the proof of Claim 1.11 and use Exercise 3.5(b).)

Exercise 3.7   Prove that Theorem 3.3 holds also after removing the restrictions about the curves being irreducible and distinct. (Hint: Do not change the proof of Theorem 3.3. Apply this theorem as is.)

Exercise 3.8   Let $\mathcal{P}$ be a set of m points and let $\mathcal{L}$ be a set of n lines, both in $\mathbb{R}^2$. Prove that there exist $\mathcal{P}' \subset \mathcal{P}$ and $\mathcal{L}' \subset \mathcal{L}$ such that $|\mathcal{P}'| = \Theta(m), |\mathcal{L}'| = \Theta(n)$, and $I(\mathcal{P}', \mathcal{L}') = 0$. (Hint: Consider an r-partitioning polynomial, where r is a large constant. Handle separately the case where most points are on the partition.)

Exercise 3.9   Revise the proof of Lemma 3.4 so that the resulting bound shows the asymptotic dependency in t. In other words, the bound should be of the form $O_s(\cdot)$, rather than $O_{s,t}(\cdot)$.

Exercise 3.10   Revise the proof of Theorem 3.3 so that the resulting bound shows the asymptotic dependency in t. In other words, the bound should be of the form $O_{s,k}(\cdot)$. Most of the proof can remain unchanged – focus on the few parts that need to be revised.

Exercise 3.11   Let $\mathcal{P}$ be a set of n points in $\mathbb{R}^2$. Let v_2 be the Veronese map as defined in the proof of Theorem 3.6. Prove that the set $v_2(\mathcal{P}) = \{v_2(p) : p \in \mathcal{P}\}$ is in convex position. (Recall that a point set is in convex position if every $p \in \mathcal{P}$ can be separated from $\mathcal{P}\setminus\{p\}$ by a hyperplane.)

Exercise 3.12   *Radon's theorem* states that any set of $d + 2$ points in $\mathbb{R}^d$ can be partitioned into two disjoint subsets whose convex hulls intersect. A polynomial $f \in \mathbb{R}[x_1, \ldots, x_d]$ *separates* two sets $\mathcal{P}_1, \mathcal{P}_2 \subset \mathbb{R}^d$ if $f(p) > 0$ for every $p \in \mathcal{P}_1$ and $f(q) < 0$ for every $q \in \mathcal{P}_2$. The convex hulls of two finite point sets are disjoint if and only if no hyperplane separates them.

Prove that every sufficiently large finite $\mathcal{P} \subset \mathbb{R}^d$ can be partitioned into two disjoint subsets $\mathcal{P}_1, \mathcal{P}_2 \subset \mathcal{P}$ with the following property: Every polynomial that separates $\mathcal{P}_1$ and $\mathcal{P}_2$ is of degree $\Omega_d(n^{1/d})$. You may rely on the statements from the preceding paragraph without proving them.

Exercise 3.13   Show that the polynomial partitioning theorem (Theorem 3.1) remains valid after adding the following restriction: No monomial of f contains a variable with a power larger than $r/2$.

Exercise 3.14   As in Section 3.4, let $\mathcal{G}$ be a $\sqrt{n} \times \sqrt{n}$ section of the integer lattice in $\mathbb{R}^2$. Let $\gamma \subset \mathbb{R}^2$ be a curve of degree k. Use Bézout's theorem to prove that γ contains at most $2k\sqrt{n}$ points of $\mathcal{G}$. You may not use Theorem 3.3 or similar incidence bounds. Note that γ may be reducible.

Exercise 3.15   Improve the bound of Claim 3.7 to $O_k(n^{1/3})$. The following paragraph provides some directions.

Let $\mathcal{G}'$ and Γ be as in the proof of Claim 3.7. We rely on the following variant of the point-line duality from Section 1.10. Recall that every curve $\gamma_j \in \Gamma$ is a translation of γ. Each translation can be described as a translation of distance s in the x-direction and a translation of distance t in the y-direction. The dual of γ_j is the point $(-s, -t) \in \mathbb{R}^2$, where s and t describe the translation of γ that led to γ_j. The dual of a point $p \in \mathcal{G}'$ is the set S_p of points (s, t) that describe translations of γ that are incident to p. Find what S_p looks like and then study the incidence graph in the dual plane.

Exercise 3.16 Let $\gamma \subset \mathbb{R}^2$ be an irreducible curve of degree $k \geq 2$. Prove that, for any two points $p, q \in \mathbb{R}^2$, at most k^2 translations of γ are incident to both p and q. (Hint: Use the duality that was presented in Exercise 3.15.)

Exercise 3.17 Let $\mathcal{P}$ be a set of m points and let $\mathcal{L}$ be a set of n lines, both in $\mathbb{R}^d$. By Theorem 3.1, there exists a partition with $O(r^d)$ cells, each containing at most m/r^d points of $\mathcal{P}$ (for some $1 < r < m$). Show that we can further partition the existing $O(r^d)$ cells so that

(i) Every new cell is intersected by at most n/r^{d-1} lines of $\mathcal{L}$.
(ii) The number of cells remains $O(r^d)$.
(iii) Every incidence that was in one of the original cells is in exactly one of the new cells.

In particular, partition each cell C into several abstract subcells, all corresponding to the same geometric region. Each subcell of C consists of the same set of points as C but only of a subset of the lines.

3.6 Open Problems

Theorem 3.3 provides a general point-curve incidence bound in $\mathbb{R}^2$. In Chapter 1, we saw that this bound is asymptotically tight for the case of lines. However, we also know that this bound is not tight in many other cases. The following is a common conjecture.

Conjecture 3.8 *Let $\mathcal{P}$ be a set of n points and let Γ be a set of n curves of degree at most k, both in $\mathbb{R}^2$. If the incidence graph of $\mathcal{P} \times \Gamma$ contains no $K_{s,t}$, then*

$$I(\mathcal{P}, \Gamma) = O_{s,t,k}(n^{4/3}).$$

Conjecture 3.8 is false when the number of curves is significantly larger than the number of points. For example, there exists a set of m points and a set of n parabolas with $\Theta(m^{1/2}n^{5/6})$ incidences (see Exercise 1.1). This expression is asymptotically larger than $m^{2/3}n^{2/3}$ when n is asymptotically larger than m. For $n = \Theta(m^2/\sqrt{\log m})$, there exists a configuration of m points, n circles, and $\Theta(m^{2/3}n^{2/3}\log^{1/3} m)$ incidences (see Pach and Sharir, 2004).

Conjecture 1.7 suggests that the number of incidences between n points and n unit circles is $O(n^{1+\varepsilon})$, for every $\varepsilon > 0$. This may also be the case for other variants, such as the degenerate hyperbolas that are described in Exercise 1.5. The current best bound for all these problems is $O(n^{4/3})$.

The following theorem contains the current best bound for point-curve incidences in $\mathbb{R}^2$. The proof combines polynomial partitioning with the *lens cutting* method. We do not discuss lens cutting in this book.

Consider the polynomial $f(x, y) = (x - c_x)^2 + (y - c_y)^2 - r^2$ with parameters $c_x, c_y, r \in \mathbb{R}$. Every circle in $\mathbb{R}^2$ is defined by $f(x, y)$ after setting specific values for c_x, c_y, and r. We say that the family of all circles in $\mathbb{R}^2$ is defined by $f(x, y)$ and the parameters c_x, c_y, and r. More generally, an infinite *family of curves* in $\mathbb{R}^2$ is defined by a polynomial $f \in \mathbb{R}[x, y]$ whose coefficients depend on parameters $s_1, \ldots, s_k \in \mathbb{R}$. Each curve in the family is defined by $f(x, y)$ after setting some values to the parameters. The *dimension* of a family of curves is the number of parameters. For example, the family of circles in $\mathbb{R}^2$ is three-dimensional.

Theorem 3.9 (Sharir and Zahl, 2017) *Let $\mathcal{P}$ be a set of m points and let Γ be a set of n irreducible curves, both in $\mathbb{R}^2$. Assume that the curves of Γ belong to an s-dimensional family of curves of degree at most k. Then for every $\varepsilon > 0$ we have that*

$$I(\mathcal{P}, \Gamma) = O_{k,s,\varepsilon} \left(m^{\frac{2s}{5s-4}+\varepsilon} n^{\frac{5s-6}{5s-4}} + m^{2/3} n^{2/3} + m + n \right).$$

Theorems 3.9 and 3.3 lead to the same bounds for incidences with lines and with unit circles. However, Theorem 3.9 is stronger in many other cases. For incidences with arbitrary circles, Theorem 3.3 leads to the bound $O(m^{3/5} n^{4/5} + m + n)$ and Theorem 3.9 leads to $O(m^{6/11+\varepsilon} n^{9/11} + m^{2/3} n^{2/3} + m + n)$. When $m = n$, the former is $O(n^{7/5})$ while the latter is $O(n^{15/11+\varepsilon})$. For a detailed discussion about families of varieties and their applications, see Chapter 14.

Figure 3.10 is an important recap of everything that we've learned in the past two chapters.

Figure 3.10 A drawing by Zachary Chase.

4

Basic Real Algebraic Geometry in $\mathbb{R}^d$

Every field has its taboos. In algebraic geometry the taboos are (1) writing a draft that can be followed by anyone but two or three of one's closest friends, (2) claiming that a result has applications, (3) mentioning the word "combinatorial," and (4) claiming that algebraic geometry existed before Grothendieck.

Gian-Carlo Rota (2008).

In Chapter 2 we studied basic properties of varieties in $\mathbb{R}^2$. We now generalize these properties to $\mathbb{R}^d$, and introduce new ones. Some of the definitions in $\mathbb{R}^d$ require the notion of a polynomial ideal.

4.1 Ideals

Varieties are the basic geometric objects of this book. Polynomial ideals are some of the basic *algebraic* objects of this book. A set of polynomials $J \subseteq \mathbb{R}[x_1, \ldots, x_d]$ is an *ideal* if it satisfies:

- $0 \in J$.
- If $f, g \in J$ then $f + g \in J$.
- If $f \in J$ and $h \in \mathbb{R}[x_1, \ldots, x_d]$, then $f \cdot h \in J$.

For any polynomial $f \in \mathbb{R}[x_1, \ldots, x_d]$, the set $\{f \cdot h : h \in \mathbb{R}[x_1, \ldots, x_d]\}$ is an ideal. More generally, the ideal *generated* by $f_1, \ldots, f_k \in \mathbb{R}[x_1, \ldots, x_d]$ is

$$\langle f_1, \ldots, f_k \rangle = \left\{ \sum_{j=1}^{k} f_j \cdot h_j : h_1, \ldots, h_k \in \mathbb{R}[x_1, \ldots, x_d] \right\}.$$

We also say that $\{f_1, \ldots, f_k\}$ is a *basis* of this ideal.

In this book we are specifically interested in ideals of varieties. Given a variety $U \subset \mathbb{R}^d$, the ideal of U is

$$\mathbf{I}(U) = \{f \in \mathbb{R}[x_1, \ldots, x_d] : f(a) = 0 \text{ for every } a \in U\}.$$

51

It is not difficult to verify that $\mathbf{I}(U)$ is indeed an ideal. Given $f_1, \ldots, f_k \in \mathbb{R}[x_1, \ldots, x_d]$, is it always the case that $\langle f_1, \ldots, f_k \rangle = \mathbf{I}(\mathbf{V}(f_1, \ldots, f_k))$?

Claim 4.1 *Given $f_1, \ldots, f_k \in \mathbb{R}[x_1, \ldots, x_d]$, we have $\langle f_1, \ldots, f_k \rangle \subseteq \mathbf{I}(\mathbf{V}(f_1, \ldots, f_k))$ although equality need not occur.*

Proof We set $U = \mathbf{V}(f_1, \ldots, f_k)$. If $g \in \langle f_1, \ldots, f_k \rangle$ then g vanishes on every point of U, and is thus in $\mathbf{I}(\mathbf{V}(f_1, \ldots, f_k))$. Thus, the containment relation holds.

To see that equality does not always hold, we consider $f(x, y) = x^2 + y^2$. We have that $\mathbf{V}(f) = \{(0, 0)\} \subset \mathbb{R}^2$. Note that $x \in \mathbf{I}(\mathbf{V}(f))$ but $x \notin \langle f \rangle$. □

When defining a variety U, it is often useful to use a basis of $\mathbf{I}(U)$ rather than an arbitrary set of polynomials that define U. Later in this chapter, we see that we must use a basis of $\mathbf{I}(U)$ to study singular points of U. We now observe another connection between ideals and varieties.

Claim 4.2 *Let $U, W \subset \mathbb{R}^d$ be varieties. Then*
(a) $U \subsetneq W$ if and only if $\mathbf{I}(W) \subsetneq \mathbf{I}(U)$.
(b) $U = W$ if and only if $\mathbf{I}(W) = \mathbf{I}(U)$.

Proof We only prove part (a); part (b) is proved in a similar manner.

First assume that $U \subsetneq W$ and consider a polynomial $f \in \mathbf{I}(W)$. By definition, f vanishes on every point of W. Since $U \subset W$, we get that f also vanishes on every point of U. This in turn implies that $f \in \mathbf{I}(U)$, so $\mathbf{I}(W) \subset \mathbf{I}(U)$. Since $U \neq W$, there exists a polynomial that vanishes on U but not on W.

Next, assume that $\mathbf{I}(W) \subsetneq \mathbf{I}(U)$ and consider a point $p \in U$. Every polynomial of $\mathbf{I}(U)$ vanishes on p, which in turn implies that every polynomial of $\mathbf{I}(W)$ vanishes on p. This implies that $p \in W$, so $W \subset U$. Since $\mathbf{I}(W) \neq \mathbf{I}(U)$, there exists a point $q \in W \backslash U$. □

The following result is an *ascending chain condition*.

Theorem 4.3 *Let $I_1 \subseteq I_2 \subseteq I_3 \subseteq \cdots$ be an infinite chain of ideals in $\mathbb{R}[x_1, \ldots, x_d]$. Then there exists an integer $n \geq 1$ such that $I_n = I_{n+1} = I_{n+2} = \cdots$.*

4.2 Dimension

Consider an irreducible variety $U \subset \mathbb{R}^d$. The *dimension* of U is the maximum integer d' such that there exists a sequence

$$U_0 \subsetneq U_1 \subsetneq \cdots \subsetneq U_{d'} = U,$$

where each U_j is a nonempty irreducible variety. We also write $\dim U = d'$. When $U \subset \mathbb{R}^d$ is a reducible variety with irreducible components $U_1, \ldots, U_k$, we set

$$\dim U = \max_j \{\dim U_j\}.$$

The above definition of a dimension aligns with the way that we intuitively think of a dimension. For example, lines and circles are one-dimensional, spheres and planes are two-dimensional, and finite point sets are zero-dimensional. The following claim is a warm-up for working with dimensions of varieties.

Claim 4.4 *Let U and U' be varieties of dimension d' in $\mathbb{R}^d$ that do not have any common irreducible component. Then $\dim(U \cap U') < d'$.*

Proof We first consider the case where both U and U' are irreducible. Assume for contradiction that $\dim(U \cap U') \geq d'$. Let W be a component of $U \cap V$ of dimension at least d'. By definition, there exists a chain

$$W_0 \subsetneq W_1 \subsetneq \cdots \subsetneq W_{d'-1} \subsetneq W,$$

where each W_j is a nonempty irreducible variety. Since U and U' have no common components, W is a proper subset of U. We thus have the chain

$$W_0 \subsetneq W_1 \subsetneq \cdots \subsetneq W_{d'-1} \subsetneq W \subsetneq U.$$

This contradicts the dimension of U being d', which implies that $\dim(U \cap U') < d'$.

Next, we consider the case where U and U' may be reducible. Let $U_1, \ldots, U_k$ be the components of U, and let $U'_1, \ldots, U'_\ell$ be the components of U'. We note that $U \cap U' = \bigcup_{\substack{1 \leq j \leq k \\ 1 \leq j' \leq \ell}} \left(U_j \cap U'_{j'} \right)$. By the above, each $U_j \cap U_{j'}$ is of dimension smaller than d'. We conclude that

$$\dim(U \cap U') = \max_{j,j'} \left\{ \dim(U_j \cap U'_{j'} \right\} < d'. \qquad \square$$

In Chapter 14 we study more advanced arguments that rely on dimensions of varieties. A few dimension-based definitions:

- A *curve* is a variety that consists only of one-dimensional components. (This generalizes the definition of a curve in $\mathbb{R}^2$ from Chapter 2.)
- A *k-flat* is a translation of a k-dimensional linear space.
- A *hypersurface* in $\mathbb{R}^d$ is a variety that consists only of $(d-1)$-dimensional components. Similarly, a *hyperplane* in $\mathbb{R}^d$ is a $(d-1)$-flat and a *hypersphere* is a $(d-1)$-dimensional sphere.

Consider an irreducible variety $U \subset \mathbb{R}^d$ of dimension d'. One might expect U to look like a d'-dimensional set in a small neighborhood around every point $p \in U$. We already saw one counterexample to this claim in Chapter 2: The curve $\mathbf{V}(y^2 - x^3 + x^2) \subset \mathbb{R}^2$ is irreducible but includes an isolated point (see Figure 2.2). For a higher-dimensional counterexample, consider the *Whitney Umbrella* $\mathbf{V}(x^2 - y^2 z) \subset \mathbb{R}^3$. See Figure 4.1. This is an irreducible two-dimensional variety that contains the z-axis. In the half-space defined by $z < 0$, the Whitney Umbrella consists only of the one-dimensional z-axis. When removing this axis, we obtain a set that is not a variety.

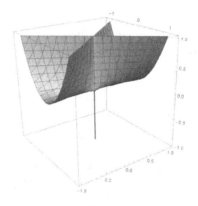

Figure 4.1 The Whitney Umbrella is irreducible but includes the z-axis.

Hypersurfaces are often easier to study than lower-dimensional varieties. We already saw this phenomena in Chapter 2: Results such as Lemma 2.6 are valid for curves, but not for arbitrary varieties in $\mathbb{R}^2$. The following lemma states a main property that makes hypersurfaces easier to study. A *principal ideal* is an ideal that is generated by a single element. In other words, an ideal $J \subset \mathbb{R}[x_1, \ldots, x_d]$ is principal if there exists $f \in \mathbb{R}[x_1, \ldots, x_d]$ such that $J = \langle f \rangle$.

Lemma 4.5 *For every hypersurface $U \subset \mathbb{R}^d$, the ideal $\mathbf{I}(U)$ is principal.*

You are asked to partially prove Lemma 4.5 in Exercise 4.3. For a complete proof, see Zahl (2013). As a first example of the usefulness of Lemma 4.5, it implies the following property of hypersurfaces.

Corollary 4.6 *Consider a hypersurface $U \subset \mathbb{R}^d$ and let $f \in \mathbb{R}[x_1, \ldots, x_d]$ be a minimum-degree polynomial that satisfies $U = \mathbf{V}(f)$. Then there is a bijection between the factors of f and the components of U, such that each factor generates the ideal of the corresponding component.*

Corollary 4.6 is false for varieties that are not hypersurfaces. For example, for an integer $n > 2$, consider the polynomial $f \in \mathbb{R}[x, y, z]$ defined as

$$f(x, y, z) = \prod_{j=1}^{n}(x - j)^2 + \prod_{k=1}^{n}(y - k)^2. \tag{4.1}$$

Note that $\mathbf{V}(f)$ is a set of n^2 lines in $\mathbb{R}^3$, all parallel to the z-axis. Since $\deg f = 2n$ and $\mathbf{V}(f)$ consists of n^2 components, there cannot be a bijection between the factors and the components.

Consider an irreducible variety $U \subset \mathbb{R}^d$ of dimension $d' < d - 1$. In other words, U is not a hypersurface. It is tempting to conjecture that $\mathbf{I}(U)$ is generated by $d - d'$ polynomials. For example, when $U \subset \mathbb{R}^3$ is the z-axis, we have that $\mathbf{I}(U) = \langle x, y \rangle$. To see that the conjecture is false, consider the *twisted cubic*

$$U = \left\{ (t, t^2, t^3) \in \mathbb{R}^3 : t \in \mathbb{R} \right\}.$$

See Figure 4.2. This is an irreducible curve in $\mathbb{R}^3$, but the ideal $\mathbf{I}(U)$ is generated by the three polynomials $x_1 x_3 - x_2^2$, $x_2 - x_1^2$, and $x_3 - x_1 x_2$.

Figure 4.2 The twisted cubic $\left\{ (t, t^2, t^3) \in \mathbb{R}^3 : t \in \mathbb{R} \right\}$.

4.3 Tangent Spaces and Singular Points

Consider a variety $U \subset \mathbb{R}^d$ and a point $p \in U$. Intuitively, the *tangent space* to U at p is the union of all lines that are tangent to U at p, followed by a translation that takes p to the origin. Because of this translation, the tangent space contains the origin and is a linear subspace. We denote this linear subspace as $T_p U$.

For a rigorous definition of a tangent space, we require the following definition. Given a polynomial $f \in \mathbb{R}[x_1, \ldots, x_d]$, the *gradient* of f is

$$\nabla f = \left(\frac{\partial f}{\partial x_1}, \frac{\partial f}{\partial x_2}, \ldots, \frac{\partial f}{\partial x_d} \right).$$

Consider a variety $U \subset \mathbb{R}^d$ and let $f_1, \ldots, f_k \in \mathbb{R}[x_1, \ldots, x_d]$ generate $\mathbf{I}(U)$. For $p \in U$, the tangent space $T_p U$ is the subspace of vectors of $\mathbb{R}^d$ that are orthogonal to all of $\nabla f_1(p), \ldots, \nabla f_k(p)$.

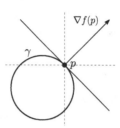

Figure 4.3 The circle γ is defined by $f(x, y) = (x + 1)^2 + (y + 1)^2 - 1$. Since $\nabla f(x, y) = (2x + 2, 2y + 2)$, we have that $\nabla f(0, 0) = (2, 2)$. This vector is orthogonal to the line tangent to γ at $p = (0, 0)$.

For a first intuition, consider a curve $\gamma \subset \mathbb{R}^2$ and a point $p \in \gamma$. By Lemma 4.5, there exists $f \in \mathbb{R}[x, y]$ such that $\langle f \rangle = \mathbf{I}(\gamma)$. Note that $\nabla f(p)$ is the direction of the normal of γ at p. Thus, the set of vectors orthogonal to $\nabla f(p)$ is indeed the line tangent to γ at p, after a translation that takes p to the origin. See Figure 4.3 for an example.

The case of hypersurfaces in $\mathbb{R}^d$ is the same as for curves in $\mathbb{R}^2$: The tangent at p is the hyperplane orthogonal to $\nabla f(p)$. The case of a lower dimensional variety $U \subset \mathbb{R}^d$ is less intuitive. In this case, one might imagine several hypersurfaces whose intersection is U. The tangent space at $p \in U$ is the set of vectors that are orthogonal to the normals of all these hypersurfaces at p.

A tangent space may not be well defined at a point p of a variety U. For example, consider the apex of a circular conical surface (see Figure 4.4). The tangent space may also have a different dimension than U. For example, recall the z-axis of the Whitney Umbrella (Figure 4.1). We now study these cases.

Singular points: Consider a variety $U \subset \mathbb{R}^d$ and a point $p \in U$. Intuitively (and with exceptions), p is a *singular point* of U if one of the following holds:

- The tangent space $T_p U$ is not a well-defined linear subspace.

- The point p is contained in more than one irreducible component of U. For example, consider the union of two spheres in $\mathbb{R}^3$ that have the same tangent

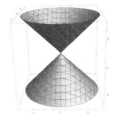

Figure 4.4 The conical surface $\mathbf{V}(x^2 + y^2 - z^2)$. The apex is a singular point.

plane at a common point p. One might say that the tangent plane is well defined at p, but p is still a singular point.

- The dimension of the tangent space $T_p U$ is different from the dimension of an irreducible component that contains p (such as on the line of the Whitney Umbrella).

A point that satisfies one of the three above cases is always singular, but there are singular points that do not fit any of these cases. For an example, see the discussion surrounding Figure 2.3 in Section 2.2. A point of U that is not singular is a *regular* point of U. We denote the set of singular points of U as U_{sing}, and the set of regular points of U as U_{reg}.

We now study a rigorous definition of a singular point. We begin with the simpler case of a hypersurface $U \subset \mathbb{R}^d$. By Lemma 4.5, there exists $f \in \mathbb{R}[x_1, \ldots, x_d]$ such that $\mathbf{I}(U) = \langle f \rangle$. A point $p \in U$ is singular if and only if $\nabla f(p)$ consists of d zeros. For a discussion about why the condition $\mathbf{I}(U) = \langle f \rangle$ is necessary, see Section 2.2 after Figure 2.2.

Recall that a polynomial $f \in \mathbb{R}[x_1, \ldots, x_d]$ is *square-free* if the factorization of f into irreducible factors does not contain any factor more than once. Let $f \in \mathbb{R}[x_1, \ldots, x_d]$ be a square-free polynomial and let g be a component of f that depends on x_j. Then $\frac{\partial f}{\partial x_j}$ is not divisible by g, as shown in Section 2.2.

Claim 4.7 *Every hypersurface $U \subset \mathbb{R}^d$ contains a regular point.*

Proof Let C be an irreducible component of U. By Lemma 4.5 there exist $f, g \in \mathbb{R}[x_1, \ldots, x_d]$ such that $\mathbf{I}(U) = \langle f \rangle$ and $\mathbf{I}(C) = \langle g \rangle$. Corollary 4.6 implies that g is a factor of f. By definition, both f and g are square-free.

Without loss of generality, assume that x_1 appears in g. This implies that $f_1 = \frac{\partial f}{\partial x_1}$ is not identically zero. Since f is square-free, f_1 is not divisible by g. That is, $\mathbf{V}(f_1)$ and C do not have any common components. By Claim 4.4, the intersection $C \cap \mathbf{V}(f_1)$ is of dimension at most $d - 2$. Since C is of dimension $d - 1$ and $C \cap \mathbf{V}(f_1)$ is of dimension at most $d - 2$, there exists $p \in C$ that satisfies $f_1(p) \neq 0$. By definition, p is a regular point of U. □

To define singular points of varieties that are not hypersurfaces, we require a generalization of the gradient. Given $f_1, \ldots, f_k \in \mathbb{R}[x_1, \ldots, x_d]$, the *Jacobian matrix* of $f_1, \ldots, f_k$ is

$$
\mathbf{J}_{f_1, \ldots, f_k} = \begin{pmatrix} \frac{\partial f_1}{\partial x_1} & \frac{\partial f_1}{\partial x_2} & \cdots & \frac{\partial f_1}{\partial x_d} \\ \frac{\partial f_2}{\partial x_1} & \frac{\partial f_2}{\partial x_2} & \cdots & \frac{\partial f_2}{\partial x_d} \\ \vdots & \vdots & \vdots & \vdots \\ \frac{\partial f_k}{\partial x_1} & \frac{\partial f_k}{\partial x_2} & \cdots & \frac{\partial f_k}{\partial x_d} \end{pmatrix}.
$$

Let $U \subset \mathbb{R}^d$ be a variety of dimension d', and consider $f_1, \ldots, f_k \in \mathbb{R}[x_1, \ldots, x_d]$ such that $\mathbf{I}(U) = \langle f_1, \ldots, f_k \rangle$. Then $p \in U$ is a singular point of U if and only if $\operatorname{rank}(\mathbf{J}_{f_1, \ldots, f_k}(p)) < d - d'$. Basic linear algebra states that

$$
\operatorname{rank}\left(\mathbf{J}_{f_1, \ldots, f_k}(p)\right) + \ker\left(\mathbf{J}_{f_1, \ldots, f_k}(p)\right) = d.
$$

Since the tangent space $T_p U$ is orthogonal to the vectors $\nabla f_1, \ldots, \nabla f_k$, we can intuitively think of $d - \operatorname{rank}(\mathbf{J}_{f_1, \ldots, f_k}(p))$ as the dimension of $T_p U$. This fits with the intuition that a singular point does not have a well-defined tangent space of dimension d'.

Lemma 4.8 *Every variety contains a regular point.*

Unlike the case of Claim 4.7, we are not aware of a proof for Lemma 4.8 that is simple enough to present here. For a more advanced discussion, see Bochnak et al. (2013, Section 3.3). This is yet another instance of hypersurfaces being simpler to study.

4.4 Generic Objects

Let U be a variety in $\mathbb{R}^d$. Let X be a property that some points of U have. A *generic point* of U has property X if the points of U that do not have property X are contained in a subvariety of U of a smaller dimension. For example, let $H \subset \mathbb{R}^3$ be a plane and let $O \subset \mathbb{R}^3$ be a sphere. Then

- A generic point of $\mathbb{R}^3$ is not contained in H.
- A generic point of O is not contained in H.
- A generic point of O has a tangent space that is not parallel to H.

Let $U \subset \mathbb{R}^d$ be a hypersurface. By inspecting the proof of Claim 4.7, we note that a generic point of U is regular. In Theorem 4.11 we see that this is the case for every variety.

In Section 3.6, we defined families of curves. For example, we can describe all circles in $\mathbb{R}^2$ with the polynomial $f(x, y) = (x - c_x)^2 + (y - c_y)^2 = r^2$. We then think of the set of circles as $\mathbb{R}^3$, where the coordinates of a point in $\mathbb{R}^3$ represent the parameters c_x, c_y, r. Some of these points correspond to degenerate circles of radius zero. Every circle with a nonzero radius corresponds to two points in $\mathbb{R}^3$. The point-line duality from Section 1.10 is a similar parameterization of lines in $\mathbb{R}^2$.

We say that a *generic circle* in $\mathbb{R}^2$ has property X if the circles that do not have property X are contained in a variety in $\mathbb{R}^3$ of dimension at most two. More precisely, the points in $\mathbb{R}^3$ that correspond to these circles are contained in such a variety. For example, the circles in $\mathbb{R}^2$ that are incident to the origin correspond to the two-dimensional variety $\mathbf{V}(c_x^2 + c_y^2 - r^2)$ (this is the conical surface from Figure 4.4). Thus, a generic circle in $\mathbb{R}^2$ is not incident to the origin. On the other hand, it is not true that a generic circle in $\mathbb{R}^2$ does not intersect the x-axis.

The above considers the special case of circles in $\mathbb{R}^2$. We can similarly discuss generic hyperplanes in $\mathbb{R}^d$, generic spheres in $\mathbb{R}^2$ that are incident to the origin, and so on. In each such case, we first parameterize the objects as points in some space. We then check if the objects that do not have some property are contained in a lower-dimensional variety. For a more advanced study of parameterizations of varieties, see Chapter 14.

4.5 Degree and Complexity

Defining the degree of a real variety is similar to taking candy from a baby: Everyone thinks it should be easy, but then it turns out to be surprisingly tricky and unsatisfying. In a complex space there is a well-defined notion of degree, with many equivalent definitions. A variety $U \subset \mathbb{C}^d$ of dimension d' has degree k if it intersects a generic $(d - d')$-flat of $\mathbb{C}^d$ in exactly k points. For example, a circle in $\mathbb{C}^2$ has degree 2 because it intersects a generic line in two points. A circle in $\mathbb{C}^3$ has degree 2 because it intersects a generic plane in two points. Unfortunately this is not the case in $\mathbb{R}^2$ and $\mathbb{R}^3$.

Several nonequivalent definitions of degree are being used in $\mathbb{R}^d$. Some authors avoid referring to degrees of varieties, and instead work with degrees of polynomials. That is, such authors consider a variety defined by X polynomials of degree at most Y. This is the approach of one of the main real algebraic geometry textbooks (Bochnak et al., 2013).

For hypersurfaces in $\mathbb{R}^d$, the different definitions of degree are equivalent and well behaved. We define the *degree* of a hypersurface $U \subset \mathbb{R}^d$ as the minimum degree of a polynomial $f \in \mathbb{R}[x_1, \ldots, x_d]$ that satisfies $U = \mathbf{V}(f)$.[1]

---

[1] Other equivalent definitions for hypersurfaces: (i) The degree of a polynomial $f \in \mathbb{R}[x_1, \ldots, x_d]$ that satisfies $\mathbf{I}(U) = \langle f \rangle$. (ii) The minimum degree of a variety in $\mathbb{C}^d$ that contains the embedding of U into $\mathbb{C}^d$.

This generalizes the definition of degree for curves in $\mathbb{R}^2$, as presented in Chapter 2.

We do not define the degree of a variety in $\mathbb{R}^d$ that is not a hypersurface. Instead, we rely on the notion of complexity, which is gradually becoming more common (for example, see Breuillard et al., 2011; Solymosi and Tao, 2012). The *complexity* of a variety $U \subset \mathbb{R}^d$ is the minimum integer k such that U can be defined by at most k polynomial of degree at most k. We write comp $U = k$. Sometimes we only need to state that a variety is defined by a constant number of polynomials of constant degree. In such cases, relying on the notion of complexity simplifies many proofs and tools.

Like all the degree definitions of varieties in $\mathbb{R}^d$, the notion of complexity has disadvantages. For example, it is counterintuitive that the complexity of a line in $\mathbb{R}^3$ is 2. Indeed, we can define such a line either with one quadratic polynomial or with two linear polynomials. When using complexity, Bézout's theorem (Theorem 2.5) does not generalize to dimensions $d \geq 3$. That is, when two varieties have a finite intersection, the number of points in this intersection may be significantly larger than the product of their complexities. For example, consider the varieties

$$U = \mathbf{V}((x-1)(x-2)(x-3),(y-1)(y-2)(y-3)) \subset \mathbb{R}^3,$$
$$W = \mathbf{V}(z) \subset \mathbb{R}^3.$$

We note that W is a plane and U is a set of nine lines parallel to the z-axis. We also note that $U \cap W$ is a set of nine points. Since comp $U = 3$ and comp $W = 1$, we get that $|U \cap W| > $ comp $U \cdot$ comp W.

Bézout's theorem similarly fails for the other common degree definitions. Even worse, the above example still holds over $\mathbb{C}^3$. Instead of Bézout's theorem, we can use the following celebrated result of Milnor (1964), Thom (2015), and others.

Theorem 4.9 (**Milnor–Thom theorem**) *Let* $f_1, \ldots, f_m \in \mathbb{R}[x_1, \ldots, x_d]$ *be of degree at most* k. *Then the number of connected components of* $\mathbf{V}(f_1, \ldots, f_m)$ *is at most*

$$k(2k-1)^{d-1}.$$

Theorem 4.9 does not require the notion of complexity, since the number of polynomials does not affect the bound of the theorem. When possible, we avoid the use of complexity. We rely on complexity in three cases:

- When discussing projections of varieties.
- When discussing upper bounds for the number of irreducible components of a nonhypersurface variety.
- When discussing the number of connected components in sets of the form $U \setminus W$, where U and W are varieties.

By Corollary 4.6, a hypersurface in $\mathbb{R}^d$ of degree k has at most k irreducible components. As shown in Equation (4.1), this is false for arbitrary varieties in $\mathbb{R}^d$. We can rely on complexity to bound the number of irreducible components of arbitrary varieties.

Lemma 4.10 *The number of irreducible components of a variety of degree k in $\mathbb{R}^d$ is $O_{d,k}(1)$.*

The bound $O_{k,d}(1)$ in Lemma 4.10 may seem unusual. It is a way of saying that the number of components depends only on the dimension d and the complexity k. In other words, the number of components cannot be arbitrarily large. The proof of Theorem 4.11 demonstrates how we use such bounds.

Theorem 4.11 *Let $U \subset \mathbb{R}^d$ be a variety of complexity k and dimension d_U. Then U_{sing} is a variety of dimension smaller than d_U and of complexity at most $k^d d^{d_U}$.*

Proof By definition, there exist $\ell \leq k$ and polynomials $f_1, \ldots, f_\ell$ of degree at most k such that $\langle f_1, \ldots, f_\ell \rangle = \mathbf{I}(U)$. By the definition of singular points,

$$U_{\text{sing}} = \left\{ p \in U : \text{rank}\left(\mathbf{J}_{f_1,\ldots,f_\ell}(p)\right) < d - d_U \right\}.$$

We recall from linear algebra that rank $\left(\mathbf{J}_{f_1,\ldots,f_\ell}(p)\right) < d - d_U$ if and only if every $(d - d_U) \times (d - d_U)$ minor of $\mathbf{J}_{f_1,\ldots,f_\ell}(p)$ is zero. Such a minor is an equation of degree at most $k(d - d_U)$ in the coordinates of p. Since $\mathbf{J}_{f_1,\ldots,f_\ell}(p)$ is an $\ell \times d$ matrix, the number of $(d - d_U) \times (d - d_U)$ minors is

$$\binom{\ell}{d - d_U} \cdot \binom{d}{d - d_U} \leq \binom{k}{d - d_U} \cdot \binom{d}{d_U} \leq k^d d^{d_U}.$$

Thus, U_{sing} is defined by at most $k^d d^{d_U}$ polynomials of degree at most $k(d - d_U)$. We conclude that U_{sing} is a variety of complexity at most $k^d d^{d_U}$.

It remains to prove that U_{sing} is of dimension smaller than d_U. First, assume that U is irreducible. Lemma 4.8 states that U contains a regular point, so $U_{\text{sing}} \subsetneq U$. Since U is irreducible, U_{sing} must have a smaller dimension.

Next, assume that U is reducible. By the preceding paragraph, the set of singular points of a single component of U is a variety of dimension smaller than d_U. By Lemma 4.10, the are $O_{d,k}(1)$ such components. The union of a finite number of varieties of dimension smaller than d_U is a variety of dimension smaller than d_U.

It remains to consider points that are not singular points of any component of U but are contained in more than one such component. By Claim 4.4, the intersection of two components of dimension at most d_U is a variety of dimension smaller than d_U. By Lemma 4.10, there are $O_{d,k}(1)$ pairs of intersecting components of U. Thus, their union is a variety of dimension smaller than d_U. $\qquad\square$

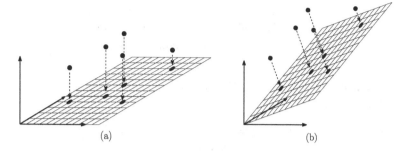

(a)   (b)

Figure 4.5   (a) A projection from $\mathbb{R}^3$ to the plane $\mathbf{V}(z)$. (b) A projection from $\mathbb{R}^3$ to a different plane.

Projections of varieties: The standard projection from $\mathbb{R}^3$ to $\mathbb{R}^2$ is $\pi(x, y, z) = (x, y)$. We can think of $\pi(x, y, z)$ as a projection on the plane $\mathbf{V}(z)$. An example is depicted in Figure 4.5(a). We may also consider a projection on a different plane in $\mathbb{R}^3$. To project a point $p \in \mathbb{R}^3$ on a plane $H \subset \mathbb{R}^3$, we consider the line $\ell_{p,H}$ that is incident to p and orthogonal to H. The projection is the intersection point $\ell_{p,H} \cap H$. An example is depicted in Figure 4.5(b). Every such projection is equivalent to applying a rotation $R \colon \mathbb{R}^3 \to \mathbb{R}^3$ and then applying the projection $\pi(x, y, z) = (x, y)$. Specifically, this is the projection on the plane H that satisfies $R(H) = \mathbf{V}(z)$.

For integers $0 < d' < d$, we can consider a projection on any d'-flat in $\mathbb{R}^d$. Such a projection can be thought of as a rotation $R \colon \mathbb{R}^d \to \mathbb{R}^d$ followed by discarding the last $d - d'$ coordinates of $\mathbb{R}^d$. Since a rotation consists of d linear polynomials, such a projection consists of d' linear polynomials in the coordinates of $\mathbb{R}^d$. We can also adapt the line intersection approach from the preceding paragraph, as follows. To project a point $p \in \mathbb{R}^d$ on a d'-flat H, we consider the $(d - d')$-flat $\ell_{p,H}$ that is incident to p and orthogonal to H. The projection is the intersection point $\ell_{p,H} \cap H$.

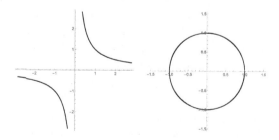

Figure 4.6   (Left) The hyperbola $\mathbf{V}(xy - 1)$ does not contain a point with a zero x-coordinate. (Right) The x-coordinates of the points on the circle $\mathbf{V}(x^2 + y^2 - 1)$ form the interval $[-1, 1]$.

We consider a set $S \subset \mathbb{R}^d$ and a projection $\pi \colon \mathbb{R}^d \to \mathbb{R}^e$. Then the projection of S is $\pi(S) = \{\pi(p) : p \in S\}$. The projection of a variety in $\mathbb{R}^d$ may not be a variety. For example, let γ be the hyperbola $\mathbf{V}(xy - 1)$ (see the left part of Figure 4.6). Consider the projection $\pi(x, y) = x$. Then $\pi(\gamma)$ is $\mathbb{R}\setminus\{0\}$, which is not a variety. Similarly, let C be the circle $\mathbf{V}(x^2 + y^2 - 1)$. Then $\pi(C)$ is the interval $[-1, 1]$, which is not a variety.[2] Lemma 4.12 is a weaker property of projections.

Lemma 4.12 *Let $U \subset \mathbb{R}^d$ be a variety of dimension d_U and complexity k. Let $\pi \colon \mathbb{R}^d \to \mathbb{R}^{d'}$ be a projection. Then $\pi(U)$ is contained in a variety of dimension at most d_U and complexity $O_{k,d}(1)$.*

Consider a set $S \subset \mathbb{R}^d$. The *Zariski closure* of S, denoted $\overline{S}$, is the smallest variety in $\mathbb{R}^d$ that contains S. If a variety $U \subset \mathbb{R}^d$ contains S, then U also contains $\overline{S}$. For example, the Zariski closure of the interval $\{(k, k) : 0 \le k \le 1\} \subset \mathbb{R}^2$ is the line $\mathbf{V}(x - y)$. We can rephrase Lemma 4.12 as follows.

Corollary 4.13 *Let $U \subset \mathbb{R}^d$ be a variety of dimension d_U and complexity k. Let $\pi \colon \mathbb{R}^d \to \mathbb{R}^{d'}$ be a projection. Then $\overline{\pi(U)}$ is a variety of dimension at most d_U and complexity $O_{k,d}(1)$.*

4.6 Polynomial Partitioning in $\mathbb{R}^d$

The polynomial partitioning theorem that was presented in Chapter 3 (Theorem 3.1) holds in $\mathbb{R}^d$ for every $d \ge 2$. When using this theorem in dimension $d \ge 3$, it would be helpful to have the following slightly revised variant.

Theorem 4.14 (Zahl, 2013) *Let $\mathcal{P}$ be a set of m points in $\mathbb{R}^d$. Then for every $1 < r \le m$, there exists an r-partitioning polynomial $f \in \mathbb{R}[x_1, \ldots, x_d]$ of degree $O(r)$. Moreover, we may assume that $\mathbf{V}(f)$ is a hypersurface.*

The novelty of Theorem 4.14 is that the partition is a hypersurface. To prove Theorem 4.14, one might suggest applying the original polynomial partitioning theorem (Theorem 3.1) and remove the lower-dimensional components from the partition. Removing the lower-dimensional components cannot merge cells of the partition and cannot increase the degree of the partitioning polynomial. There is a more subtle issue: A lower dimensional component may contain points of $\mathcal{P}$. When removing the component, these points may get added to some of the cells. This may in turn lead to a cell that contains too many points.

---

[2] A projection of a variety in $\mathbb{R}^d$ is a semi-algebraic set. Such sets are outside the scope of this book. For more information, see for example Bochnak et al. (2013).

The proof of Theorem 4.14 replaces each lower dimensional component with a hypersurface that contains this component. To do so without increasing the degree of the partitioning polynomial, we require more tools from algebraic geometry. We do not discuss these tools here. For the full details, see Zahl (2013).

When applying polynomial partitioning in $\mathbb{R}^2$, we use Bézout's theorem to bound the number of cells that are intersected by a curve. In dimension $d \geq 3$, we instead rely on the following result by Solymosi and Tao (2012) (see also similar work of Barone and Basu, 2012).

Theorem 4.15   *Let* $U \subset \mathbb{R}^d$ *be a variety of dimension* d_U *and complexity* k_U. *Let* $f \in \mathbb{R}[x_1, \dots, x_d]$ *be a polynomial of degree* k_f. *Then the number of connected components of* $U \backslash \mathbf{V}(f)$ *is* $O_{d, k_U}(k_f^{d_U})$.

When applying Theorem 4.15, we need the exact dependency in the degree of f. We assume that the other parameters are constant, which is why the exact dependency in d, k_U, and d_U is not needed.

4.7 Exercises

Exercise 4.1   Use Theorem 4.3 to prove that every variety can be decomposed into a finite number of irreducible components. You may not apply Lemma 4.10.

Exercise 4.2   Prove part (b) of Claim 4.2.

Exercise 4.3   Prove Lemma 4.5. Use the following claim without proving it: Consider $f, g \in \mathbb{R}[x_1, \dots, x_d]$ such that $\mathbf{V}(f)$ and $\mathbf{V}(g)$ contain the same $(d-1)$-dimensional component. Then f and g have a common factor. (Hint: Start with the case where U is irreducible.)

Exercise 4.4   Use Lemma 4.5 to prove Corollary 4.6.

Exercise 4.5   (a) Prove that a generic sphere in $\mathbb{R}^3$ is not tangent to the plane $\mathbf{V}(z)$. (Hint: Define the family of spheres in $\mathbb{R}^3$ and check the dimension of the subset of spheres tangent to $\mathbf{V}(z)$.)
(b) True or false? A generic sphere in $\mathbb{R}^3$ does not intersect the plane $\mathbf{V}(z)$.

Exercise 4.6   Let ℓ be a line in $\mathbb{R}^3$ such that every two points in ℓ have different z-coordinates. Prove that a generic plane in $\mathbb{R}^3$ does not contain ℓ.
   Instructions: In Section 1.10 we studied point-line duality in $\mathbb{R}^2$. A similar point-plane duality in $\mathbb{R}^3$:

$$(a, b, c) \in \mathbb{R} \qquad \Longleftrightarrow \qquad \mathbf{V}(z - ax - by + c).$$

This duality does not include planes that are parallel to the xy-plane. What happens to the set of planes that contain ℓ under this duality?

Exercise 4.7 (a) Consider two polynomials $f, g \in \mathbb{R}[x_1, x_2, x_3]$ of degrees k_f and k_g. Prove that if f and g have no common factors then $\mathbf{V}(f) \cap \mathbf{V}(g)$ contains at most $k_f k_g$ lines. (Hint: consider a generic plane in $\mathbb{R}^3$.)

(b) Consider two polynomials $f, g \in \mathbb{R}[x_1, \ldots, x_d]$ of degrees k_f and k_g. Prove that if f and g have no common factors then $\mathbf{V}(f) \cap \mathbf{V}(g)$ contains at most $k_f k_g$ flats of dimension $d - 2$.

Exercise 4.8 Consider $f \in \mathbb{R}[x_1, x_2, x_3]$ of degrees k such that $\mathbf{V}(f)$ is one-dimensional. Prove that $\mathbf{V}(f)$ contains at most $k(2k - 1)$ lines. (Hint: First solve Exercise 4.7(a).)

Exercise 4.9 The following result is by Green and Tao (2013).

Theorem. *Let $\mathcal{P}$ be a set of n points in $\mathbb{R}^2$. Then at most $n^2/6 - n/2 + 1$ lines contain three points of $\mathcal{P}$.*

Use this theorem to prove that the same result holds also in $\mathbb{R}^d$ for every $d \geq 3$. (Hint: First explain why a generic projection of a line results in a line.)

Exercise 4.10 (a) Show that Warren's theorem (Theorem 3.2) is asymptotically tight: For every $d \geq 2$ and k, prove that there exists a polynomial $f \in \mathbb{R}[x_1, \ldots, x_d]$ of degree k such that $\mathbb{R}^d \backslash \mathbf{V}(f)$ has $\Theta(k^d)$ connected components. (Hint: You might like to first consider the case of $d = 2$.)

(b) Show that the Milnor–Thom theorem (Theorem 4.9) is asymptotically tight: For every $d \geq 2$ and k, prove that there exists a polynomial $f \in \mathbb{R}[x_1, \ldots, x_d]$ such that $\mathbf{V}(f)$ has $\Theta(k^d)$ connected components.

Exercise 4.11 (a) Prove Warren's theorem (Theorem 3.2) by using the Milnor–Thom theorem (Theorem 4.9). (Hint: Change the polynomial by an $\varepsilon > 0$.)

(b) Let $f \in \mathbb{R}[x_1, \ldots, x_d]$ be a polynomial of degree k in $\mathbb{R}^d$. Prove that $\mathbf{V}(f)$ has $O_d(k^d)$ connected components by using Warren's theorem. (Hint: Change the polynomial by an $\varepsilon > 0$.)

Exercise 4.12 In this problem you are asked to prove a special case of Theorem 4.15: Let $U \subset \mathbb{R}^d$ be a hyperplane and let $f \in \mathbb{R}[x_1, \ldots, x_d]$ be a polynomial of degree k. Prove that the number of connected components of $U \backslash \mathbf{V}(f)$ is $O(k^{d-1})$. You are not allowed to rely on Theorem 4.15, but you may use similar results that we encountered.

5

The Joints Problem and Degree Reduction

The situation is a little bit like the spread of a disease in a population. If each member of a population is exposed to many other members of the population, then a fairly small outbreak can become an epidemic.

Larry Guth (2016) explaining a degree reduction argument before epidemics were in style.

Before their seminal distinct distances paper, Guth and Katz wrote a paper that introduced new polynomial methods (Guth and Katz, 2010). In this chapter we study one of the two problems that were resolved in that paper: *the joints problem*. The solution to this problem relies on a simple polynomial technique, which is a good warm-up for working in dimensions $d \geq 3$. We apply variants of this technique for two additional problems. A large part of this chapter is based on ideas from Guth (2016).

5.1 The Joints Problem

Let $\mathcal{L}$ be a set of lines in $\mathbb{R}^3$. A *joint* of $\mathcal{L}$ is a point of $\mathbb{R}^3$ incident to three lines of $\mathcal{L}$ that are not contained in the same plane. In other words, the directions of the three lines are linearly independent. The joint may be incident to additional lines of $\mathcal{L}$, including ones that lie in the same plane. The joints problem asks for the maximum number of joints in a set of n lines in $\mathbb{R}^3$.

Claim 5.1 *There exists a set of n lines in $\mathbb{R}^3$ that spans $\Theta(n^{3/2})$ joints.*

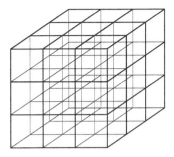

Figure 5.1 A construction with $n/3$ lines parallel to each axis and $\Theta(n^{3/2})$ joints.

Proof Consider $n/3$ lines in the direction of the x-axis, $n/3$ lines in the direction of the y-axis, and $n/3$ lines in the direction of the z-axis:

$$\mathcal{L} = \left\{ \mathbf{V}(x - a, y - b) \subset \mathbb{R}^3 : a, b \in \mathbb{N} \text{ and } 1 \le a, b \le \sqrt{n/3} \right\}$$

$$\bigcup \left\{ \mathbf{V}(x - a, z - c) \subset \mathbb{R}^3 : a, c \in \mathbb{N} \text{ and } 1 \le a, c \le \sqrt{n/3} \right\}$$

$$\bigcup \left\{ \mathbf{V}(y - b, z - c) \subset \mathbb{R}^3 : b, c \in \mathbb{N} \text{ and } 1 \le b, c \le \sqrt{n/3} \right\} .$$

An example is depicted in Figure 5.1. The joints spanned by this set are

$$\left\{ (a, b, c) \in \mathbb{N}^3 : 1 \le a, b, c \le \sqrt{n/3} \right\} . \qquad \square$$

The construction of Claim 5.1 has many structural properties. For example, every two lines lines are either parallel or orthogonal. Also, many planes contain $\Theta(n^{2/3})$ lines. The following construction leads to the same asymptotic number of joints, with less structure.

Claim 5.2 *There exists a set of n lines in $\mathbb{R}^3$ that spans $\Theta(n^{3/2})$ joints, with no two lines being parallel or orthogonal and with every plane containing $O(n^{1/2})$ lines.*

Proof Let H be a set of m generic planes in $\mathbb{R}^3$ for a parameter m that is set below. By generic planes, we mean that no two planes are parallel, no three intersect in a line, and no four intersect in a point. Set

$$\mathcal{L} = \{ \Pi \cap \Pi' : \Pi, \Pi' \in H \text{ and } \Pi \ne \Pi' \}.$$

Since no three planes intersect in a line, $\mathcal{L}$ is a set of $\binom{m}{2}$ distinct lines. We may also assume that no two lines of $\mathcal{L}$ are parallel or orthogonal. We fix the value of m so that $|\mathcal{L}| = n$, and note that $m = \Theta(\sqrt{n})$. For any three distinct planes Π, Π', Π'', the three lines $\Pi \cap \Pi'$, $\Pi \cap \Pi''$, and $\Pi' \cap \Pi''$ form a distinct joint. See Figure 5.2. Thus, the number of joints is $\binom{m}{3} = \Theta(n^{3/2})$. $\qquad \square$

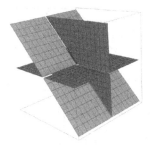

Figure 5.2 Every two planes intersect in a line. The three lines form a joint.

The difficult part of the joints problem is to prove that every set of n lines in $\mathbb{R}^3$ spans $O(n^{3/2})$ joints. This started as a discrete geometry problem (for example, see Chazelle et al., 1992). Then, Wolff (1999) observed a connection between the joints problem and the Kakeya problem. This led harmonic analysts to also study the joints problem. After a sequence of increasingly better bounds, the problem was completely solved by Guth and Katz.

Theorem 5.3 (Guth and Katz, 2010) *The maximum number of joints in a set of n lines in $\mathbb{R}^3$ is $\Theta(n^{3/2})$.*

In Chapter 3, we relied on polynomials that partition the space into well-behaved cells. We now use a different approach, considering polynomials that vanish on a point set.

Lemma 5.4 *Consider a set $\mathcal{P}$ of m points in $\mathbb{R}^d$ and a positive integer D, such that $\binom{d+D}{d} > m$. Then there exists $f \in \mathbb{R}[x_1, \ldots, x_d] \setminus \{0\}$ of degree at most D that satisfies $\mathcal{P} \subset \mathbf{V}(f)$.*

Proof As we say in the proof of the polynomial ham sandwich theorem (Theorem 3.6), the number of distinct monomials in $x_1, \ldots, x_d$ of degree at most D is $\binom{D+d}{d}$. We set $k = \binom{D+d}{d}$ and note that $k > m$.

For a polynomial $f \in \mathbb{R}[x_1, \ldots, x_d]$ of degree D, we denote the coefficients of the monomials of f as $c_1, \ldots, c_k$. Asking f to vanish on a point $p \in \mathcal{P}$ gives a linear homogeneous equation in $c_1, \ldots, c_k$. Asking f to vanish on every point of $\mathcal{P}$ gives a system of m linear homogeneous equations in k variables. Since $k > m$, this system has a nontrivial solution. This solution is a set of values for the coefficients $c_1, \ldots, c_k$ such that f vanishes on $\mathcal{P}$ and not all values are zero. $\qquad\square$

The following proof is our first use of the polynomial method that relies on Lemma 5.4.

Lemma 5.5 *Consider a set $\mathcal{L}$ of lines in $\mathbb{R}^3$ and let $\mathcal{J}$ be the set of joints of $\mathcal{L}$. Then there exists a line of $\mathcal{L}$ that is incident to at most $3|\mathcal{J}|^{1/3}$ of the joints.*

Proof We assume for contradiction that every line of $\mathcal{L}$ is incident to more than $3|\mathcal{J}|^{1/3}$ joints. Let $f \in \mathbb{R}[x, y, z]$ be a minimum degree nonzero polynomial that vanishes on $\mathcal{J}$. Since $\binom{3+3|\mathcal{J}|^{1/3}}{3} > |\mathcal{J}|$, Lemma 5.4 implies that $\deg f \le 3|\mathcal{J}|^{1/3}$.

We consider a line $\ell \in \mathcal{L}$ and choose a plane Π that contains ℓ. We may also assume that $\Pi \not\subseteq \mathbf{V}(f)$, since this is the case for a generic plane that contains ℓ. This implies that $\gamma = \mathbf{V}(f) \cap \Pi$ is a variety of dimension at most one. We think of Π as $\mathbb{R}^2$ and of γ as a variety in $\mathbb{R}^2$. Then γ is defined by a polynomial of degree at most $3|\mathcal{J}|^{1/3}$. By applying Bézout's theorem (Theorem 2.5) inside of Π, we get that either $\ell \subseteq \gamma$ or $|\gamma \cap \ell| \le 3|\mathcal{J}|^{1/3}$. By assumption, ℓ contains more than $3|\mathcal{J}|^{1/3}$ joints, so we have that $\ell \subseteq \gamma \subset \mathbf{V}(f)$. Since this holds for every line of $\mathcal{L}$, we conclude that $\mathbf{V}(f)$ contains all lines of $\mathcal{L}$.

Consider a point $p \in \mathcal{J}$, and let $\ell_1, \ell_2, \ell_3 \in \mathcal{L}$ be three lines that are incident to p and are not contained in a common plane. By the above, all three lines are contained in $\mathbf{V}(f)$. Let $\tau \colon \mathbb{R}^3 \to \mathbb{R}^3$ be the translation that takes p to the origin. We set $\ell_1' = \tau(\ell_1)$, $\ell_2' = \tau(\ell_2)$, and $\ell_3' = \tau(\ell_3)$. If p is a regular point of $\mathbf{V}(f)$, then the tangent plane $T_p\mathbf{V}(f)$ contains ℓ_1', ℓ_2', and ℓ_3'. Since these three lines are not contained in the same plane by assumption, p is a singular point of $\mathbf{V}(f)$. We conclude that every joint of $\mathcal{J}$ is a singular point of $\mathbf{V}(f)$. In other words, $\nabla f(p) = 0$ for every $p \in \mathcal{J}$.

Since f is not identically zero, it includes at least one of the three variables x, y, z. Without loss of generality, we assume that f includes x. This implies that the first partial derivative $f_x = \frac{\partial f}{\partial x}$ is not identically zero. Since ∇f vanishes on every joint of $\mathcal{J}$, so does f_x. This contradicts f being a minimum degree polynomial that vanishes on $\mathcal{J}$, and completes the proof of the lemma.

A pedantic reader might have noticed a missing case in the above proof. It is possible that $\mathbf{V}(f)$ is one-dimensional around a joint $p \in \mathcal{J}$, so $T_p\mathbf{V}(f)$ is not a plane. In this case, each of the lines ℓ_1, ℓ_2, and ℓ_3 is a component of $\mathbf{V}(f)$. Since p is contained in more than one component, it is again a singular point of $\mathbf{V}(f)$. $\quad\square$

After deriving Lemma 5.5, proving Theorem 5.3 is straightforward.

Proof of Theorem 5.3 Let $\mathcal{L}$ be a set of n lines in $\mathbb{R}^3$, and let $\mathcal{J}$ be the set of joints of $\mathcal{L}$. We set $x = |\mathcal{J}|$. We repeatedly consider a line that is incident to at most $3x^{1/3}$ joints of $\mathcal{J}$, remove this line from $\mathcal{L}$, and update $\mathcal{J}$ accordingly (the value of x remains fixed during this process). By Lemma 5.5, such a line exists at every step. Since every line removal destroys at most $3x^{1/3}$ joints, and eventually all x joints are destroyed, we have that

$$x \le n \cdot 3x^{1/3}.$$

The assertion of the theorem is obtained by tidying up this equation.   □

5.2 Additional Applications of the Polynomial Argument

To solve the joints problem, we relied on a polynomial method that is based on Lemma 5.4. We now study two additional applications of this method.

Reguli: Let ℓ_1, ℓ_2, ℓ_3 be parallel lines in $\mathbb{R}^3$ and consider the union of all lines that intersect ℓ_1, ℓ_2, and ℓ_3. When ℓ_1, ℓ_2, ℓ_3 are contained in a common plane Π, the union of all lines that intersect them is Π. When ℓ_1, ℓ_2, ℓ_3 are not contained in a common plane, no line intersects all three and the union is empty.

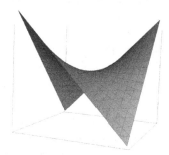

Figure 5.3 The hyperbolic paraboloid $\mathbf{V}(z - xy)$ contains two infinite families of pairwise-skew lines.

Two lines in $\mathbb{R}^3$ are *skew* if no plane contains both lines. Equivalently, two lines are skew if they are neither parallel nor intersecting. A *regulus* is the union of all lines in $\mathbb{R}^3$ that intersect three *pairwise-skew* lines ℓ_1, ℓ_2, ℓ_3. One example of a regulus is the hyperbolic paraboloid $\mathbf{V}(z - xy)$, depicted in Figure 5.3. This variety is the union of lines of the form $\mathbf{V}(x - c, z - cy)$ where $c \in \mathbb{R}$. It is also the union of the lines of the form $\mathbf{V}(y - c, z - cx)$. The lines in each of these families are pairwise-skew. We obtain $\mathbf{V}(z - xy)$ by taking the union of all lines that intersect three lines of the form $\mathbf{V}(y - c, z - cx)$. This is the family of lines of the form $\mathbf{V}(x - c, z - cy)$. Similarly, we obtain $\mathbf{V}(z - xy)$ by taking the union of all lines that intersect three lines of the form $\mathbf{V}(x - c, z - cy)$.

We are now ready for our second application of Lemma 5.4.

Lemma 5.6 *Every regulus in* $\mathbb{R}^3$ *is contained in an irreducible variety of dimension two and degree two.*

Proof Consider a regulus $S \subset \mathbb{R}^3$ that is defined by the three pairwise-skew lines ℓ_1, ℓ_2, and ℓ_3. Let $\mathcal{P}$ be a set of nine points that is obtained by arbitrarily choosing three points from each of ℓ_1, ℓ_2, and ℓ_3. Since $\binom{3+2}{3} = 10 > 9$, Lemma 5.4 implies that there exists a nonzero $f \in \mathbb{R}[x, y, z]$ of degree at most two that vanishes on $\mathcal{P}$. We apply Bézout's theorem in a plane that contains ℓ_1 and is not contained in $\mathbf{V}(f)$, like we did in the proof of Lemma 5.5. Since ℓ_1 intersects $\mathbf{V}(f)$ in at least three points, Bézout's theorem implies that $\ell_1 \subset \mathbf{V}(f)$. For the same reason, we also have that $\ell_2, \ell_3 \subset \mathbf{V}(f)$. If $\mathbf{V}(f)$ is a plane or a union of two planes, then at least two of ℓ_1, ℓ_2, ℓ_3 are contained in a common plane. This would contradict the assumption that the three lines are pairwise-skew. We conclude that $\mathbf{V}(f)$ is an irreducible variety of degree two.

Consider a line ℓ' that intersects all three lines ℓ_1, ℓ_2, ℓ_3. The three intersection points are distinct since ℓ_1, ℓ_2, ℓ_3 are disjoint. Since ℓ' intersects $\mathbf{V}(f)$ in at least three points, Bézout's theorem implies that $\ell' \subset \mathbf{V}(f)$. Since S is the union of all lines that intersect ℓ_1, ℓ_2, ℓ_3, we get that $S \subseteq \mathbf{V}(f)$. $\square$

With some more work, we can show that the reguli in $\mathbb{R}^3$ are hyperbolic paraboloids and hyperboloids of one sheet. For more information, see Section 14.5.

Degree reduction: In the proofs of Lemma 5.5 and Lemma 5.6, we argued that various lines are contained in a variety $\mathbf{V}(f)$. In both cases, we achieved this by applying Bézout's theorem in a generic plane that contains the line. The following claim is another variant of this approach.

Claim 5.7 *Let* $\mathcal{L}$ *be a set of n lines in* $\mathbb{R}^3$. *Then there exists a nonzero* $f \in \mathbb{R}[x, y, z]$ *of degree at most* $3\sqrt{n}$ *such that* $\mathbf{V}(f)$ *contains all lines of* $\mathcal{L}$.

Proof We create a point set $\mathcal{P}$ by arbitrarily choosing $4\sqrt{n}$ points from every line of $\mathcal{L}$. For simplicity, we do not choose points that are incident to more than one line of $\mathcal{L}$. This implies that $|\mathcal{P}| = 4n^{3/2}$. Since $\binom{3+3\sqrt{n}}{3} > 4n^{3/2}$, Lemma 5.4 implies that there exists a nonzero $f \in \mathbb{R}[x, y, z]$ of degree at most $3\sqrt{n}$ that vanishes on $\mathcal{P}$. We consider a line $\ell \in \mathcal{L}$ and recall that $\mathbf{V}(f)$ is incident to at least $4\sqrt{n}$ points of ℓ. By applying Bézout's theorem in a plane that contains ℓ and is not contained in $\mathbf{V}(f)$, we get that $\ell \subset \mathbf{V}(f)$. Thus, $\mathbf{V}(f)$ contains all lines of $\mathcal{L}$. $\square$

When every line of $\mathcal{L}$ contains many intersection points with other lines of $\mathcal{L}$, we can improve the bound of Claim 5.7. This idea is called *degree reduction*.

Lemma 5.8 *Consider an integer k that may depend on n. Let $\mathcal{L}$ be a set of n lines in $\mathbb{R}^3$, such that each $\ell \in \mathcal{L}$ contains at least k points where ℓ intersects other lines of $\mathcal{L}$. Then there exists a nonzero $f \in \mathbb{R}[x, y, z]$ of degree at most $O(n^{3/4}/\sqrt{k})$ such that $\mathbf{V}(f)$ contains all lines of $\mathcal{L}$.*

Note that the condition of Lemma 5.8 is stronger than asking for every line of $\mathcal{L}$ to intersect many other lines of $\mathcal{L}$. It also asks for every line to have many different intersection points with the other lines.

Proof of Lemma 5.8 When $k \le 50^2\sqrt{n}$, the lemma is immediately implied by Claim 5.7. Indeed, in this case $3\sqrt{n} = O(n^{3/4}/\sqrt{k})$. We may thus assume that $k > 50^2\sqrt{n}$. We may also assume that n is larger than some sufficiently large constant C. To handle the case of $n \le C$, we take the constant hidden by the $O(\cdot)$-notation in the lemma to be sufficiently large.

We set $p = 100\sqrt{n}/k$ and note that $0 < p < 1$. We create a subset $\mathcal{L}' \subset \mathcal{L}$ by taking every line of $\mathcal{L}$ with probability p. With positive probability, $|\mathcal{L}'| < 200n^{3/2}/k$ and every line of $\mathcal{L}\backslash\mathcal{L}'$ has at least $\sqrt{n}$ intersection points with lines of $\mathcal{L}'$. The details of this standard probabilistic argument are postponed to the optional Section 5.3. Since this scenario occurs with positive probability, there exists a subset $\mathcal{L}'$ that satisfies the above properties. We fix such a subset $\mathcal{L}'$.

By Claim 5.7, there exists a nonzero $f \in \mathbb{R}[x, y, z]$ that vanishes on every line of $\mathcal{L}'$ and is of degree at most

$$3 \cdot \left(200n^{3/2}/k\right)^{1/2} < 45n^{3/4}/\sqrt{k} < \sqrt{n}.$$

The last transition relies on the assumption that $k > 50^2\sqrt{n}$.

We consider a line $\ell \in \mathcal{L}\backslash\mathcal{L}'$. By the above, $\mathbf{V}(f)$ contains at least $\sqrt{n}$ points of ℓ. We apply Bézout's theorem in a plane that contains ℓ and is not contained in $\mathbf{V}(f)$. This implies that $\ell \subset \mathbf{V}(f)$. We conclude that $\mathbf{V}(f)$ contains every line of $\mathcal{L}$. $\qquad\square$

5.3 (Optional) The Probabilistic Argument

The proof of Lemma 5.8 relies on a probabilistic statement without proving it. In this section we prove this probabilistic statement using a standard probabilistic argument. An uninterested reader can safely skip this section.

We first recall some basic probability. Let $B(n, p)$ denote a random variable with a *binomial distribution*: the number of coin flips that landed on heads when performing n independent flips, each with probability p for heads. We write $X \sim B(n, p)$ to state that the random variable X has the distribution $B(n, p)$. For a proof of the following lemma, see for example Alon and Spencer (2004, Theorem A.1.15).

Lemma 5.9 (Chernoff bounds) *Let $X \sim B(n, p)$ with $0 < p < 1$, and let $\delta > 0$. Then*

$$\Pr[X \geq (1 + \delta)np] \leq \left(\frac{e^\delta}{(1 + \delta)^{1+\delta}} \right)^{np},$$

$$\Pr[X \leq (1 - \delta)np] \leq e^{-pn\delta^2/2}.$$

We are now ready to prove the probabilistic statement. To recall what we need to prove, we phrase it as a lemma.

Lemma 5.10 *Let $\mathcal{L}$ be a set of n lines, for a sufficiently large n. For $50^2 \sqrt{n} < k < n$, let $p = 100\sqrt{n}/k$. Assume that each line of $\mathcal{L}$ contains at least k intersection points with other lines of $\mathcal{L}$. Let $\mathcal{L}'$ be a subset of $\mathcal{L}$ obtained by taking every line of $\mathcal{L}$ with probability p. Then with positive probability. $|\mathcal{L}'| < 200n^{3/2}/k$ and every line of $\mathcal{L} \backslash \mathcal{L}'$ contains at least $\sqrt{n}$ intersection points with lines of $\mathcal{L}'$.*

Proof We note that $|\mathcal{L}'| \sim B(n, p)$. By Lemma 5.9 with $\delta = 1$, we get that

$$\Pr\left[|\mathcal{L}'| \geq 200n^{3/2}/k \right] \leq \left(\frac{e}{4} \right)^{100n^{3/2}/k} < \left(\frac{3}{4} \right)^{100\sqrt{n}}.$$

We fix a line $\ell \in \mathcal{L}$. We ignore all but k of the intersection points on ℓ with other lines. Out of these k intersection points, let X_ℓ denote the number of intersection points that ℓ has with lines of $\mathcal{L}'$. Note that $X_\ell \sim B(k, p)$. By Lemma 5.9 with $\delta = 1/2$, we have

$$\Pr\left[X_\ell < \sqrt{n} \right] < \Pr\left[X_\ell \leq 50\sqrt{n} \right] = \Pr\left[X_\ell \leq kp/2 \right] \leq e^{-100\sqrt{n}/8} < e^{-10\sqrt{n}}.$$

We recall the *union bound* principle: Given a set of potential events, the probability of at least one event occurring is at most the sum of the probabilities of the individual events. Thus, the probability that $|\mathcal{L}'| \geq 200n^{3/2}/k$ or that at least one line of $\mathcal{L} \backslash \mathcal{L}'$ has fewer than $\sqrt{n}$ intersection points with $\mathcal{L}'$, is smaller than

$$\left(\frac{3}{4} \right)^{100\sqrt{n}} + |\mathcal{L} \backslash \mathcal{L}'| \cdot e^{-10\sqrt{n}} < \left(\frac{3}{4} \right)^{100\sqrt{n}} + n \cdot e^{-10\sqrt{n}}.$$

When n is sufficiently large, this probability is smaller than 0.5. Thus, with probability larger than 0.5, we have that $|\mathcal{L}'| < 200n^{3/2}/k$ and every line of $\mathcal{L} \backslash \mathcal{L}'$ has at least $\sqrt{n}$ intersection points with $\mathcal{L}'$. □

5.4 Exercises

Exercise 5.1   In Lemma 5.4, how does the condition on D change in each of the following cases:
(a) In every monomial of f, the degree of each variable must be even.
(b) In every monomial of f, the degree of x_1 must be at least one.
(c) The degree of every monomial of f must be even.

Exercise 5.2   Let $\mathcal{P}$ be a set of m points in $\mathbb{R}^3$. Prove that there exists a one-dimensional variety of complexity $O(m^{1/2})$ that contains $\mathcal{P}$. (Hint: Use Lemma 5.4 and projections.)

Exercise 5.3   Let $\mathcal{L}$ be a set of n lines in $\mathbb{R}^d$. A *joint* of $\mathcal{L}$ is a point of $\mathbb{R}^d$ incident to d lines of $\mathcal{L}$ that are not all contained in the same hyperplane. A joint may be incident to additional lines of $\mathcal{L}$ that are contained in the same hyperplane. Derive an upper bound for the maximum number of joints of $\mathcal{L}$, and prove that this bound is asymptotically tight.

Exercise 5.4 (Guth, 2016)   Consider the lattice

$$\mathcal{P} = \{(a, b, c) \in \mathbb{R}^3 : 1 \le a \le n \text{ and } 1 \le b, c \le 2n\}.$$

(a) Find a nonzero $f \in \mathbb{R}[x, y, z]$ of degree n such that $\mathcal{P} \subset \mathbf{V}(f)$.
(b) Prove that no nonzero $f \in \mathbb{R}[x, y, z]$ of degree smaller than n satisfies $\mathcal{P} \subset \mathbf{V}(f)$. (Hint: Use Exercise 4.7(a).)

Exercise 5.5   Prove the following claim with x as small as possible: Let $\mathcal{L}$ be a set of n lines in $\mathbb{R}^d$. Then there exists a polynomial $f \in \mathbb{R}[x_1, \ldots, x_d]$ of degree $O(n^x)$ such that $\mathbf{V}(f)$ contains all lines of $\mathcal{L}$.

Exercise 5.6   A *planar curve in* $\mathbb{R}^3$ is a curve that is contained in some plane in $\mathbb{R}^3$.
(a) Prove Claim 5.7 when the lines are replaced with irreducible planar curves of degree d. That is, each curve is the intersection of a plane with a variety defined by a polynomial of degree d. The bound of the claim should change from $3\sqrt{n}$ to $O(\sqrt{n} \cdot d^{3/2})$.
(b) Prove Lemma 5.8 when the lines are replaced with irreducible planar curves of degree d. The bound of the lemma should change from $O(n^{3/4}/\sqrt{k})$ to $O(n^{3/4}d^{11/4}/\sqrt{k})$. If you prefer not to delve into the probabilistic calculations of Section 5.3, you can instead assume that random events behave like their expectation.

Exercise 5.7

(a) Let $\mathcal{P}$ be a set of m points in $\mathbb{R}^2$. Use Theorem 5.4 to prove that the number of ellipses in $\mathbb{R}^2$ that are incident to more than $4\sqrt{m}$ points of $\mathcal{P}$ is $O(\sqrt{m})$.

(b) For any $0 < \alpha < 1/2$, use part (a) to prove that the number of ellipses that contain $\Omega(m^{\alpha+1/2})$ points of $\mathcal{P}$ is $O(n^{1/2-\alpha})$. (Hint: this requires familiarity with the probabilistic method.)

5.5 Open Problems

The joints problem has been generalized to the case of d lines in $\mathbb{R}^d$ (Kaplan et al., 2010; Quilodrán, 2010) (see also Exercise 5.3). The $O(\cdot)$-notation in the joints bound has been replaced with an exact constant (Yu and Zhao, 2019). Many variants of the problem have been suggested and resolved: joints formed by higher-dimensional flats, joints formed by varieties, joints in spaces over other fields, and more (for example, see Carbery and Iliopoulou, 2020; Tidor et al., 2020; Zhang, 2020). Thus, one might say that no *major* open problems remain open.

One might also say that the structural joints problems are major open problems. For example, the problem of characterizing all configurations of n lines in $\mathbb{R}^3$ that span $\Theta(n^{3/2})$ joints. One can also consider the nonasymptotic variant: Characterize all configurations of n lines in $\mathbb{R}^3$ that span $\sqrt{2} \cdot n^{3/2}/3 + o(n^{3/2})$ joints.[1] The construction from Claim 5.2 has this number of joints. The following is a special case of a more general conjecture by Yu and Zhao (2019).

Conjecture 5.11 *The only sets of n lines in $\mathbb{R}^3$ that have $\sqrt{2} \cdot n^{3/2}/3 + o(n^{3/2})$ joints are the ones described in Claim 5.2.*

---

[1] The notation $o(n^{3/2})$ means asymptotically strictly smaller than $n^{3/2}$.

6

Polynomial Methods in Finite Fields

In this chapter, we use polynomial methods to study incidence-related problems in spaces over finite fields. We focus on two breakthroughs: A solution to the finite field Kakeya problem and the cap set problem. The proofs of these results are short, elegant, and require mostly elementary tools. In Chapter 13, we study point-line incidences in spaces over finite fields, which requires more involved arguments.

6.1 Finite Fields Preliminaries

We begin by recalling a few basic definitions. A field is *finite* if it contains finitely many elements. The *order* of a finite field is the number of elements in that field. There is a finite field of order $q \in \mathbb{N}\setminus\{0\}$ if and only if $q = p^r$ for some prime p and positive integer r. We say that such a q is a *prime power*. For every such q there is a unique field of that order, up to isomorphisms. We denote this finite field as $\mathbb{F}_q$ and as $\mathbb{F}_{p^r}$.

For a prime p, we can think of $\mathbb{F}_p$ as the set of integers $\{0, 1, \ldots, p-1\}$ under addition and multiplication mod p. To describe a field $\mathbb{F}_{p^r}$ with prime p and integer $r > 1$, we consider an irreducible polynomial $f \in \mathbb{F}_p[x]$ of degree r. We think of $\mathbb{F}_{p^r}$ as the set of polynomials in $\mathbb{F}_p[x]$ under addition and multiplication modulo f. That is, when multiplying two polynomials in $\mathbb{F}_p[x]$ we first perform the standard polynomial multiplication, then replace each coefficient with its value mod p, and finally divide by f and take the remainder. For example, by setting $f = x^2 + 1$ we get that $\mathbb{F}_9 = \{0, 1, 2, x, x+1, x+2, 2x, 2x+1, 2x+2\}$. The multiplicative group of $\mathbb{F}_{p^r}$ is cyclic, and the additive group of $\mathbb{F}_{p^r}$ is the direct product of r cyclic groups of order p.

In this chapter we work in the vector space $\mathbb{F}_q^d$, for some prime power q and integer $d \geq 2$. We borrow the standard geometric notation from $\mathbb{R}^d$.

76

Figure 6.1 The line in $\mathbb{F}_5^2$ defined by $y \equiv x + 2$.

For example, we refer to $\mathbb{F}_q^2$ as a finite plane and to an element of $\mathbb{F}_q^2$ as a point in the plane. Let 0_d be a vector of d zeros. In other words, 0_d is the origin of a d-dimensional space.

A line in $\mathbb{F}_q^2$ is the zero set of a linear polynomial in $\mathbb{F}_q[x, y]$. Figure 6.1 shows that such a line might have a nonintuitive behavior. As in $\mathbb{R}^2$, an equivalent definition of a line is $\{u + tv : t \in \mathbb{F}_q\}$ where $u, v \in \mathbb{F}_q^2$ and $v \neq 0_d$ (see Exercise 6.1). This equivalent definition of a line is often easier to work with in a space $\mathbb{F}_q^d$ of dimension $d > 2$.

When working in $\mathbb{F}_{p^r}$, our arithmetic is as described above: We first fix an irreducible polynomial $f \in \mathbb{F}_p[x_1, \ldots, x_d]$ of degree r. After each arithmetic operation, we take each coefficient mod p and the entire expression modulo f. To distinguish this operation from standard additions and multiplications, we use the notation $x \equiv y$ (for brevity we do not add the 'mod' afterwards).

Let $a = (a_1, \ldots, a_d)$ and $b = (b_1, \ldots, b_d)$ be points in $\mathbb{F}_q^d$. We define point addition in $\mathbb{F}_q^d$ as $a + b \equiv (a_1 + b_1, \ldots, a_d + b_d)$. The addition operations in each coordinate are with respect to an irreducible polynomial, as described above.

6.2 The Finite Field Kakeya Problem

A set $S \subset \mathbb{R}^d$ is a *Kakeya set* if S contains a line segment of length 1 in every direction. For example, the disk $\{(p_x, p_y) \in \mathbb{R}^2 : p_x^2 + p_y^2 \leq 1/4\}$ is a Kakeya set in $\mathbb{R}^2$. See Figure 6.2. Surprisingly, there exist Kakeya sets in $\mathbb{R}^d$ of measure zero, for any $d \geq 2$. This led to the study of the minimum dimension that a Kakeya set can have, under various definitions of dimension. This is the *Kakeya problem*, which is a main open problem in harmonic analysis. Since this problem is far from being resolved, Wolff (1999) suggested that it might be useful to study the following finite fields variant.

Consider a finite field $\mathbb{F}_q$ and $d > 1$. Every line $\ell \subset \mathbb{F}_q^d$ can be written as $\{u + tv : t \in \mathbb{F}_q\}$, where $u \in \mathbb{F}_q^d$ is a point on ℓ and $v \neq 0_d$ is the direction vector of ℓ. A set $\mathcal{P} \subset \mathbb{F}_q^d$ is a *Kakeya set* if, for every $u \in \mathbb{F}_q^d \setminus \{0_2\}$, at least one line with direction u is contained in $\mathcal{P}$. As an example, while $\mathbb{F}_q^2 \setminus \{0_d\}$ consists of

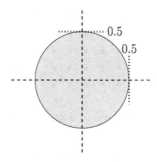

Figure 6.2 The disk $\{(p_x, p_y) \in \mathbb{R}^2 : p_x^2 + p_y^2 \leq 1/4\}$ contains a line segment of length 1 in every direction.

$q^2 - 1$ vectors, the lines in $\mathbb{F}_q^2$ only have $q + 1$ directions: $(1, 0), (1, 1), (1, 2), \ldots,$ $(1, q - 1), (0, 1)$. Thus, a Kakeya set in $\mathbb{F}_q^2$ contains $q + 1$ lines with distinct directions.

Dvir (2009) solved the finite field Kakeya problem by introducing a simple and elegant polynomial method. Dvir's proof is often considered as the work that began the new era of polynomial methods.

Theorem 6.1 (Dvir, 2009) *Let $\mathcal{P} \subset \mathbb{F}_q^d$ be a Kakeya set. Then*

$$|\mathcal{P}| > \frac{(q - 1)^d}{d!}.$$

The space $\mathbb{F}_q^d$ consists of q^d points. Thus, Theorem 6.1 states that any Kakeya set in $\mathbb{F}_q^d$ contains a constant fraction of the points of $\mathbb{F}_q^d$. To prove Theorem 6.1, we first need to introduce a couple of tools.

The only polynomial from $\mathbb{R}[x_1, \ldots, x_d]$ that vanishes on all the points of $\mathbb{R}^d$ is 0. When working in $\mathbb{F}_q[x_1, \ldots, x_d]$, this claim is no longer true. As a simple example, note that $x_1^2 - x_1 \in \mathbb{F}_2[x_1, \ldots, x_d]$ vanishes on $\mathbb{F}_2^d$. The following result implies that 0 is the only polynomial of degree smaller than q that vanishes on $\mathbb{F}_q^d$.

Lemma 6.2 (Schwartz–Zippel lemma, Schwartz, 1980; Zippel, 1979) *Let $f \in \mathbb{F}_q[x_1, \ldots, x_d] \backslash \{0\}$ be of degree k. Then f vanishes on at most kq^{d-1} points of $\mathbb{F}_q^d$.*

The bound of Lemma 6.2 is tight. Indeed, the polynomial $(x_1 - 1)(x_1 - 2) \cdots (x_1 - k)$ is of degree k and vanishes on kq^{d-1} points of $\mathbb{F}_q^d$. To prove Lemma 6.2, we first recall the *factor theorem*.

Theorem 6.3 (Factor theorem) *Let $\mathbb{F}$ be a field or a commutative ring. For every $f \in \mathbb{F}[t]$ and $a \in \mathbb{F}$, if $g(a) = 0$ then $g(t) = (t - a) \cdot h(t)$ for some $h \in \mathbb{F}[t]$.*

Proof of Lemma 6.2 We prove the lemma by induction on d. For the induction basis, consider the case of $d = 1$. Since f is a product of at most k factors, Theorem 6.3 implies that f vanishes on at most k elements of $\mathbb{F}_q$.

For the induction step, we consider $d > 1$ and assume that the lemma holds for $d - 1$. We define

$$A = \{a \in \mathbb{F}_q \; : \; f(a, x_2, \ldots, x_d) \equiv 0\}.$$

That is, when plugging the value of any $a \in A$ into x_1, the polynomial f becomes identically zero. Let $p \in \mathbb{F}_q^d$ satisfy that $f(p) \equiv 0$. We say that p is a *type I zero* of f if the x_1-coordinate of p is in A. Otherwise, we say that p is a *type II zero* of f.

We consider f as a polynomial in x_1 with coefficients from the polynomial ring $\mathbb{F}_q[x_2, \ldots, x_d]$. Then, for every $a \in A$, Theorem 6.3 implies that f is divisible by $x_1 - a$. We set $r = |A|$ and write $A = \{a_1, \ldots, a_r\}$. This leads to

$$f(x_1, \ldots, x_d) = g(x_1, \ldots, x_d) \cdot (x_1 - a_1)^{m_1} \cdots (x_1 - a_r)^{m_r},$$

where $g \in \mathbb{F}_q[x_1, \ldots, x_d]$ and $m_1, \ldots, m_r \geq 1$. We note that

$$\deg g = k - (m_1 + \cdots + m_r) \leq k - r.$$

For $b \in \mathbb{F}_q \backslash A$, we set $g_b = g(b, x_2, \ldots, x_d)$. By definition, $g_b \in \mathbb{F}_q[x_2, \ldots, x_d]$ is of degree at most $k - r$ and is not identically zero. By the induction hypothesis, g_b vanishes on at most $(k - r) \cdot q^{d-2}$ points of $\mathbb{F}_q^{d-2}$. We sum this bound over all $b \in \mathbb{F}_q \backslash A$, to obtain that f has at most $(k - r) \cdot q^{d-1}$ type II zeros. Since $|A| = r$, we get that f has $r \cdot q^{d-1}$ type I zeros. In total, f has at most $k \cdot q^{d-1}$ zeros, which completes the induction step. □

The following lemma is a finite field variant of Lemma 5.4. The proof is identical to the proof of Lemma 5.4, so we do not repeat it here.

Lemma 6.4 *Consider a set $\mathcal{P}$ of m points in $\mathbb{F}_q^d$ and a positive integer k, such that $\binom{d+k}{d} > m$. Then there exists $f \in \mathbb{F}_q[x_1, \ldots, x_d] \backslash \{0\}$ of degree at most k that vanishes on every point of $\mathcal{P}$.*

We are now ready for Dvir's analysis of finite field Kakeya sets.

Proof of Theorem 6.1 Assume for contradiction that there exists a Kakeya set $\mathcal{P} \subset \mathbb{F}_q^d$ such that $|\mathcal{P}| \leq (q - 1)^d/d!$. By Lemma 6.4, there exists $f \in \mathbb{F}_q[x_1, \ldots, x_d]$ of degree $1 \leq k \leq q - 1$ that vanishes on $\mathcal{P}$. We write $f = \sum_{j=0}^k f_j$, where f_j is a homogeneous polynomial of degree j. By the definition of k, we have that $f_k \not\equiv 0$.

We consider a vector $v \in \mathbb{F}_q^d \backslash \{0_d\}$. Since $\mathcal{P}$ is a Kakeya set, it contains a line with direction v. Thus, there exists $u \in \mathcal{P}$ such that $\{u + tv : t \in \mathbb{F}_q\} \subset \mathcal{P}$. Since f vanishes on $\mathcal{P}$, we get that $f(u + tv)$ vanishes on every $t \in \mathbb{F}_q$. We set

$g(t) = f(u + tv)$ and note that $g(t)$ is a polynomial of degree k in $\mathbb{F}_q[t]$. Since $k < q$, Lemma 6.2 implies that $g(t) \equiv 0$. Equivalently, $f(u + tv) \equiv 0$, which implies that the coefficient of t^k in $f(u + tv)$ is zero. The coefficient of t^k in $f(u + tv)$ is $f_k(v)$, so $f_k(v) \equiv 0$.

By the preceding paragraph, $f_k(v) \equiv 0$ for every $v \neq 0_d$. Since $\deg f_k < q$, Lemma 6.2 implies that $f_k \equiv 0$. This contradicts f being of degree k, and completes the proof.   □

One may briefly describe the above proof as follows. We consider a minimum degree nonzero polynomial that vanishes on the point set. We show that the degree of the polynomial is bounded from above. We also show that this polynomial must vanish on many other points. This leads to a contradiction, since a nonzero polynomial of such a degree cannot vanish on so many points.

In the following section, we amplify the approach that is described in the preceding paragraph. This optional section can be safely skipped.

6.3 (Optional) The Method of Multiplicities

In Theorem 6.1, the asymptotic dependency in q is tight: Every Kakeya set in $\mathbb{F}_q^d$ is of size $\Theta_d(q^d)$. However, the factor $1/d!$ is far from tight. By using the *method of multiplicities*, Dvir et al. (2013) proved that every Kakeya set in $\mathbb{F}_q^d$ is of size at least $q^d/2^d$. When q is arbitrarily large, this is known to be tight up to a factor of 2.

In this section, we introduce the method of multiplicities. We use this method to prove a slightly weaker Kakeya result from Saraf and Sudan (2008).

Theorem 6.5   *There exists a constant $c \in \mathbb{R}$ that satisfies the following. If $\mathcal{P} \subset \mathbb{F}_q^d$ is a Kakeya set, then*

$$|\mathcal{P}| > \frac{c \cdot q^d}{2.6^d}.$$

The stronger lower bound $q^d/2^d$ also relies on the method of multiplicities. However, it requires a more involved proof. We do not present this proof here so as not to disrupt the sequence of short and elementary proofs that form this chapter. Yu and Zhao (2019) use a variant of the method of multiplicities to derive the correct constant for the joints problem (see Section 5.1).

Vanishing with multiplicity: For a field $\mathbb{F}$ that is not necessarily finite, consider a univariate polynomial $h \in \mathbb{F}[t]$. We recall that h vanishes on $p \in \mathbb{F}$ with *multiplicity k* if h is divisible by $(t - p)^k$; in other words, if we can write $h(t) = (t - p)^k \cdot h^*(t)$ for some $h^* \in \mathbb{R}[t]$. We now study a generalization of this notion to polynomials with any number of variables.

We write $x = (x_1, \ldots, x_d)$, where $x_1, \ldots, x_d$ are the coordinates of $\mathbb{F}_q^d$. The *multiplicity* of $g \in \mathbb{F}_q[x_1, \ldots, x_d]$ at $a \in \mathbb{F}_q^d$ is the largest integer k that satisfies: The polynomial $g(x+a) \in \mathbb{F}_q[x_1, \ldots, x_d]$ consists only of monomials of degree at least k; in other words, all the monomials of degree smaller than k in $g(x + a)$ have the coefficient 0. To see that this definition generalizes the univariate case, consider $h \in \mathbb{F}_q[t]$ that vanishes on $p \in \mathbb{F}$ with multiplicity k. Then $h(t) = (t - p)^k \cdot h^*(t)$ for some $h^* \in \mathbb{R}[t]$. We note that $h(t + p) = t^k \cdot h^*(t + p)$, so all the monomials of $h(t + p)$ are of degree at least k. For more information about this notion of multiplicity, see Tao (2014, Section 2).

For $g \in \mathbb{F}_q[x_1, \ldots, x_d]$ and $a \in \mathbb{F}_q^d$, the terms of degree 0 in $g(x + a)$ sum up to $g(a)$. Thus, $g(a) \equiv 0$ if and only if the multiplicity of g at a is at least 1. The following claim introduces another nice property of the multiplicity of g. Let $\ell \subset \mathbb{F}_q^d$ be a line. Then $\ell = \{u + tv : t \in \mathbb{F}_q\}$ for some $u, v \in \mathbb{F}_q^d$ with $v \neq 0_d$. The restriction of g to ℓ leads to the univariate function $g_{u,v}(t) = g(u + tv)$. Note that $g_{u,v}(t) \in \mathbb{F}_q[t]$.

Claim 6.6 *Consider a line $\ell = \{u + tv : t \in \mathbb{F}_q\} \subset \mathbb{F}_q^d$ for $u, v \in \mathbb{F}_q^d$ with $v \neq 0_d$. If $g \in \mathbb{F}_q[x_1, \ldots, x_d]$ has multiplicity k at $u + t_0 v$ (where $t_0 \in \mathbb{F}$), then $g_{u,v}$ has multiplicity at least k at t_0.*

Proof By definition, the lowest degree monomials of $g(x + (u + t_0 v))$ are of degree k. When setting $x = tv$, we get that $g(tv + (u + t_0 v)) \in \mathbb{F}[t]$ has no monomials of degree smaller than k. We note that $g(tv + (u + t_0 v)) = g_{u,v}(t + t_0)$. Thus, $g_{u,v}(t + t_0)$ has no monomials of degree smaller than k. This implies that $g_{u,v}(t + t_0)$ is divisible by t^k, which in turn implies that $g_{u,v}(t)$ is divisible by $(t - t_0)^k$. We conclude that $g_{u,v}$ has multiplicity at least k at t_0. $\square$

The polynomial method with multiplicities: We now strengthen the proof of Theorem 6.1 by also relying on multiplicities. To do that, we require a variant of Lemma 6.4 where the polynomial has a high multiplicity at each point. When working with polynomials from $\mathbb{F}_q[x_1, \ldots, x_d]$, we say that a monomial is *proper* if every variable x_j has a power of at most $q - 1$ (this is not a standard definition). Let $N_q(d, k)$ denote the number of proper monomials in $x_1, \ldots, x_d$ of degree smaller than kq.

Lemma 6.7 *Consider a set $\mathcal{P}$ of m points in $\mathbb{F}_q^d$ and a positive integer k, such that $\binom{d+k-1}{d} \cdot m < N_q(d, k)$. Then there exists $g \in \mathbb{F}_q[x_1, \ldots, x_d] \backslash \{0\}$ of degree smaller than kq that consists of proper monomials and has multiplicity at least k at each point of $\mathcal{P}$.*

Proof We adapt the proof of Lemma 5.4. By definition, there are $N_q(d, k)$ proper monomials in $x_1, \ldots, x_d$ of degree smaller than kq. Thus, g is defined by $N_q(d, k)$ coefficients from $\mathbb{F}_q$. Asking for a monomial of g to vanish at a point of $\mathcal{P}$ is a linear homogeneous equation in the coefficients of g.

As explained in the proof of Theorem 3.6, the number of monomials in $x_1, \ldots, x_d$ of degree at most $k - 1$ is $\binom{d+k-1}{d}$. The number of such proper monomials is at most $\binom{d+k-1}{d}$. Thus, each point of $\mathcal{P}$ leads to at most $\binom{d+k-1}{d}$ linear homogeneous equations in the coefficients of g. By the assumption $\binom{d+k-1}{d} \cdot m < N_q(d, k)$, we have a system of linear homogeneous equations with more variables than equations. Such a system has a nontrivial solution, which leads to the asserted g.   □

Instead of using Lemma 6.2, we rely on the following claim. For the proof of this claim, see Exercise 6.3.

Claim 6.8   *If $g \in \mathbb{F}_q[x_1, \ldots, x_d]$ consists of proper monomials and vanishes on every point of $\mathbb{F}_q^d$, then $g \equiv 0$.*

We are now ready to strengthen the proof of Theorem 6.1, by relying on multiplicities.

Proof of Theorem 6.5   Assume for contradiction that there exists a Kakeya set $\mathcal{P} \subset \mathbb{F}_q^d$ such that $|\mathcal{P}| < N_q(d, k) / \binom{d+k-1}{d}$, for some $0 < k < q$. By Lemma 6.7, there exists $g \in \mathbb{F}_q[x_1, \ldots, x_d] \backslash \{0\}$ of degree $D < kq$ that consists of proper monomials and has multiplicity at least k at each point of $\mathcal{P}$. We write $g = \sum_{j=0}^{D} g_j$, where g_j is a homogeneous polynomial of degree j. By the definition of D, we have that $g_D \not\equiv 0$.

We consider a vector $v \in \mathbb{F}_q^d \backslash \{0_d\}$. Since $\mathcal{P}$ is a Kakeya set, it contains a line with direction v. Thus, there exists $u \in \mathcal{P}$ such that $\{u + tv : t \in \mathbb{F}_q\} \subset \mathcal{P}$. As before, we consider the restriction of g to this line, setting $g_{u,v}(t) = g(u + tv)$. We note that $g_{u,v} \in \mathbb{F}_q[t]$ is of degree at most D. By Claim 6.6, $g_{u,v}$ has multiplicity at least k in each point. That is, $g_{u,v}$ has multiplicity at least k in q points. A nonzero polynomial of degree $D < kq$ cannot have kq factors, so $g_{u,v}(t) \equiv 0$. Equivalently, $g(u + tv) \equiv 0$, which implies that the coefficient of t^D in $g(u + tv)$ is zero. The coefficient of t^D in $g(u + tv)$ is $g_D(v)$, so $g_D(v) \equiv 0$.

By the preceding paragraph, $g_D(v) \equiv 0$ for every $v \neq 0_d$. Since g_D is homogeneous of degree $D > 0$, we also have that $g_D(0_d) \equiv 0$. Since g_D consists of proper monomials, Claim 6.8 implies that $g_D \equiv 0$. This contradicts g being of degree D. We conclude that every Kakeya set $\mathcal{P} \subset \mathbb{F}_q^d$ satisfies $|\mathcal{P}| \geq N_q(d, k) / \binom{d+k-1}{d}$.

Simplifying the bound: The above holds for every $0 < k < q$. We now set $k = \lceil d/2 \rceil$. We claim that, in this case, $N_q(d, k) \geq q^d/2$. Indeed, there are q^d proper monomials in $x_1, \ldots, x_d$. If the degree of such a monomial M is at least kq, then the degree of $x_1^{q-1} x_2^{q-1} \cdots x_d^{q-1}/M$ is smaller than kq. Since this quotient is a bijection from the set of proper monomials to itself, at least half of these q^d monomials have degrees smaller than kq.

Stirling's approximation states that $m! = \Theta(\sqrt{m} \cdot (m/e)^m)$. We consider the case where d is even. In this case $k = d/2$, so $\binom{d+k-1}{d} < \binom{3d/2}{d}$. Combining this with the above leads to

$$|\mathcal{P}| \geq N_q(d, k) \Big/ \binom{d+k-1}{d} > \frac{q^d}{2} \Big/ \binom{3d/2}{d} = \frac{q^d}{2} \cdot \frac{d!(d/2)!}{(3d/2)!}$$

$$= \Omega\left(q^d \cdot \frac{\sqrt{d} \cdot (d/e)^d \cdot \sqrt{d/2} \cdot ((d/2)/e)^{d/2}}{\sqrt{3d/2} \cdot ((3d/2)/)^{3d/2}}\right)$$

$$= \Omega\left(q^d \cdot \frac{\sqrt{d^2/2}}{\sqrt{3d/2}} \cdot \frac{(1/2)^{d/2}}{(3/2)^{3d/2}}\right) = \Omega\left(q^d \cdot \frac{2^d}{3^{3d/2}}\right) = \Omega\left(\frac{q^d}{2.6^d}\right).$$

The case where d is odd is handled symmetrically. □

By choosing the value of k more carefully in the above proof, one can improve 2.6^d to about 2.46^d. This is still far from the correct value 2^d. To improve the bound of Theorem 6.5 to $|\mathcal{P}| > q^d/2^d$, we need to study additional properties of multiplicities. In particular, we require a variant of the Schwartz–Zippel Lemma (Lemma 6.2) that also depends on multiplicities.

6.4 The Cap Set Problem

The following result was cited as one of the two reasons for Klaus Roth's Fields Medal.

Theorem 6.9 (Roth, 1953) *There exists a constant $c > 0$ that satisfies the following. For every integer n, each set $A \subset \{1, 2, \ldots, n\}$ with $|A| \geq cn/ \log \log n$ contains a 3-term arithmetic progression.*

Many works improved Theorem 6.9, generalized it, and studied variants of it. We now consider a finite fields variant of this problem, called *the cap set problem*.

We say that three distinct points $p, q, r \in \mathbb{F}_3^d$ form a *3-term arithmetic progression* if there exists $s \in \mathbb{F}_3^d$ such that $q \equiv p + s$ and $r \equiv p + 2s$. This is equivalent to $p + r - 2q \equiv 0_d$ and thus to $p + q + r \equiv 0_d$. We say that a set $A \subset \mathbb{F}_3^d$ is a *cap set* if no three points of A form a 3-term arithmetic progression. The cap set problem asks to find the largest cap set in $\mathbb{F}_3^d$.

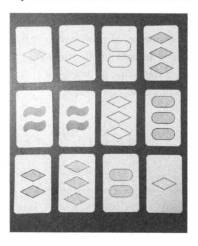

Figure 6.3 The SET card game corresponds to $\mathbb{F}_3^4$.

If you know the SET® card game, then you can also think about the cap set problem as such a game.[1] In this game, every card has four properties: shape, color, number, and shading (see Figure 6.3). The cap set problem is obtained by setting every card to have d properties, and asking for the maximum number of cards that do not contain a *SET*.

We note that $\{0, 1\}^d \subset \mathbb{F}_3^d$ is a cap set of size 2^d. In $\mathbb{F}_3^2$ there is no cap set of size larger than four. However, for every $d \geq 3$, the space $\mathbb{F}_3^d$ contains a cap set of size larger than 2^d. Edel (2004) proved that $\mathbb{F}_3^d$ contains a cap set of size $\Omega(2.217^d)$.

By adapting Roth's argument, Meshulam (1995) proved that any capset in $\mathbb{F}_3^d$ is of size $O(3^d/d)$. Meshulam's proof is a simple and elegant Fourier transform argument. Bateman and Katz (2012) introduced a long and involved analysis that further pushed Meshulam's argument. This sophisticated proof improved the bound to $O(3^d/d^{1+\varepsilon})$, for some small $\varepsilon > 0$.

For years, experts disagreed about whether the size of the largest cap set has a number smaller than 3 at the base of the exponent. Ellenberg and Gijswijt (2017) settled this question by proving the following results.

Theorem 6.10 *Let A be a cap set in $\mathbb{F}_3^d$. Then $|A| = O(2.756^d)$.*

The proof of Theorem 6.10 is surprisingly short and elegant. Rather than using Fourier analysis, the proof relies on a simple polynomial technique. This technique is based on a previous result of Croot et al. (2017).

---

[1] SET is a registered trademark of Cannei, LLC. The distinctive SET symbols and cards are copyrights of Cannei, LLC. All rights reserved. Used with permission from PlayMonster, LLC.

In Sections 6.6 and 6.7 we present an elegant variant of the cap set proof by Tao (2016). Both the original proof and Tao's variant rely on the linear algebra notions of dimension and rank. As a warm-up, in Section 6.5 we solve two other combinatorial problems by using these linear algebra concepts.

6.5 Warmups: Two Distances and Odd Towns

The two distances problem: What is the maximum size of a set $\mathcal{P} \subset \mathbb{R}^d$ such that the distance between every two points of $\mathcal{P}$ is one? By taking the vertices of a d-dimensional simplex with side length 1 in $\mathbb{R}^d$, we obtain $d + 1$ points that span only the distance 1. It is not difficult to show that every $d + 2$ points in $\mathbb{R}^d$ span more than a single distance (Exercise 6.5).

A point set $\mathcal{P}$ is a *two-distance set* if there exist $r, s \in \mathbb{R}$ such that the distance between every pair of points of $\mathcal{P}$ is either r or s. What is the maximum size of a two-distance set in $\mathbb{R}^d$?

Consider the set of points in $\mathbb{R}^d$ that have two coordinates with value 1 and the other $d - 2$ coordinates with value 0. There are $\binom{d}{2}$ such points, and the distance between every pair of those is either $\sqrt{2}$ or 2. Larman et al. (1977) showed that the above example is not far from being tight.

Theorem 6.11 *Every two-distance set in $\mathbb{R}^d$ has size at most $\binom{d}{2} + 3d + 2$.*

Proof Let $\mathcal{P} = \{p_1, p_2, \ldots, p_m\}$ be a two-distance set in $\mathbb{R}^d$, and denote the two distances as $r, s \in \mathbb{R}$. Recall that, for points $a, b \in \mathbb{R}^d$, we denote the distance between the points as $|ab|$. We write $x = (x_1, \ldots, x_d)$, where $x_1, \ldots, x_d$ are the coordinates of $\mathbb{R}^d$. For $1 \le j \le m$, we define the polynomial $f_j \in \mathbb{R}[x_1, \ldots, x_d]$ as

$$f_j(x) = \left(|xp_j|^2 - r^2 \right) \cdot \left(|xp_j|^2 - s^2 \right).$$

Note that f_j vanishes on $q \in \mathbb{R}^d$ if and only if the distance between p_j and q is either r or s. Every $f_j(x)$ is a linear combination of the polynomials

$$\left(\sum_{j=1}^{d} x_j^2 \right)^2, \qquad x_k \sum_{j=1}^{d} x_j^2, \qquad x_k, \qquad x_\ell x_k, \qquad 1,$$

where $1 \le \ell, k \le d$. Note that the above also includes the case of $x_\ell x_k$ with $\ell = k$. Since every f_j is a linear combination of $t = \binom{d}{2} + 3d + 2$ polynomials, we can represent f_j as a vector in $\mathbb{R}^t$. That is, the vector $(v_1, \ldots, v_t)$ corresponds to the polynomial

$$v_1 \cdot \left(\sum_{j=1}^{d} x_j^2 \right)^2 + v_2 \cdot x_1 \sum_{j=1}^{d} x_j^2 + v_3 \cdot x_2 \sum_{j=1}^{d} x_j^2 + \cdots + v_t \cdot 1.$$

For $1 \leq j \leq m$, let V_j be the vector corresponding to $f_j(x)$. Consider $\alpha_1, \ldots, \alpha_m \in \mathbb{R}$ such that $\sum_{j=1}^{m} \alpha_j V_j = 0$. We set

$$g(x) = \sum_{j=1}^{m} \alpha_j f_j(x) = 0.$$

For every $j \neq k$, we have that $f_j(p_k) = 0$ and that $f_j(p_j) = r^2 s^2$. This leads to

$$g(p_k) = \sum_{j=1}^{m} \alpha_j f_j(p_k) = \alpha_k r^2 s^2.$$

Since $g(x) = 0$, we have that $\alpha_k = 0$. Thus, the only solution $\sum_{j=1}^{m} \alpha_j V_j = 0$ is $\alpha_1 = \cdots = \alpha_m = 0$. This implies that the vectors $V_1, \ldots, V_m$ are linearly independent. Since these vectors are in $\mathbb{R}^t$, we conclude that $m \leq t = \binom{d}{2} + 3d + 2$. □

In the proof of Theorem 6.11, we think of a space of polynomials as a vector space and study the dimension of this space. The following problem can be solved in a similar manner. However, we present the proof in slightly different way to involve the rank of a matrix. This is a step towards the cap set proof in the following sections.

The odd town problem: The odd town problem studies sets of odd size that have even-sized intersections.

Theorem 6.12 *In the town of Liouville there are n people and m clubs. Every club has an odd number of members. For every two clubs, the number of people who are in both is even. Then $m \leq n$.*

Proof Let $P = \{p_1, \ldots, p_n\}$ be the set of people and let $C = \{c_1, \ldots, c_m\}$ be the set of clubs. Let Z be an $m \times m$ matrix with entries in $\mathbb{F}_2$. The value of $Z_{j,k}$ is the number of people who are in both c_j and c_k modulo 2. By the assumptions of the theorem, Z has ones on the main diagonal and zeros everywhere else. This implies that $\mathrm{rk}(Z) = m$.

For $1 \leq j \leq n$, let W^j be an $m \times m$ matrix with entries in $\mathbb{F}_2$. We set $W^j_{k,\ell} \equiv 1$ if and only if person p_j is a member of both c_k and c_ℓ. Then $Z = \sum_{j=1}^{n} W^j$, under coordinate-wise addition modulo 2.

We now study W^j for a fixed $1 \leq j \leq n$. For $1 \leq k \leq m$, if person p_j is not a member of club c_k then the kth row of W^j is all zeros. Also, all the nonzero rows of W^j are identical: There are ones at the indices of the clubs that include p_j. This implies that $\mathrm{rk}(W^j) \leq 1$. Since Z is a sum of n matrices of rank at most one, we have $\mathrm{rk}(Z) \leq n$. Recalling that $\mathrm{rk}(Z) = m$, we conclude that $m \leq n$. □

6.6 Tensors and Slice Rank

Tensors: In the two distances problem we studied the dimension of a vector space. In the odd town problem we studied ranks of matrices. In the cap set problem, we further generalize this approach to three-dimensional objects. These three-dimensional objects are tensors of order three. See Figure 6.4. For brevity, we write *tensors* and omit "of order three."

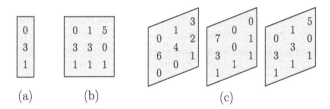

Figure 6.4 (a) A vector of size 3. (b) A 3×3 matrix. (c) A $3 \times 3 \times 3$ tensor.

We recall that matrices are used to represent *bilinear forms*: functions that receive two points and are linear in the coordinates of each. For an example, see Figure 6.5.

$$(x_1, x_2, x_3) \begin{bmatrix} 0 & 1 & 5 \\ 3 & 3 & 0 \\ 0 & 0 & 1 \end{bmatrix} \begin{pmatrix} y_1 \\ y_2 \\ y_3 \end{pmatrix} = x_1 y_3 + 5 x_1 y_3 + 3 x_2 y_1 + 3 x_2 y_2 + x_3 y_3.$$

Figure 6.5 A matrix that describes a bilinear form for $x, y \in \mathbb{R}^3$.

Tensors can be similarly used to represent *trilinear forms*: functions that receive three points and are linear in the coordinates of each. For example, the tensor in Figure 6.6 corresponds to the trilinear form

$$x_1 y_1 z_2 + 6 x_1 y_2 z_1 + 4 x_1 y_2 z_2 + x_1 y_1 z_3 + 7 x_2 y_1 z_1$$

$$+ 3 x_2 y_3 z_2 + x_3 y_1 z_2 + 3 x_3 y_2 z_1 + x_3 y_3 z_3.$$

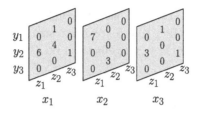

Figure 6.6 A $3 \times 3 \times 3$ tensor corresponds to a trilinear form for $x, y, z \in \mathbb{R}^3$.

Slice rank: A matrix has rank one if and only if it is the outer product of two nonzero vectors. For example, consider the rank one matrix

$$M = \begin{pmatrix} 0 & 1 & 2 \\ 0 & 2 & 4 \\ 0 & 3 & 6 \end{pmatrix} = \begin{pmatrix} 1 \\ 2 \\ 3 \end{pmatrix} \begin{pmatrix} 0 & 1 & 2 \end{pmatrix}.$$

Since M has rank one, this is also the rank of the corresponding bilinear form.

Generalizing the above to tensors, we say that a tensor has a *slice rank* of one if it is the outer product of a matrix and a vector. For an example, see Figure 6.7.

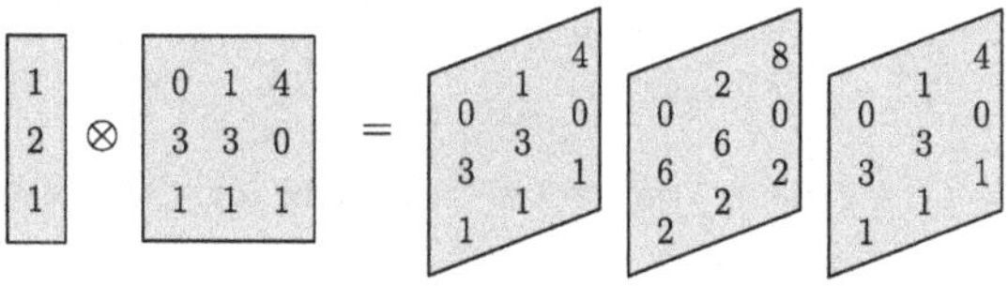

$$\begin{bmatrix} 1 \\ 2 \\ 1 \end{bmatrix} \otimes \begin{bmatrix} 0 & 1 & 4 \\ 3 & 3 & 0 \\ 1 & 1 & 1 \end{bmatrix} = \cdots$$

Figure 6.7 An outer product leads to a tensor with slice rank one.

We say that a trilinear form $f(x, y, z)$ has slice rank one if it corresponds to a tensor with slice rank one. In other words, $f(x, y, z)$ has slice rank one if we can write

$$f(x, y, z) = g(x) \cdot h(y, z) \quad \text{or} \quad f(x, y, z) = g(y) \cdot h(x, z)$$
$$\text{or} \quad f(x, y, z) = g(z) \cdot h(x, y),$$

where g is linear and h is bilinear.

We now return to $\mathbb{F}_3^d$ and generalize the above to arbitrary polynomials. Consider a set $A \subset \mathbb{F}_3^d$ and a polynomial $f \colon A^3 \to \mathbb{F}_3$. We say that $f(x, y, z)$ has a *slice rank* of one if we can write

$$f(x, y, z) = g(x) \cdot h(y, z) \quad \text{or} \quad f(x, y, z) = g(y) \cdot h(x, z)$$
$$\text{or} \quad f(x, y, z) = g(z) \cdot h(x, y),$$

for some polynomials $g \colon A \to \mathbb{F}_3$ and $h \colon A^2 \to \mathbb{F}_3$. In this case, g is not necessarily linear and h is not necessarily bilinear.

The slice rank of a polynomial $f : A^3 \to \mathbb{F}_3$ is the smallest k that satisfies the following: There exist k polynomials $f_1, \ldots, f_k : A^3 \to \mathbb{F}_3$ of slice rank 1 such that $f(x, y, z) = \sum_{j=1}^{k} f_j(x, y, z)$. We write $\mathrm{sr}(f) = k$. By definition, a single monomial has slice rank one. Thus, the slice rank of f is at most the number of monomials in f. As another example, when $A = \mathbb{F}_3$ we have $\mathrm{sr}(x + y + z) = 2$.

We consider $g, h : A \to \mathbb{F}_3$ as the same function if $g(a) = h(a)$ for every $a \in A$. That is, g and h may be considered distinct when their domain is $\mathbb{F}_3^d$, but identical when their domain is restricted to A. The heart of the cap set proof lies in the following property of slice rank.

Lemma 6.13 *Consider $A \subset \mathbb{F}_3^d$ and $f : A^3 \to \mathbb{F}_3$ such that $f(x, y, z) \not\equiv 0$ if and only if $x = y = z$. Then* $\mathrm{sr}(f) = |A|$.

Note the similarity to the odd town proof, where $c_j \cap c_k \not\equiv 0 \mod 2$ if and only if $j = k$.

Proof For $a \in A$, we define the function $\mathbf{1}_a : A \to \mathbb{F}_3$ as

$$\mathbf{1}_a(x) = \prod_{j=1}^{n} \left(1 - (a_j - x_j)^2 \right).$$

For $b, c \in \mathbb{F}_3$, we note that $1 - (b - c)^2 \equiv 0$ unless $b = c$. Thus, $\mathbf{1}_a(x) \equiv 1$ when $a = x$, and otherwise $\mathbf{1}_a(x) \equiv 0$.

We have that

$$f(x, y, z) = \sum_{a \in A} \mathbf{1}_a(x) f(a, y, z).$$

Since this is a sum of $|A|$ polynomials of slice rank one, we get that $\mathrm{sr}(f) \leq |A|$. This bound holds for every polynomial $f : A^3 \to \mathbb{F}_3$.

By the definition of $\mathrm{sr}(f)$, there exist polynomials $f_j : A \to \mathbb{F}_3$ and $g_j : A^2 \to \mathbb{F}_3$ such that

$$f(x, y, z) = \sum_{j=1}^{s} f_j(x) g_j(y, z) + \sum_{j=s+1}^{t} f_j(y) g_j(x, z) + \sum_{j=t+1}^{\mathrm{sr}(f)} f_j(z) g_j(x, y). \quad (6.1)$$

Note that Equation (6.1) implicitly defines the variables s and t.

Let P be the space of functions from A to $\mathbb{F}_3$. For any function $h : A \to \mathbb{F}_3$ we have $h(x) = \sum_{a \in A} h(a) \cdot \mathbf{1}_a(x)$. Thus, P is spanned by $|A|$ polynomials. These polynomials are linearly independent, so $\dim P = |A|$.

Let P' be the set of polynomials $h \in P$ that satisfy the following: For every $1 \leq j \leq s$ we have

$$\sum_{a \in A} f_j(a) h(a) \equiv 0.$$

This is a set of s homogeneous equations in elements of P, so $\dim P' \geq |A| - s$.

The *support* of a polynomial $h \in P'$ is $S_h = \{a \in A : h(a) \neq 0\}$. We fix $h \in P'$ with maximal support. Assume for contradiction that $|S_h| < \dim P'$. Asking for $h' \in P'$ to vanish on every point of S_h leads to a system of linear homogeneous equations with more variables than equations. Thus, there exists such nonzero $h' \in P'$. We note that $h + h'$ is in P' and has a larger support than h, contradicting our choice of h. We conclude that $|S_h| \geq \dim P' \geq |A| - s$.

We define

$$F(y, z) = \sum_{a \in A} h(a) \cdot f(a, y, z).$$

Combining this with Equation (6.1) leads to

$$F(y, z) = \sum_{a \in A} h(a) \left(\sum_{j=1}^{s} f_j(a) g_j(y, z) + \sum_{j=s+1}^{t} f_j(y) g_j(a, z) + \sum_{j=t+1}^{\mathrm{sr}(f)} f_j(z) g_j(a, y) \right)$$

$$= \sum_{j=1}^{s} \left(g_j(y, z) \sum_{a \in A} f_j(a) h(a) \right) + \sum_{j=s+1}^{t} \left(f_j(y) \sum_{a \in A} g_j(a, z) h(a) \right)$$

$$+ \sum_{j=t+1}^{\mathrm{sr}(f)} \left(f_j(z) \sum_{a \in A} g_j(a, y) h(a) \right).$$

By the definition of P', we have that $\sum_{a \in A} f_j(a) h(a) = 0$ and we may remove the first term in the above expression. Rewriting the above, there exist polynomials $\varphi_j : A \to \mathbb{F}_3$ such that

$$F(y, z) = \sum_{j=s+1}^{t} f_j(y) \varphi_j(z) + \sum_{j=t+1}^{\mathrm{sr}(f)} f_j(z) \varphi_j(y). \tag{6.2}$$

Imitating the proof of Theorem 6.12, we complete the current proof by double counting the rank of a matrix. Write $A = \{a_1, a_2, \ldots, a_{|A|}\}$. Let Z be an $|A| \times |A|$ matrix such that $Z_{j,k} = F(a_j, a_k)$. By the assumption of the theorem, we have that $f(x, y, z) = 0$ whenever $y \neq z$. This implies that $F(a', a'') = \sum_{a \in A} h(a) \cdot f(a, a', a'') \equiv 0$ whenever $a' \neq a''$. Thus, Z is a diagonal matrix. Similarly, the assumption implies that $F(a', a') = h(a') \cdot f(a', a', a')$. This in turn implies that $F(a', a') \not\equiv 0$ for every $a' \in S_h$. Recalling that $|S_h| \geq |A| - s$, we get that $\mathrm{rk}(Z) \geq |A| - s$.

For $s < j \leq t$, let W^j be the $|A| \times |A|$ matrix defined by $W^j_{k,\ell} = f_j(a_k) \cdot \varphi_j(a_\ell)$. For $t < j \leq \mathrm{sr}(f)$, let W^j be the $|A| \times |A|$ matrix defined by $W^j_{k,\ell} = f_j(a_\ell) \cdot \varphi_j(a_k)$. For every $s < j \leq \mathrm{sr}(f)$, every two rows of W^j are linearly dependent. Thus, the rank of every W^j is at most one.

By inspecting Equation (6.2), we note that $Z = \sum_{j=s+1}^{\mathrm{sr}(f)} W_j$. Since Z is a sum of $\mathrm{sr}(f) - s$ matrices of rank at most one, we get that $\mathrm{rk} Z \leq \mathrm{sr}(f) - s$. Combining this with $\mathrm{rk}(Z) \geq |A| - s$ implies $|A| \leq \mathrm{sr}(f)$, which completes the proof. $\square$

6.7 A Polynomial Method with Slice Rank

We now prove the cap set result. The proof is mostly an application of Lemma 6.13. We first recall the statement of this theorem.

Theorem 6.10 Let A be a cap set in $\mathbb{F}_3^d$. Then $|A| = O(2.756^d)$.

Proof Since A is a cap set, we have that $a, b, c \in A$ satisfy $a + b + c = 0_d$ if and only if $a = b = c$. Consider the polynomial $f \colon A^3 \to \mathbb{F}_3$ defined as

$$f(x, y, z) = \prod_{j=1}^{d} \left(1 - (x_j + y_j + z_j)^2 \right). \tag{6.3}$$

We note that $f(x, y, z) \equiv 1$ if $x + y + z = 0_d$, and otherwise $f(x, y, z) \equiv 0$. That is, for $a, b, c \in A$ we have that $f(a, b, c) \equiv 1$ if and only if $a = b = c$. We may thus apply Lemma 6.13 with A and $f(x, y, z)$ to obtain that $\mathrm{sr}(f) = |A|$.

For $x = (x_1, \ldots, x_d)$ and $p = (p_1, \ldots, p_d)$, we write $x^p = x_1^{p_1} x_2^{p_2} \cdots x_d^{p_d}$. We also write $|p| = p_1 + p_2 + \cdots + p_d$ with *summation over* $\mathbb{R}$. The degree of a monomial of $f(x, y, z)$ *in* x is the sum of the degrees of $x_1, \ldots, x_d$ in this monomial (over $\mathbb{R}$). Since $\deg f(x, y, z) \leq 2d$, each monomial of $f(x, y, z)$ has degree at most $2d/3$ in at least one of x, y, and z. We can thus rewrite Equation (6.3) as

$$f(x, y, z) = \sum_{\substack{p \in \{0,1,2\}^d \\ |p| \leq 2d/3}} x^p g_{x,p}(y, z) + \sum_{\substack{p \in \{0,1,2\}^d \\ |p| \leq 2d/3}} y^p g_{y,p}(x, z) + \sum_{\substack{p \in \{0,1,2\}^d \\ |p| \leq 2d/3}} z^p g_{z,p}(x, y),$$

$$\tag{6.4}$$

where $g_{x,p}, g_{y,p}, g_{z,p} \colon A^2 \to \mathbb{F}_3$ are polynomials of degree at most $2d$.

Set $r = \left| \{ p \in \mathbb{F}_3^d : |p| \leq 2d/3 \} \right|$. By Equation (6.4) we have that $\mathrm{sr}(f) \leq 3r$. Recalling that $\mathrm{sr}(f) = |A|$ gives $|A| \leq 3r$. It remains to derive an upper bound for r. This is a rather standard calculation that relies on elementary combinatorial tools.

For a fixed $p \in \{0, 1, 2\}^d$ and $0 \leq j \leq 2$, we denote by m_j the number of coordinates of p that are equal to j. Using this notation, we get that

$$r = \sum_{\substack{m_0+m_1+m_2=d \\ m_1+2m_2 \leq 2d/3}} \frac{d!}{m_0! m_1! m_2!}.$$

The multinomial theorem (for example, see Chapter 4 of Bóna, 2006) implies

$$\left(1 + x + x^2 \right)^d = \sum_{m_0+m_1+m_2=d} \frac{d!}{m_0! m_1! m_2!} \cdot x^{m_1+2m_2}.$$

When assuming that $0 < x < 1$, the above leads to

$$x^{-2d/3}\left(1 + x + x^2\right)^d = \sum_{m_0+m_1+m_2=d} \frac{d!}{m_0!m_1!m_2!} \cdot x^{m_1+2m_2-2d/3}$$

$$> \sum_{\substack{m_0+m_1+m_2=d \\ m_1+2m_2 \leq 2d/3}} \frac{d!}{m_0!m_1!m_2!} \cdot x^{m_1+2m_2-2d/3}$$

$$> \sum_{\substack{m_0+m_1+m_2=d \\ m_1+2m_2 \leq 2d/3}} \frac{d!}{m_0!m_1!m_2!} = r.$$

Set $g(x) = x^{-2/3}(1 + x + x^2)$. The minimum of $g(x)$ when $0 < x < 1$ is obtained when $x = (\sqrt{33} - 1)/8$. This minimum value satisfies $g(x) < 2.76$, implying that $|A| \leq 3r = O(2.76^d)$. $\qquad\square$

6.8 Exercises

Exercise 6.1 Let $q > 1$ be a prime power.
(a) Let $\ell \subset \mathbb{F}_q^2$ be a line that is defined by $y \equiv ax + b$, for some $a, b \in \mathbb{F}_q$. Find $u, v \in \mathbb{F}_q^2$ such that $\ell = \{u + tv : t \in \mathbb{F}_q\}$.
(b) Let $\ell = \{u + tv : t \in \mathbb{F}_q\} \subset \mathbb{F}_q^2$ where $u, v \in \mathbb{F}_q^2$ and the x-coordinate of v is nonzero. Prove that there exist $a, b \in \mathbb{F}_q$ such that ℓ is defined by $y \equiv ax + b$.

Exercise 6.2 Let $0 < \alpha < 1$. A set $\mathcal{P} \subset \mathbb{F}_q^d$ is a q^α-*Kakeya set* if, for every $u \in \mathbb{F}_q^d \setminus \{0_d\}$, there exists a line with direction u that contains at least q^α points of $\mathcal{P}$. Adapt the proof of the finite field Kakeya theorem to obtain a lower bound for the minimum size of a q^α-Kakeya set.

Exercise 6.3 Prove Claim 6.8. (Hint: It is easier to consider the contrapositive: For every nonzero $f \in \mathbb{F}_q[x_1, \ldots, x_d]$ that consists of proper monomials, there exists $p \in \mathbb{F}_q^d$ such that $f(p) \not\equiv 0$. Find the coordinates of p one by one.)

Exercise 6.4 (Moshkovitz, 2010) We proved Lemma 6.2 by induction on d. We now prove Lemma 6.2 in a different way.

Let f_k be the homogeneous component of degree k of f. By Claim 6.8, there exists $p \in \mathbb{F}_q^d$ such that $f_k(p) \neq 0$.

- Explain why $p \not\equiv 0_d$.
- Partition $\mathbb{F}_d^q$ into q^{d-1} lines with direction p.

- Show that f is never identically zero when restricted to such a line.
- Complete the proof.

Exercise 6.5 Prove that every $d + 2$ points in $\mathbb{R}^d$ span at least two distinct distances.

Exercise 6.6 Construct a two-distance set in $\mathbb{R}^d$ of size $\binom{d}{2} + 1$. (Hint: Start with the construction from Section 6.5.)

Exercise 6.7 A library contains n books and has $n+1$ members. Every member read at least one book from the library. Prove that there exist two disjoint sets of members that read exactly the same set of books. It does not matter how many people from the same set read a book. Some members may not be in either set.

Exercise 6.8 The *even town problem*: There are n people and m clubs. Every club has an even number of members. Every two clubs have an odd number of common members.
(a) Use a simple reduction to the odd town problem, to argue that $m \leq n + 1$.
(b) Assuming that m is even, prove that $m \leq n$.

Exercise 6.9 Consider a set $A \subset \mathbb{F}_3^d$ and a $f: A^3 \to \mathbb{F}_3$. We say that $f(x, y, z)$ has a *tensor rank* of one if we can write $f(x, y, z) = g_x(x)g_y(y)g_z(z)$, for some polynomials $g_x, g_y, g_z: A \to \mathbb{F}_3$. The tensor rank of a polynomial $f: A^3 \to \mathbb{F}_3$ is the smallest k that satisfies the following: There exist k polynomials $f_1, \ldots, f_k: A^3 \to \mathbb{F}_3$ of tensor rank one such that $f(x, y, z) = \sum_{j=1}^k f_j(x, y, z)$.

(a) Prove that the tensor rank of a polynomial f is at most $\operatorname{sr}(f) \cdot |A|$.

(b) Find a set $A \subset \mathbb{F}_3^d$ and $f: A^3 \to \mathbb{F}_3$ such that $|A| = 2d$, $\operatorname{sr}(f) = 1$, and the tensor rank of f is $|A| \cdot \operatorname{sr}(f) = 2d$.

Exercise 6.10 (Naslund and Sawin, 2017) Three sets form a *3-sunflower* if every two sets have the same intersection. For example, the sets $\{1, 2, 3\}$, $\{3, 4, 5, 6, 7, 8\}$, and $\{3, 20\}$ form a 3-sunflower, since the intersection of each two sets is $\{3\}$.

Let C be a set whose elements are subsets of $\{1, \ldots, n\}$, with no three of the subsets forming a 3-sunflower. Imitate the cap set proof to derive the bound $|C| = O(1.89^n)$. Instructions: Work in $\mathbb{F}_2^n$ and consider the polynomial

$$f(x, y, z) = \prod_{j=1}^n (2 - (x_j + y_j + z_j)).$$

Separately consider sets of different sizes.

6.9 Open Problems

Theorem 6.10 settled the question about the existence of a cap set whose size has 3 in the base of the exponent. However, the cap set problem remains wide open. Edel (2004) proved that $\mathbb{F}_3^d$ contains a cap set of size $\Omega(2.217^d)$. Theorem 6.10 states that every cap set in $\mathbb{F}_3^d$ is of size $O(2.756^d)$. Eliminating the big gap between these two bounds is a main open problem.

Open Problem 6.14   *Find the size of the largest cap set in $\mathbb{F}_3^d$.*

The finite field Kakeya problem is solved, up to a factor of 2. However, an interesting generalization of the problem remains wide open: *Kakeya sets over finite rings*. Hickman and Wright (2018) recently made the following conjecture.

Conjecture 6.15   *For every $\varepsilon > 0$ and integer d, there exists a constant $C_{d,\varepsilon}$ that satisfies the following. Each Kakeya set $A \subset (\mathbb{Z}/N\mathbb{Z})^d$ satisfies*

$$|A| \geq C_{d,\varepsilon} \cdot N^{d-\varepsilon}.$$

Dhar and Dvir (2020) proved Conjecture 6.15 in the case where N is square-free.

7

The Elekes–Sharir–Guth–Katz Framework

By the way, in case of something unexpected happens to me (car accident, plane crash, a brick on the top of my skull) I definitely ask you to publish anything we have, at your will.

György Elekes, in an email to Micha Sharir, a few years before he passed away (Sharir, 2009).

We started to study the distinct distances problem in Section 1.6. The mathematicians Elekes and Sharir used to discuss this problem. Around the turn of the millennium, Elekes discovered a reduction from this problem to a problem about intersections of helices in $\mathbb{R}^3$. Later on, Elekes sent Sharir the above quote.

Elekes passed away in 2008 and, as requested, Sharir then published their ideas. Before publishing, Sharir simplified the reduction so that it led to a problem about intersections of *parabolas* in $\mathbb{R}^3$. Sharing the reduction with the general community had surprising consequences. Hardly any time had passed before Guth and Katz managed to apply the reduction to almost completely solve the distinct distances problem.

Theorem 7.1 (Guth and Katz, 2015) *Every set of n points in $\mathbb{R}^2$ determines $\Omega(n/\log n)$ distinct distances.*

Recall from Section 1.6 that the current best upper bound for the problem is $O(n/\sqrt{\log n})$. Thus, the problem is now solved up to a small gap of $\sqrt{\log n}$. To obtain Theorem 7.1, Guth and Katz further revised the reduction so that it led to a problem about intersections of *lines* in $\mathbb{R}^3$. We thus refer to this reduction as the *Elekes–Sharir–Guth–Katz framework* (or the *ESGK framework*, for short). In the current chapter, we study the ESGK framework.

7.1 Warmup: Distances between Points on Two Lines

Before we study the ESGK framework, we consider a simpler distinct distances problem. This problem can be easily reduced to an incidence problem. While the ESGK reduction is more involved, the two reductions have several ideas in common.

In a *bipartite* distinct distances problem we have two point sets $\mathcal{P}_1$ and $\mathcal{P}_2$. We denote by $D(\mathcal{P}_1, \mathcal{P}_2)$ the number of distinct distances spanned by pairs from $\mathcal{P}_1 \times \mathcal{P}_2$. That is, we ignore pairs of points from the same set. In other words,

$$D(\mathcal{P}_1, \mathcal{P}_2) = \left| \{ |pq| \ : \ p \in \mathcal{P}_1, q \in \mathcal{P}_2 \} \right|$$

(where $|pq|$ denotes the distance between the points p and q).

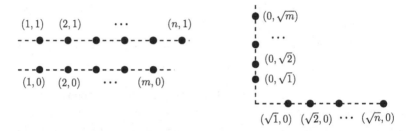

Figure 7.1 When the lines are either parallel or orthogonal, the points can be arranged so that $D(\mathcal{P}_1, \mathcal{P}_2) = \Theta(n)$. (Left) The distances in the parallel case are $\{2, 3, 4, \ldots, n\}$. (Right) The distances in the orthogonal case are $\{2, 3, 4, \ldots, n + m\}$.

We consider the case where $\mathcal{P}_1$ is a set of m points on a line ℓ_1 and $\mathcal{P}_2$ is a set of n points on a different line ℓ_2. Without loss of generality, we assume that $n \geq m$. When the two lines are parallel or orthogonal, the points can be arranged so that $D(\mathcal{P}_1, \mathcal{P}_2) = \Theta(n)$. Such constructions are depicted in Figure 7.1. It is not difficult to show that an asymptotically smaller number of distances is not possible for this problem.

When ℓ_1 and ℓ_2 are neither parallel nor orthogonal, the problem significantly changes. The current best construction for minimizing the number of distances gives $D(\mathcal{P}_1, \mathcal{P}_2) = \Theta(n^2/\sqrt{\log n})$. In this construction, we take ℓ_1 to be the x-axis and $\ell_2 = \mathbf{V}(y - x)$ (the line of slope one that is incident to the origin). We set

$$\mathcal{P}_1 = \{(j, 0) \ : \ 1 \leq j \leq m\} \quad \text{and} \quad \mathcal{P}_2 = \{(j, j) \ : \ 1 \leq j \leq n\}.$$

By Theorem 1.8, the number of integers in $\{1, 2, \ldots, n\}$ that are the sum of two integer squares is $\Theta(n/\sqrt{\log n})$. In our case, every distance in $\mathcal{P}_1 \times \mathcal{P}_2$ is of the form $\sqrt{d_x^2 + d_y^2}$, where d_x and d_y are integers between $-n$ and $m - 1$.

We may consider the squares of the distances since this does not change the number of distinct values. Theorem 1.8 implies that $D(\mathcal{P}_1, \mathcal{P}_2) = O(n^2/\sqrt{\log n})$.

We now derive the current best lower bound for distinct distances between points on two lines. Note the huge gap between the current best upper and lower bounds.

Theorem 7.2 (Sharir et al., 2013) *Let ℓ_1 and ℓ_2 be lines in $\mathbb{R}^2$ that are neither parallel nor orthogonal. Let $\mathcal{P}_1$ be a set of m points on ℓ_1 and let $\mathcal{P}_2$ be a set of n points on ℓ_2. Then*

$$D(\mathcal{P}_1, \mathcal{P}_2) = \Omega\left(\min\{m^{2/3}n^{2/3}, n^2, m^2\}\right).$$

Proof We first simplify the problem. We rotate the plane so that ℓ_1 becomes the x-axis. We then translate the plane so that the intersection point $\ell_1 \cap \ell_2$ becomes the origin. These transformations do not change $D(\mathcal{P}_1, \mathcal{P}_2)$. Let s be the slope of ℓ_2 after the rotation. Since ℓ_1 and ℓ_2 are neither parallel nor orthogonal, s is nonzero and finite. If the origin is in $\mathcal{P}_1$, then we remove it from this set. Such a removal cannot increase $D(\mathcal{P}_1, \mathcal{P}_2)$.

We set $D = D(\mathcal{P}_1, \mathcal{P}_2)$ and denote the D distinct distances in $\mathcal{P}_1 \times \mathcal{P}_2$ as $\delta_1, \ldots, \delta_D$. We define the set of quadruples

$$Q = \{(a, p, b, q) \in \mathcal{P}_1 \times \mathcal{P}_2 \times \mathcal{P}_1 \times \mathcal{P}_2 \ : \ |ap| = |bq|\}.$$

The quadruples of Q are ordered. That is, (a, p, b, q) and (b, q, a, p) are considered as two distinct elements of Q. Figure 7.2 depicts such a quadruple.

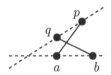

Figure 7.2 The quadruple (a, p, b, q) is in Q since $|ap| = |bq|$.

We prove the theorem by double counting $|Q|$. For every $1 \leq j \leq D$, we set

$$E_j = \{(a, p) \in \mathcal{P}_1 \times \mathcal{P}_2 \ : \ |ap| = \delta_j\}.$$

Since every pair of $\mathcal{P}_1 \times \mathcal{P}_2$ is in exactly one set E_j, we have that

$$\sum_{j=1}^{D} |E_j| = |\mathcal{P}_1| \cdot |\mathcal{P}_2| = mn.$$

The number of quadruples $(a, p, b, q) \in Q$ that satisfy $|ap| = |bq| = \delta_j$ is $|E_j|^2$. This implies that $|Q| = \sum_{j=1}^{D} |E_j|^2$. By the Cauchy–Schwarz inequality (Theorem A.1),

$$\left(\sum_{j=1}^{D} |E_j| \right)^2 \leq \left(\sum_{j=1}^{D} |E_j|^2 \right) \left(\sum_{j=1}^{D} 1 \right) = D \sum_{j=1}^{D} |E_j|^2.$$

By combining the above, we get that

$$|Q| = \sum_{j=1}^{D} |E_j|^2 \geq \frac{1}{D} \left(\sum_{j=1}^{D} |E_j| \right)^2 = \frac{m^2 n^2}{D}. \tag{7.1}$$

Deriving an upper bound for $|Q|$: We consider a quadruple $(a, p, b, q) \in \mathcal{P}_1 \times \mathcal{P}_2 \times \mathcal{P}_1 \times \mathcal{P}_2$, and write $a = (a_x, 0), b = (b_x, 0), p = (p_x, sp_x)$, and $q = (q_x, sq_x)$. Recall that this quadruple is in Q if and only if $|ap| = |bq|$, or equivalently

$$(a_x - p_x)^2 + s^2 p_x^2 = (b_x - q_x)^2 + s^2 q_x^2. \tag{7.2}$$

For a pair of points $(p, q) \in \mathcal{P}_2^2$, we define the point $v_{pq} = (p_x, q_x) \in \mathbb{R}^2$. We also define $\mathcal{P}' = \{v_{pq} : (p, q) \in \mathcal{P}_2^2\}$. It might be easier to think of the points of $\mathcal{P}'$ as living in a different plane from the points of $\mathcal{P}_1$ and $\mathcal{P}_2$. We refer to this new plane as the *dual plane*. That is, $\mathcal{P}'$ is a set of n^2 distinct points in the dual plane. For a pair of points $(a, b) \in \mathcal{P}_1^2$, we define the hyperbola

$$\gamma_{ab} = \mathbf{V} \left(a_x^2 - b_x^2 - 2a_x x + 2b_x y + x^2 (1 + s^2) - y^2 (1 + s^2) \right).$$

Finally, we set $\mathcal{H} = \left\{ \gamma_{ab} : (a, b) \in \mathcal{P}_1^2 \right\}$. We think of $\mathcal{H}$ as a set of m^2 distinct hyperbolas in the dual plane.

For a quadruple (a, p, b, q), Condition (7.2) is satisfied if and only if the point v_{pq} is incident to the hyperbola γ_{ab}. Indeed, when placing the coordinates of v_{pq} in the polynomial defining γ_{ab}, we get Condition (7.2). Thus, to obtain an upper bound for $|Q|$ it suffices to obtain an upper bound for $I(\mathcal{P}', \mathcal{H})$.

To apply an incidence theorem to $\mathcal{P}' \times \mathcal{H}$, we first check that elements of $\mathcal{H}$ do not have common components. Given an equation that defines a hyperbola γ_{ab}, we can recover the coordinates of a and b from the coefficients of x and y. This implies that no two elements of $\mathcal{H}$ are identical. Distinct hyperbolas may have common components when these are degenerate hyperbolas (products of two lines). However, since $s \neq 0$ and the origin is not in $\mathcal{P}_1$, it is not difficult to verify that no hyperbola in $\mathcal{H}$ is degenerate. We conclude that no two elements of $\mathcal{H}$ have a common component.

By Bézout's theorem (Theorem 2.5), two nondegenerate hyperbolas intersect in at most four points. Thus, there is no $K_{5,2}$ in the incidence graph of $\mathcal{P}' \times \mathcal{H}$.

This is a rather weak restriction, but we can improve it as follows. In the above proof, we replace the roles of $(\ell_1, \mathcal{P}_1)$ and $(\ell_2, \mathcal{P}_2)$: We take ℓ_2 to be the x-axis, have the pairs of $\mathcal{P}_1^2$ form the point set $\overline{\mathcal{P}}'$, and have the pairs of $\mathcal{P}_2^2$ form the set of hyperbolas $\overline{\mathcal{H}}$ (in the same way we formed $\mathcal{P}'$ and $\mathcal{H}$). As before, Bézout's theorem implies that the incidence graph of $\overline{\mathcal{P}}' \times \overline{\mathcal{H}}$ contains no copy of $K_{5,2}$. The incidence graph of $\overline{\mathcal{P}}' \times \overline{\mathcal{H}}$ is identical to the incidence graph of $\mathcal{P}' \times \mathcal{H}$, but with the two sides switched. This occurs because the edges of both graphs correspond to the quadruples of Q.

Since the incidence graph of $\overline{\mathcal{P}}' \times \overline{\mathcal{H}}$ contains no $K_{5,2}$, the incidence graph of $\mathcal{P}' \times \mathcal{H}$ contains no $K_{2,5}$. We apply our point-curve incidence bound (Theorem 3.3) on $\mathcal{P}' \times \mathcal{H}$ with $s = 2$ and $t = 5$, to obtain

$$|Q| = I(\mathcal{P}', \mathcal{H}) = O\left(|\mathcal{P}'|^{2/3}|\mathcal{H}|^{2/3} + |\mathcal{P}'| + |\mathcal{H}|\right) = O\left(m^{4/3}n^{4/3} + m^2 + n^2\right).$$

Combining this upper bound with the lower bound in Condition (7.1) leads to

$$m^2 n^2 / D = O\left(m^{4/3}n^{4/3} + m^2 + n^2\right).$$

We split the analysis into three cases, according to the term that dominates the right-hand side. Each case leads to a different term in the bound of the theorem.   □

7.2 The ESGK Framework

We are now ready to discuss the ESGK Framework. Consider a set $\mathcal{P}$ of n points in $\mathbb{R}^2$. We let x denote the number of nonzero distinct distances that are determined by pairs of points from $\mathcal{P}$. The problem asks us to derive a lower bound for x. Adapting the proof of Theorem 7.2, we consider the set

$$Q = \left\{(a, p, b, q) \in \mathcal{P}^4 \ : \ |ap| = |bq|\right\}. \tag{7.3}$$

The quadruples in Q are ordered. That is, the quadruples (a, p, b, q), (b, q, a, p), and (p, a, q, b) are considered as three distinct elements of Q. Some of the points of a quadruple may be identical, and we also allow the case of $a = p = b = q$. As before, the reduction is based on double counting $|Q|$. We first derive a lower bound for $|Q|$. We denote the distances that are determined by $\mathcal{P} \times \mathcal{P}$ as $\delta_1, \ldots, \delta_x$ (including the distance 0). For every $1 \le j \le x$, we set

$$E_j = \left\{(p, q) \in \mathcal{P}^2 \ : \ |pq| = \delta_j\right\}.$$

We consider (p, q) and (q, p) as two distinct pairs of E_j.

Since every ordered pair of points of $\mathcal{P}^2$ is contained in a unique set E_j, we have that $\sum_{j=1}^{x} |E_j| = n^2$. The number of quadruples $(a, p, b, q) \in Q$ that satisfy $|ap| = |bq| = \delta_j$ is $|E_j|^2$. By summing this over every j, we get that $|Q| = \sum_{j=1}^{x} |E_j|^2$. The Cauchy–Schwarz inequality leads to

$$|Q| = \sum_{j=1}^{x} |E_j|^2 \geq \frac{1}{x}\left(\sum_{j=1}^{x} |E_j|\right)^2 = \frac{n^4}{x}. \tag{7.4}$$

It remains to obtain an upper bound for $|Q|$. Guth and Katz prove Theorem 7.1 by obtaining the bound $|Q| = O(n^3 \log n)$ and combining it with Equation (7.4).

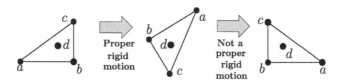

Figure 7.3 The second transformation is not a proper rigid motion: The sequence abc forms a left turn before the transformation and a right turn after.

From quadruples to rigid motions: A transformation of $\mathbb{R}^2$ is a *rigid motion* if it preserves distances between points. The rigid motions of $\mathbb{R}^2$ are rotations, translations, reflections, and their combinations. A *proper rigid motion* also preserves orientation: An ordered triple of points $(a, b, c) \in \mathbb{R}^2$ forms a left turn after applying the transformation if and only if it formed a left turn before the transformation. See Figure 7.3 for an example.

The only proper rigid motions of $\mathbb{R}^2$ are rotations and translations. Any combination of rotations and translations results in a single translation or in a single rotation (more details can be found in Stillwell (2008, Section 1.5) and in the exercises following it).

Consider a pair of points $a, b \in \mathcal{P}$. The *perpendicular bisector* of a and b is the set of points of $\mathbb{R}^2$ that are equidistant from a and b. Equivalently, the perpendicular bisector is the line incident to the midpoint of a and b, and orthogonal to the line through a and b. See Figure 7.4 for several examples. For brevity, we often refer to the *bisector* of a and b. The center of a rotation that takes a to b is equidistant from a and b, so it is on the bisector. Conversely, every point on the bisector of a and b is the origin of a rotation that takes a to b. An example is depicted in Figure 7.4(a).

We consider a quadruple $(a, p, b, q) \in Q$ and recall that $|ap| = |bq|$. For brevity, we write *a transformation that takes ap to bq* instead of a transformation that takes a to b and p to q. There always exists a proper rigid motion that takes ap to bq: A translation that takes a to b, followed by a rotation around the new position of a that takes p to q. The composition of a translation with a rotation

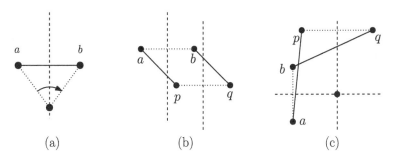

(a) (b) (c)

Figure 7.4 (a) The origin of the rotation is on the bisector. (b) When the bisectors are parallel, there is a translation that takes ap to bq. (c) When the bisectors intersect, there is a rotation that takes ap to bq. The center of the rotation is the intersection point of the bisectors.

is either a translation or a rotation (the composition is a translation only when the rotation is the identity).

We claim that there is a *unique* proper rigid motion that takes ap to bq. Indeed, let ℓ_1 be the bisector of a and b and let ℓ_2 be the bisector of p and q. If ℓ_1 and ℓ_2 are parallel, then there is a unique translation that takes ap to bq, and no rotation. The direction of the translation is orthogonal to the direction of the two bisectors. See Figure 7.4(b). If ℓ_1 and ℓ_2 intersect, then there is a unique rotation taking ap to bq, and no translations. This rotation is centered at the intersection point $\ell_1 \cap \ell_2$. The angle of rotation that takes a to b is equal to the angle that takes p to q, since $|ap| = |bq|$. See Figure 7.4(c).

In the preceding paragraph, we ignored the cases where $a = b$ or $p = q$. If both $a = b$ and $p = q$, then the unique proper rigid motion that takes ap to bq is the identity. If $a = b$ and $p \neq q$, then the unique proper rigid motion that takes ap to bq is a rotation around a. If $a \neq b$ and $p = q$, then the unique proper rigid motion that takes ap to bq is a rotation around p.

The above leads to an equivalent definition for Q: A quadruple $(a, p, b, q) \in \mathcal{P}^4$ is in Q if and only if there exists a proper rigid motion τ that takes ap to bq. We say that the quadruple (a, p, b, q) *corresponds* to τ. Thus, to derive an upper bound for $|Q|$, it suffices to derive an upper bound for the number of quadruples that correspond to a proper rigid motion.

We first bound the number of quadruples in $\mathcal{P}^4$ that correspond to a translation. When fixing the first three points of a quadruple $(a, p, b, ?)$, at most one point of $\mathcal{P}$ completes it to a quadruple that corresponds to a translation. Indeed, there is a unique translation τ that takes a to p, and the fourth point must be $\tau(b)$. We conclude that $O(n^3)$ quadruples in $\mathcal{P}^4$ correspond to a translation.

Obtaining the required upper bound for the number of quadruples in $\mathcal{P}^4$ that correspond to a rotation is significantly more difficult. We arbitrarily set all rotations to be clockwise. A rotation of $\mathbb{R}^2$ can be described using three parameters: two coordinates for the center of rotation and another for the angle of rotation. Given a rotation with center (o_x, o_y) and angle α, Guth and Katz parameterized it as $(o_x, o_y, \cot(\alpha/2)) \in \mathbb{R}^3$. With this parameterization, the rotations that take a fixed point $a \in \mathbb{R}^2$ to a fixed point $b \in \mathbb{R}^2$ form the following line in $\mathbb{R}^3$:

$$\ell_{ab} = \left\{ \left(\frac{a_x + b_x}{2}, \frac{a_y + b_y}{2}, 0 \right) + t \left(\frac{b_y - a_y}{2}, \frac{a_x - b_x}{2}, 1 \right) \ : \ t \in \mathbb{R} \right\}. \quad (7.5)$$

Showing that ℓ_{ab} corresponds to the rotations that take a to b is a standard technical calculation, so we postpone it to the optional Section 7.3.

The projection of ℓ_{ab} on the xy-plane is a line with slope $(a_x - b_x)/(b_y - a_y)$. We note that this slope is orthogonal to the slope of the line incident to a and b. By setting $t = 0$, we also note that ℓ_{ab} contains the midpoint of a and b. Thus, the projection of ℓ_{ab} on the xy-plane is the bisector of a and b. By inspecting the coefficient of t in Equation (7.5), we note that, when moving a distance of 1 in the z-direction, on the projection of ℓ_{ab} on the xy-plane we move

$$\sqrt{\left(\frac{b_y - a_y}{2} \right)^2 + \left(\frac{a_x - b_x}{2} \right)^2} = \sqrt{\frac{|ab|^2}{4}} = \frac{|ab|}{2}.$$

In other words, ℓ_{ab} is obtained by giving the bisector of a and b a slope of $2/|ab|$ in the z-direction.

Consider a quadruple $(a, p, b, q) \in \mathcal{P}^4$ and the corresponding lines ℓ_{ab} and ℓ_{pq} in $\mathbb{R}^3$. If the intersection point $\ell_{ab} \cap \ell_{pq}$ exists, then it parameterizes a rotation of $\mathbb{R}^2$ that takes ap to bq. Thus, (a, p, b, q) corresponds to a rotation if and only if the lines ℓ_{ab} and ℓ_{pq} intersect. For an example, see Figure 7.5.

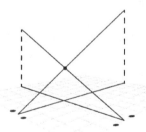

Figure 7.5 A quadruple of points in $\mathbb{R}^2$, two bisectors, and their lifting to $\mathbb{R}^3$. Since the lifted lines intersect, the quadruple is in Q.

Let

$$\mathcal{L} = \left\{ \ell_{ab} \ : \ (a, b) \in \mathcal{P}^2 \right\}.$$

We note that $\mathcal{L}$ is a set of n^2 distinct lines in $\mathbb{R}^3$. By the above, there is a bijection between quadruples of Q that correspond to a rotation and pairs of intersecting lines from $\mathcal{L}^2$. Guth and Katz obtained an upper bound of $O(n^3 \log n)$ for the number of pairs of intersecting lines, which in turn implies that $|Q| = O(n^3 \log n)$, as required. One can easily find sets of n^2 lines in $\mathbb{R}^3$ where every pair of lines intersect. However, the set $\mathcal{L}$ is not an arbitrary set of lines. We study properties of this set in Chapter 9.

7.3 (Optional) Lines in the Parametric Space $\mathbb{R}^3$

Section 7.2 assumes, without proof, that the rotations that take point a to point b correspond to the line parameterized in Equation (7.5). We now go over a technical proof of this claim, which can be safely skipped.

We fix points $a, b \in \mathbb{R}^2$ and a rotation of angle α around $o \in \mathbb{R}^2$ that takes a to b. Recall that we parameterize such a rotation as $\left(o_x, o_y, \cot \frac{\alpha}{2}\right) \in \mathbb{R}^3$. We rewrite Equation (7.5) as

$$\ell_{ab} = \left\{ \left(\frac{a_x + b_x + t(b_y - a_y)}{2}, \frac{a_y + b_y + t(a_x - b_x)}{2}, t \right) : t \in \mathbb{R} \right\}.$$

For $\left(o_x, o_y, \cot \frac{\alpha}{2}\right)$ to be on ℓ_{ab}, we set $t = \cot \frac{\alpha}{2}$. Then, our goal is to show that

$$o_x = \frac{a_x + b_x + t(b_y - a_y)}{2} \quad \text{and} \quad o_y = \frac{a_y + b_y + t(a_x - b_x)}{2}. \quad (7.6)$$

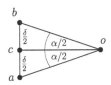

Figure 7.6 A rotation with origin o and angle α that takes a to b.

We first assume that $0 < \alpha < \pi$ and that $a_y \neq b_y$. We discuss the other cases below. The center of any rotation that takes a to b is equidistant from a and b, so o is on the bisector of a and b. We define the midpoint of a and b as $c = ((a_x + b_x)/2, (a_y + b_y)/2)$. We define the distance between a and b as $\delta = |ab| = \sqrt{(a_x - b_x)^2 + (a_y - b_y)^2}$. Figure 7.6 shows the relationships between these values.

By definition, the bisector of a and b is incident to the midpoint c and the slope of the bisector is $s = (a_x - b_x)/(b_y - a_y)$. Since $a_y \neq b_y$, we have that s is finite. Thus, the bisector is defined by the equation

$$y - \frac{a_y + b_y}{2} = s\left(x - \frac{a_x + b_x}{2}\right).$$

Since o is incident to the bisector of a and b, we obtain that

$$o_y - \frac{a_y + b_y}{2} = s\left(o_x - \frac{a_x + b_x}{2}\right). \tag{7.7}$$

We set $d_x = |o_x - c_x|$ and $d_y = |o_y - c_y|$. With this notation, Equation (7.7) becomes $d_y = |s|d_x$. This in turn implies that

$$|co| = \sqrt{d_x^2 + d_y^2} = |d_x|\sqrt{1 + s^2}$$

$$= |d_x|\sqrt{\frac{(b_y - a_y)^2 + (a_x - b_x)^2}{(b_y - a_y)^2}}$$

$$= \left|(o_x - c_x) \cdot \frac{\delta}{b_y - a_y}\right| = \delta \cdot \left|\frac{o_x - c_x}{b_y - a_y}\right|. \tag{7.8}$$

We recall that $0 < \alpha < \pi$ and that all rotations are clockwise. Thus, if $a_y < b_y$ then $c_x < o_x$, as depicted in Figure 7.6. If $a_y > b_y$ then $c_x > o_x$ (try replacing a and b in Figure 7.6 while maintaining an angle of rotation smaller than π). In either case, we can remove the absolute value from Equation (7.8).

By considering Figure 7.6, we note that $|co| = \frac{\delta}{2}\cot\frac{\alpha}{2} = \frac{\delta \cdot t}{2}$. Combining this with Equation (7.8) (without the absolute value) leads to

$$\frac{t}{2} = \frac{o_x - c_x}{b_y - a_y}, \qquad \text{or equivalently,} \qquad o_x = \frac{a_x + b_x + t(b_y - a_y)}{2}.$$

This establishes the first part of Equation (7.6). Plugging this value of o_x into Equation (7.7) gives

$$o_y = \frac{a_y + b_y}{2} + s\left(\frac{a_x + b_x + t(b_y - a_y)}{2} - \frac{a_x + b_x}{2}\right)$$

$$= \frac{a_y + b_y}{2} + s \cdot \frac{t(b_y - a_y)}{2}$$

$$= \frac{a_y + b_y}{2} + \frac{t(a_x - b_x)}{2}.$$

This establishes the second part of Equation (7.6).

When $\pi < \alpha < 2\pi$, the above holds with one change: In Figure 7.6, both $\alpha/2$ angles are replaced with $\pi - \alpha/2$. This in turn replaces $|co| = \frac{\delta \cdot t}{2}$ with $|co| = \frac{-\delta \cdot t}{2}$. The change also implies that, when removing the absolute value from $\left|\frac{o_x - c_x}{a_y - b_y}\right|$, a minus sign is introduced. The two new minus signs cancel each other, so the above analysis still holds.

When a rotation of angle $\alpha = 0$ takes a to b, we have that $a = b$. In this case, the rotations that take a to b are the rotations centered around a. We have that $\ell_{aa} = \{(a_x, a_y, t) : t \in \mathbb{R}\}$, which is indeed the set of rotations centered at a. The rotation that takes a to be b with $\alpha = \pi$ is centered at the midpoint of a and b. This rotation corresponds to the point $((a_x + b_x)/2, (a_y + b_y)/2, 0)$, which indeed lies on $\ell_{a,b}$. To recap: When $a_y \neq b_y$, every rotation of $\mathbb{R}^2$ that takes a to b corresponds to a point on ℓ_{ab}. Since the above holds for any point o on the bisector of a and b, we conclude that every point of ℓ_{ab} parameterizes a rotation that takes a to b.

It remains to consider the case where $a_y = b_y$. Since we already handled the case where $a = b$, we may assume that $a_x \neq b_x$. This case is similar to the above, and is left as Exercise 7.6.

7.4 Exercises

Exercise 7.1 (Bruner and Sharir, 2018) Let $\mathcal{P}_1$ be a set of m points on the x-axis in $\mathbb{R}^2$. Let $\mathcal{P}_2$ be a set of n points in $\mathbb{R}^2$, such that every axis-parallel line contains at most one point of $\mathcal{P}_2$. Prove that

$$D(\mathcal{P}_1, \mathcal{P}_2) = \Omega\left(\min\left\{m^{8/9}m^{2/9}, n^2, m^2\right\}\right).$$

Proof suggestion: Follow the proof of Theorem 7.2. That is, define Q as before, consider the equation $|ap| = |bq|$, turn every pair $(a, b) \in \mathcal{P}_1$ to a point, turn every pair $(p, q) \in \mathcal{P}_2$ to a curve, and show that there is no $K_{5,2}$ in the incidence graph.

Exercise 7.2 Consider the square lattice $\mathcal{P} = \left\{1, 2, 3, \ldots, \sqrt{n}\right\}^2$. Let Q be as defined in Equation (7.3). Prove that $|Q| = \Omega\left(n^3 \log n\right)$.

Proof suggestion: For a positive integer k, we denote

$$r(k) = \left|\left\{(j_1, j_2) \in \mathbb{N}^2 \: : \: j_1^2 + j_2^2 = k\right\}\right|.$$

Ramanujan (1916) proved that $\sum_{j=1}^{k} r(j)^2 = \Theta(k \log k)$. How can this result be used to obtain a lower bound for Q?

Exercise 7.3 Let $\mathcal{P}$ be a set of n points in $\mathbb{R}^2$. We apply the ESGK framework to $\mathcal{P}$, obtaining a set of n^2 lines in $\mathbb{R}^3$. Let a, a', b, b', c, c' be distinct points of $\mathcal{P}$.
(a) Assume that every pair of the lines $\ell_{aa'}, \ell_{bb'}, \ell_{cc'}$ intersect. What does this tell us about the points a, a', b, b', c, c'?
(b) Assume that the three lines $\ell_{aa'}, \ell_{bb'}, \ell_{cc'}$ intersect at a single point. What new information do we have, beyond the properties obtained in part (a)?

Exercise 7.4 Instead of having all rotations be clockwise, we set all rotations to be counterclockwise. In this case, how should we revise the line definition in Equation (7.5)?

Exercise 7.5 Set $a = (0,0) \in \mathbb{R}^2$ and $\gamma = \mathbf{V}(x^2 + y^2 - 1) \subset \mathbb{R}^2$. Find an irreducible variety of degree 2 in $\mathbb{R}^3$ that contains the lines ℓ_{ab}, for every $b \in \gamma$. (Hint: Set $b = \left(b_x, \sqrt{1 - b_x^2}\right)$ and find the parameterization of ℓ_{ab}. Then find a relation that holds between the three coordinates of the line parameterization, no matter what the value of b_x is.)

Exercise 7.6 Complete the analysis of Section 7.3. That is, prove the case where $a_y = b_y$ and $a_x \neq b_x$.

Exercise 7.7 The ESGK framework leads to a problem about the maximum number of pairs of intersecting lines in $\mathbb{R}^3$. This exercise points out an issue that we have to address when studying this problem.

Construct a set of n lines in $\mathbb{R}^3$ such that: (i) Every two lines intersect. (ii) No three lines intersect at the same point.

7.5 Open Problems

To obtain Theorem 7.1, Guth and Katz proved that every set of n points in $\mathbb{R}^2$ satisfies $|Q| = O(n^3 \log n)$ (where Q is defined as in Equation (7.3)). Recall that there remains a $\sqrt{\log n}$ gap between the current best distinct distances bounds. One may suggest that, to close this gap, we should try to derive an improved upper bound for $|Q|$. This is not possible, since there exist sets of n points for which $|Q| = \Theta(n^3 \log n)$. For example, see Exercise 7.2.

Let $\mathcal{P}$ be the point set from Exercise 7.2. In Section 1.6, we saw that $D(\mathcal{P}) = \Theta(n/\sqrt{\log n})$. Why does the above proof lead to the weaker bound $D(\mathcal{P}) = \Omega(n/\log n)$? By going over the proof, we note that the only inefficient step is the Cauchy–Schwarz argument in Equation (7.4). The Cauchy–Schwarz inequality is close to an equality when all considered quantities have a similar size. In our case, some distances repeat many times while other distances repeat only a few times. Thus, the Cauchy–Schwarz inequality is far from tight. We conclude that, to improve Theorem 7.1, one has to change the ESGK framework. Another possibility is that Theorem 7.1 is tight – that there exists a set of n points with $\Theta(n/\log n)$ distinct distances.

Open Problem 7.3 *Remove the $\sqrt{\log n}$ gap between the current best bounds for the distinct distances problem in $\mathbb{R}^2$.*

We recall the problem of distinct distances between two lines, as presented in Section 7.1. This problem is interesting not just as a warm-up for the ESGK framework. The problem is simple to understand and to study, while also having many interesting generalizations (for example, see Pach and De Zeeuw, 2017;

Raz et al., 2016a,b). In the past, improving the bounds for the two lines problem has led to improvements for its many generalizations. Quoting Hilbert (Reid, 1970):

The art of doing mathematics is finding that special case that contains all the germs of generality.

Open Problem 7.4 *Improve the current best bounds for the problem of distinct distances between points on two lines.*

For simplicity, assume that each line contains n points. In this case, the current best bounds for the minimum number of distinct distances are $\Omega(n^{4/3})$ and $O(n^2/\sqrt{\log n})$ (see Section 7.1 for the full details).

8

Constant-Degree Polynomial Partitioning and Incidences in $\mathbb{C}^2$

We completely repudiate the symbol $\sqrt{-1}$, abandoning it without regret because we do not know what this alleged symbolism signifies nor what meaning to give to it.

Augustin-Louis Cauchy, taken out of context.

In mathematics, it is usually easier to study problems over $\mathbb{C}$ than over $\mathbb{R}$. Discrete geometry problems are an exception, often being significantly simpler over $\mathbb{R}$. While there are several simple proofs of the Szemerédi–Trotter theorem over $\mathbb{R}$, we only have rather involved proofs for the complex variant of the theorem. To avoid such involved proofs, we prove a slightly weaker variant of the complex Szemerédi–Trotter theorem.

In Chapter 7, we began to prove the distinct distances theorem by studying the ESGK framework. We complete this proof in Chapter 9, by relying on the *constant-degree polynomial partitioning* technique.[1] In this chapter, we introduce this technique by studying incidences with lines in the complex plane $\mathbb{C}^2$. This is a warm-up towards Chapter 9, where we use constant-degree polynomial partitioning in more involved ways.

8.1 Introduction: Incidence Issues in $\mathbb{C}^2$ and $\mathbb{R}^d$

Varieties in $\mathbb{C}^d$ are defined in the same way as in $\mathbb{R}^d$: Given a set of polynomials $f_1, \ldots, f_k \in \mathbb{C}[x_1, \ldots, x_d]$, the variety $\mathbf{V}(f_1, \ldots, f_k)$ is defined as

$$\mathbf{V}(f_1, \ldots, f_k) = \left\{ (a_1, \ldots, a_d) \in \mathbb{C}^d \ : \ f_j(a_1, \ldots, a_d) = 0 \text{ for all } 1 \leq j \leq k \right\}.$$

We refer to the coordinates of $\mathbb{C}^2$ as z_1 and z_2, to stress that $z_1, z_2 \in \mathbb{C}$ (as opposed to $x, y \in \mathbb{R}$). As in $\mathbb{R}^2$, we define a curve in $\mathbb{C}^2$ as a variety that

---

[1] Once again, there is no standard name for this technique. Some papers refer to it as *low degree polynomial partitioning*.

108

consists only of one-dimensional components. When considering a point-curve incidence problem in $\mathbb{C}^2$, it is tempting to imitate the proof of Theorem 3.3 for incidences in $\mathbb{R}^2$. That is, to combinatorially obtain a weak incidence bound, partition $\mathbb{C}^2$ using a partitioning polynomial, and then apply the weak bound separately in every cell. Unfortunately, partitioning polynomials do not exist in $\mathbb{C}^2$, since $\mathbb{C}^2 \backslash \mathbf{V}(f)$ is connected for every $f \in \mathbb{C}[z_1, z_2]$. In other words, we cannot split $\mathbb{C}^2$ into cells by using a variety.

We think of $\mathbb{C}^2$ as $\mathbb{R}^4$, since partitioning polynomials do exist in $\mathbb{R}^4$. Thinking of $\mathbb{C}^2$ as $\mathbb{R}^4$ also makes it easier to see that $\mathbb{C}^2 \backslash \mathbf{V}(f)$ is connected for every $f \in \mathbb{C}[z_1, z_2]$. We write $z_1 = x_1 + iy_1$ and $z_2 = x_2 + iy_2$, where $x_1, x_2, y_1, y_2 \in \mathbb{R}$. We then consider x_1, x_2, y_1, y_2 as the coordinates of $\mathbb{R}^4$. We define the map $\phi \colon \mathbb{C}^2 \to \mathbb{R}^4$ as

$$\phi(x_1 + iy_1, x_2 + iy_2) = (x_1, y_1, x_2, y_2). \tag{8.1}$$

Consider a set $\mathcal{P}$ of points and a set Γ of curves, both in $\mathbb{C}^2$. We note that ϕ takes distinct points of $\mathcal{P}$ to distinct points in $\mathbb{R}^4$. Consider a curve in $\mathbb{C}^2$ that is defined by a polynomial $f \in \mathbb{C}[z_1, z_2]$. Asking f to vanish on a point $(a_1 + ib_1, a_2 + ib_2)$ is equivalent to asking both the real and the imaginary parts of $f(a_1 + ib_1, a_2 + ib_2)$ to vanish. Thus, $\phi(\mathbf{V}(f)) \subset \mathbb{R}^4$ is a variety defined by two polynomials in $\mathbb{R}[x_1, y_1, x_2, y_2]$, each of degree $\deg f$. We also note that ϕ maintains the point-curve incidences. That is, $p \in \mathcal{P}$ is incident to $\gamma \in \Gamma$ if and only if $\phi(p)$ is incident to $\phi(\gamma)$.

Figure 8.1 By having planes that contain a common line, we can create a configuration where every point is incident to every plane.

Moving from $\mathbb{C}^2$ to $\mathbb{R}^4$ allows us to use polynomial partitioning, but also introduces a new issue. This issue arises when studying incidence with varieties of dimension larger than one. For simplicity, we consider the simplest case: incidences between points and planes in $\mathbb{R}^3$. Let $\ell \subset \mathbb{R}^3$ be a line, let $\mathcal{P}$ be a set of m points on ℓ, and let H be a set of n planes that contain ℓ. For an example, see Figure 8.1. Since every point is incident to every plane, the maximum number of point-plane incidences is mn. This problem is trivial to solve and not interesting.

To turn the point-plane problem into a nontrivial one, we add additional restrictions. For example, we may ask the incidence graph of $\mathcal{P} \times H$ not to contain a copy of $K_{s,t}$, for some integers $s, t \geq 2$. This leads to an interesting family of problems, which are open for most values of s and t. Such restrictions arise naturally in many cases. For example, when no k points of $\mathcal{P}$ are collinear, the incidence graph contains no $K_{k,2}$. We also rely on such restrictions when studying incidences with curves in $\mathbb{C}^2$.

Consider the problem of point-plane incidences in $\mathbb{R}^3$ with no $K_{s,t}$ in the incidence graph. To obtain an upper bound for this problem, we wish to apply the polynomial partitioning technique as in Section 3.2. In this technique, we first derive a weak incidence bound in Lemma 3.4, and then amplify this bound by applying it separately in each cell. By inspecting the proof of Lemma 3.4, we note that it does not rely on any geometric properties of the problem. It actually derives an upper bound for the number of edges in a bipartite graph with no $K_{s,t}$. Thus, the same proof holds for any set of objects in any space over any field, as long as the graph contains no $K_{s,t}$.

Lemma 8.1   *Let* $\mathbb{F}$ *be a field and let* d *be a positive integer. Let* $\mathcal{P}$ *be a set of* m *points in* $\mathbb{F}^d$ *and let* $\mathcal{V}$ *be a set of* n *subsets of* $\mathbb{F}^d$. *If the incidence graph of* $\mathcal{P} \times \mathcal{V}$ *contains no* $K_{s,t}$, *then*

$$I(\mathcal{P}, \mathcal{V}) = O_{s,t}\left(mn^{1-\frac{1}{s}} + n \right).$$

We recall that the polynomial partitioning theorem holds in $\mathbb{R}^d$. It is straightforward to bound the number of incidences in the cells, by adapting the argument of Theorem 3.3 (see Exercise 8.2). In the proof of Theorem 3.3, it is not difficult to obtain an upper bound for the number of incidences with points on the partition. In $\mathbb{R}^d$ with $d \geq 3$, handling incidences with points on the partition becomes the most complicated part of the analysis. In $\mathbb{R}^2$, the set of singular points of the partition is zero-dimensional and a curve that is not contained in the partition has a zero-dimensional intersection with the partition. In $\mathbb{R}^d$ with $d \geq 3$, neither of these properties holds and both lead to new difficulties. Solymosi and Tao (2012) introduced the constant-degree polynomial partitioning technique to overcome these difficulties.

We study the constant-degree polynomial partitioning technique in the following sections. Before that, we prove our first incidence bound in $\mathbb{R}^d$. This warm-up proof does not rely on new partitioning techniques. Instead, we simply project the problem from $\mathbb{R}^d$ to $\mathbb{R}^2$ and apply our planar incidence bound (Theorem 3.3).

Lemma 8.2   *Consider an integer* $d \geq 2$. *Let* $\mathcal{P}$ *be a set of* m *points and let* Γ *be a set of* n *varieties, both in* $\mathbb{R}^d$. *Every variety of* Γ *is of dimension at most* *one and complexity at most k. If the incidence graph of $\mathcal{P} \times \Gamma$ contains no $K_{s,t}$, then*

$$I(\mathcal{P}, \Gamma) = O_{s,t,k,d}\left(m^{\frac{s}{2s-1}} n^{\frac{2s-2}{2s-1}} + m + n\right).$$

Proof We split every variety of Γ into irreducible components and let Γ' be the multiset of these components.[2] If an irreducible variety γ is a component of r elements of Γ, then γ appears r times in Γ'. A point that is incident to γ forms a distinct incidence with each of the r copies. By Lemma 4.10, every variety of Γ consists of $O_{k,d}(1)$ components. Thus, $|\Gamma'| = O_{k,d}(n)$.

Every zero-dimensional element of Γ' is incident to at most one point of $\mathcal{P}$. Thus, the zero-dimensional elements of Γ' form $|\Gamma'| = O_{k,d}(n)$ incidences with $\mathcal{P}$. Since the incidence graph contains no $K_{s,t}$, an element that appears at least t times in Γ' is incident to at most $s - 1$ points of $\mathcal{P}$. In total, elements that appear at least t times in Γ' form $O_{k,d,s}(n)$ incidences with $\mathcal{P}$.

It remains to bound the number of incidences with one-dimensional components that appear at most $t - 1$ times in Γ'. Let Γ'' be the set of one-dimensional elements of Γ', without multiplicity. That is, if an element appears at least once in Γ then it appears exactly once in Γ''. We note that $I(\mathcal{P}, \Gamma'') \geq (t - 1) \cdot I(\mathcal{P}, \Gamma')$. Thus, it remains to bound $I(\mathcal{P}, \Gamma'')$.

Let $\pi\colon \mathbb{R}^d \to \mathbb{R}^2$ be a generic projection with respect to $\mathcal{P}$ and Γ''. We define

$$\mathcal{P}_2 = \{\pi(p) \,:\, p \in \mathcal{P}\} \quad \text{and} \quad \Gamma_2 = \left\{\overline{\pi(\gamma)} \,:\, \gamma \in \Gamma''\right\}.$$

Since π is chosen generically, we may assume that no two points of $\mathcal{P}$ are projected to the same point of $\mathbb{R}^2$. Similarly, we may assume that no two varieties of Γ_2 have a common component and that no new incidences are introduced by the projection. This implies that $|\mathcal{P}_2| = m$, that $|\Gamma_2| = O_{k,d}(n)$, that $I(\mathcal{P}, \Gamma'') = I(\mathcal{P}_2, \Gamma_2)$, and that the incidence graph of $\mathcal{P}_2 \times \Gamma_2$ contains no $K_{s,t}$. By Corollary 4.13, every element of Γ_2 is a variety of dimension at most one and complexity $O_{d,k}(1)$. By our point-curve incidence result in $\mathbb{R}^2$ (Theorem 3.3), we obtain that

$$I(\mathcal{P}, \Gamma'') = I(\mathcal{P}_2, \Gamma_2) = O_{s,t,k,d}\left(m^{\frac{s}{2s-1}} n^{\frac{2s-2}{2s-1}} + m + n\right).$$

Technical issues: By summing up the incidences from the three different cases, we obtain the bound of the lemma and complete the proof. However, several technical issues are not addressed above. First, Theorem 3.3 holds only for curves. We handle zero-dimensional components of elements of Γ_2 as before. That is, we note that there are $O_{d,k}(n)$ such components and that every

---

[2] Recall that a multiset may contain the same element multiple times.

component contributes at most one incidence. Thus, when removing the zero-dimensional components, we decrease the number of incidences by $O_{d,k}(n)$.

We note a subtler issue: Assuming that the incidence graph of $\mathcal{P} \times \Gamma$ contains no $K_{s,t}$ does not imply that the incidence graph of $\mathcal{P} \times \Gamma'$ contains no $K_{s,t}$. For example, it is possible that t components of a variety $\gamma \in \Gamma$ contain the same s points. After splitting γ into irreducible components, we obtain t components that contain the same s points. We now address this issue.

Let $c = O_{k,d}(1)$ be the maximum number of components in an element of Γ. Instead of Γ', we create c multisets $\Gamma'_1, \Gamma'_2, \ldots, \Gamma'_c$, as follows. When considering a variety of Γ, we arbitrarily place each of its components in a different Γ_j. That is, each multiset $\Gamma'_1, \Gamma'_2, \ldots, \Gamma'_c$ contains at most one component from each variety of Γ. We have that

$$I(\mathcal{P}, \Gamma) = \sum_{j=1}^{c} I(\mathcal{P}, \Gamma'_j).$$

For every $1 \le j \le c$, the above analysis for $I(\mathcal{P}, \Gamma')$ implies that

$$I(\mathcal{P}, \Gamma'_j) = O_{s,t,k,d}\left(m^{\frac{s}{2s-1}} n^{\frac{2s-2}{2s-1}} + m + n\right).$$

Since $c = O_{k,d}(1)$, this leads to the bound of the lemma. $\square$

In Chapter 9, we derive a more sophisticated bound for incidences with lines in $\mathbb{R}^3$. In Section 11.1, we prove a similar bound for incidences with arbitrary curves in $\mathbb{R}^3$. It is not difficult to further generalize this to incidences with curves in $\mathbb{R}^d$ (see Sharir et al., 2016).

8.2 Constant-Degree Polynomial Partitioning

We now introduce the constant-degree polynomial partitioning technique. While the goal of this technique is to prove bounds in dimension $d \ge 3$, we first prove a weak Szemerédi–Trotter theorem in $\mathbb{R}^2$. This demonstrates the basic idea of the technique without the additional technicalities that arise in higher dimensions.

Theorem 8.3 *Let $\mathcal{P}$ be a set of m points and let $\mathcal{L}$ be a set of n lines, both in* $\mathbb{R}^2$. *Then for any $\varepsilon > 0$, we have $I(\mathcal{P}, \mathcal{L}) = O_\varepsilon(m^{2/3+\varepsilon} n^{2/3} + m + n)$.*

In Theorem 8.3, we may choose ε to be as small as we like. In other words, the theorem states that the number of incidences is asymptotically smaller than $m^c n^{2/3} + m + n$ for every $c > 2/3$. One might say that Theorem 8.3 is tight up to *subpolynomial factors*.

It is risky to use $O(\cdot)$-notation in a proof by induction. To see the issue, we prove the false claim $2^n = O(n)$ by induction on n. For the induction basis, the claim holds for $n \leq 10$ by taking the constant hidden by the $O(\cdot)$-notation to be sufficiently large. For the induction step, assume that $2^n = O(n)$ for some n and note that $2^{n+1} = 2 \cdot 2^n = 2 \cdot O(n) = O(n)$. This concludes the false proof. The issue is that the induction step does not actually close. We get an asymptotically correct upper bound for 2^{n+1}, but the hidden constant that we get in the $O(\cdot)$-notation is larger than the constant of the induction hypothesis. To avoid this issue, the following proof relies on the constants α_1 and α_2.

Proof of Theorem 8.3 We prove the theorem by induction on $m+n$. Specifically, we prove by induction that, for any fixed $\varepsilon > 0$, there exist constants α_1, α_2 such that

$$I(\mathcal{P}, \mathcal{L}) \leq \alpha_1 m^{2/3+\varepsilon} n^{2/3} + \alpha_2(m + n). \tag{8.2}$$

For the induction basis, consider the case of $m + n \leq 200$. In this case, Inequality (8.2) holds by taking α_1 and α_2 to be sufficiently large. For example, in this case the number of incidences is at most 10^4, so we can set $\alpha_1 = 10^4$.

For the induction step, we consider m and n that satisfy $m+n > 200$. Lemma 1.2 (or Lemma 8.1) implies that $I(\mathcal{P}, \mathcal{L}) = O(m\sqrt{n} + n)$. If $m = O(\sqrt{n})$ then $I(\mathcal{P}, \mathcal{L}) = O(n)$, which completes the proof when α_2 is sufficiently large. We may thus assume that $m = \Omega(\sqrt{n})$, or equivalently

$$n = O(m^2). \tag{8.3}$$

Partitioning the plane: We take r to be a sufficiently large constant, whose value depends on ε. The meaning of "sufficiently large" is discussed below. Let f be an r-partitioning polynomial of $\mathcal{P}$. By the polynomial partitioning theorem (Theorem 3.1), $\deg f = O(r)$ and $\mathbf{V}(f)$ partitions $\mathbb{R}^2$ into c cells, each containing at most m/r^2 points of $\mathcal{P}$. Warren's theorem (Theorem 3.2) implies that $c = O(r^2)$. We choose the constants in this proof so that[3]

$$\varepsilon^{-1} \ll r \ll \alpha_2 \ll \alpha_1.$$

We denote the cells of the partition as $C_1, \ldots, C_c$. For $1 \leq j \leq c$, we set $\mathcal{P}_j = \mathcal{P} \cap C_j$ and let $\mathcal{L}_j$ be the set of lines of $\mathcal{L}$ that intersect C_j. We let $\mathcal{L}_0$ denote the subset of lines of $\mathcal{L}$ that are contained in $\mathbf{V}(f)$, and let $\mathcal{P}_0 = \mathcal{P} \cap \mathbf{V}(f)$. Note that

$$I(\mathcal{P}, \mathcal{L}) = \sum_{j=1}^{c} I(\mathcal{P}_j, \mathcal{L}_j) + I(\mathcal{P}_0, \mathcal{L}_0) + I(\mathcal{P}_0, \mathcal{L} \backslash \mathcal{L}_0).$$

---

[3] The expression $a \ll b$ means that b is larger than some expression that depends on a. For example, we might require that $b > 3a^a$.

For any line $\ell \in \mathcal{L}\backslash\mathcal{L}_0$, Bézout's theorem (Theorem 2.5) implies that $|\ell \cap V(f)| = O(r)$. We get that

$$I(\mathcal{P}_0, \mathcal{L}\backslash\mathcal{L}_0) = O(r \cdot |\mathcal{L}\backslash\mathcal{L}_0|) = O(nr). \tag{8.4}$$

We set $m_0 = |\mathcal{P}_0|$ and $m' = m - m_0$. In other words, m' is the number of points of $\mathcal{P}$ that are in the cells of the partition. Since $\deg f = O(r)$, Corollary 4.6 implies that $V(f)$ contains $O(r)$ lines. We thus get that

$$I(\mathcal{P}_0, \mathcal{L}_0) \le |\mathcal{P}_0| \cdot |\mathcal{L}_0| = O(m_0 r). \tag{8.5}$$

It remains to bound $\sum_{j=1}^{c} I(\mathcal{P}_j, \mathcal{L}_j)$. For $1 \le j \le c$, we let $m_j = |\mathcal{P}_j|$ and $n_j = |\mathcal{L}_j|$. We note that $m' = \sum_{j=1}^{c} m_j$, and recall that $m_j \le m/r^2$ for every $1 \le j \le c$. By the induction hypothesis, we have that

$$\sum_{j=1}^{c} I(\mathcal{P}_j, \mathcal{L}_j) \le \sum_{j=1}^{c} \left(\alpha_1 m_j^{2/3+\varepsilon} n_j^{2/3} + \alpha_2 (m_j + n_j) \right)$$

$$\le \alpha_1 \left(\frac{m}{r^2} \right)^{2/3+\varepsilon} \sum_{j=1}^{c} n_j^{2/3} + \alpha_2 \left(m' + \sum_{j=1}^{c} n_j \right). \tag{8.6}$$

Recall that every line $\ell \in \mathcal{L}\backslash\mathcal{L}_0$ satisfies $|\ell \cap V(f)| = O(r)$. When traveling along ℓ, we move to a different cell only after an intersection with $V(f)$. Thus, every line of $\mathcal{L}\backslash\mathcal{L}_0$ intersects $O(r)$ cells. This implies that $\sum_{j=1}^{c} n_j = O(nr)$. Combining this with Hölder's inequality (Theorem A.3) leads to

$$\sum_{j=1}^{c} n_j^{2/3} \le \left(\sum_{j=1}^{c} n_j \right)^{2/3} \left(\sum_{j=1}^{c} 1 \right)^{1/3} = O\left((nr)^{2/3} \cdot r^{2/3} \right) = O\left(n^{2/3} r^{4/3} \right). \tag{8.7}$$

By combining Inequalities (8.6) and (8.7), we obtain that

$$\sum_{j=1}^{c} I(\mathcal{P}_j, \mathcal{L}_j) = O\left(\frac{\alpha_1 m^{2/3+\varepsilon} n^{2/3}}{r^{2\varepsilon}} + \alpha_2 nr \right) + \alpha_2 m'.$$

Combining this with Inequalities (8.4) and (8.5) gives

$$I(\mathcal{P}, \mathcal{L}) = O\left(\frac{\alpha_1 m^{2/3+\varepsilon} n^{2/3}}{r^{2\varepsilon}} + \alpha_2 nr + m_0 r \right) + \alpha_2 m'.$$

Simplifying the bound: By taking α_2 to be sufficiently large with respect to r and to the constant in the $O(\cdot)$-notation, we get that

$$I(\mathcal{P}, \mathcal{L}) = O\left(\frac{\alpha_1 m^{2/3+\varepsilon} n^{2/3}}{r^{2\varepsilon}} + \alpha_2 nr\right) + \alpha_2 (m' + m_0)$$

$$= O\left(\frac{\alpha_1 m^{2/3+\varepsilon} n^{2/3}}{r^{2\varepsilon}} + \alpha_2 nr\right) + \alpha_2 m. \tag{8.8}$$

By Equation (8.3), we have that $n = n^{2/3} n^{1/3} = O(m^{2/3} n^{2/3})$. Taking α_1 to be sufficiently large with respect to α_2, r, and to the constant in the $O(\cdot)$-notation in Equation (8.8), leads to

$$O(\alpha_2 nr) = O\left(\alpha_2 m^{2/3} n^{2/3} r\right) \le \frac{\alpha_1}{2} m^{2/3} n^{2/3}.$$

Similarly, taking r to be sufficiently large with respect to ε^{-1} and to the constant in the $O(\cdot)$-notation in Equation (8.8), gives that

$$O\left(\frac{\alpha_1 m^{2/3+\varepsilon} n^{2/3}}{r^{2\varepsilon}}\right) \le \frac{\alpha_1}{2} m^{2/3+\varepsilon} n^{2/3}.$$

Combining this with Equation (8.8) completes the induction step, and thus the proof of the theorem. □

Understanding the technique: The proofs of Theorem 3.3 and of Theorem 8.2 may seem similar. Both proofs partition $\mathbb{R}^2$ with a partitioning polynomial and then separately bound the number of incidences in every cell. While the proof of Theorem 3.3 applies the weak bound of Lemma 1.2 in every cell, the proof of Theorem 8.2 applies the induction hypothesis. The use of induction allows us to use a partitioning polynomial of degree that does not depend on m and n. In other words, we can use a constant-degree partitioning polynomial. The significantly smaller degree makes it easier to study the incidences that are on the partition. As mentioned above, studying incidences on the partition becomes difficult when applying polynomial partitioning in dimension $d \ge 3$.

When removing the ε from the bound of Theorem 8.3, the induction step fails. When we apply the induction hypothesis in every cell and sum up, the result has the correct asymptotic value but the leading constant is larger than the one we started with. This is the same issue as in the above false proof that $2^n = O(n)$. When using constant-degree polynomial partitioning, we always need to add this extra ε for the induction step to close. One might say that this extra ε is the main disadvantage of the technique.

When using constant-degree polynomial partitioning, we do not apply the weak Lemma 1.2 (or Lemma 8.1) in each cell. However, the proof still requires such a weak bound. In the proof of Theorem 8.2, we use this weak bound to obtain Equation (8.3).

8.3 The Szemerédi–Trotter Theorem in $\mathbb{C}^2$

After gaining a first understanding of constant-degree polynomial partitioning, we use it to prove the Szemerédi–Trotter theorem in $\mathbb{C}^2$. That is, we have a point set $\mathcal{P}$ and a set of lines $\mathcal{L}$, both in $\mathbb{C}^2$. We think of a complex line as the zero set of a linear polynomial in $\mathbb{C}[z_1, z_2]$.

Theorem 8.4 *Let $\mathcal{P}$ be a set of m points and let $\mathcal{L}$ be a set of n lines, both in $\mathbb{C}^2$. Then for any $\varepsilon > 0$, we have $I(\mathcal{P}, \mathcal{L}) = O_\varepsilon(m^{2/3+\varepsilon} n^{2/3} + m + n)$.*

Theorem 8.4 is a weaker variant of the Szemerédi–Trotter theorem in $\mathbb{C}^2$ because of the extra ε in the exponent. There exist proofs that remove this ε (see Tóth, 2015; Zahl, 2015). These proofs are significantly more involved and beyond the scope of this book. The following proof is by Solymosi and Tao (2012).

Proof of Theorem 8.4 We recall the map $\phi \colon \mathbb{C}^2 \to \mathbb{R}^4$ as defined in Equation (8.1). We set $\mathcal{P}' = \{\phi(p) \ : \ p \in \mathcal{P}\}$ and note that $\mathcal{P}'$ is a set of m distinct points in $\mathbb{R}^4$. We also set $\mathcal{L}' = \{\phi(\ell) \ : \ \ell \in \mathcal{L}\}$.

Fix a line $\ell \subset \mathbb{C}^2$. There exist $a, a', b, b', c, c' \in \mathbb{R}$ such that

$$\ell = \mathbf{V}((a + ia')z_1 + (b + ib')z_2 + (c + ic')).$$

A point $(x_1 + iy_1, x_2 + iy_2) \in \mathbb{C}^2$ is incident to ℓ if and only if

$$(a + ia')(x_1 + iy_1) + (b + ib')(x_2 + iy_2) + (c + ic') = 0.$$

By separating the real and imaginary parts, we get that the point is incident to ℓ if and only if

$$ax_1 - a'y_1 + bx_2 - b'y_2 + c = 0 \quad \text{and} \quad a'x_1 + ay_1 + b'x_2 + by_2 + c' = 0.$$

Thus, the variety $\phi(\ell) \subset \mathbb{R}^4$ is defined by two linear equations. Each linear equation defines a distinct hyperplane, and these two hyperplanes are not parallel. We conclude that $\phi(\ell)$ is a 2-flat in $\mathbb{R}^4$.

We say that a 2-flat $\Pi \subset \mathbb{R}^4$ is *special* if there exists a line $\ell \subset \mathbb{C}^2$ that satisfies $\phi(\ell) = \Pi$. A generic 2-flat in $\mathbb{R}^4$ is not special (this can be shown with tools from Chapter 14). Since two distinct lines in $\mathbb{C}^2$ intersect in at most one point and ϕ is a bijection, every two special 2-flats intersect in at most one point.

We note that $\mathcal{L}'$ is a set of n distinct special 2-flats in $\mathbb{R}^4$. Since ϕ is a bijection, the incidence graph of $\mathcal{P} \times \mathcal{L}$ is identical to the incidence graph of $\mathcal{P}' \times \mathcal{L}'$. This implies that $I(\mathcal{P}, \mathcal{L}) = I(\mathcal{P}', \mathcal{L}')$. Since every two special 2-flats intersect in at most one point, the incidence graph of $\mathcal{P}' \times \mathcal{L}'$ contains no $K_{2,2}$. We derive an upper bound for $I(\mathcal{P}', \mathcal{L}')$ by proving the following incidence result.

Lemma 8.5 *For every fixed* $\varepsilon > 0$ *there exist sufficiently large* $\alpha_1, \alpha_2 \in \mathbb{R}$ *that satisfy the following. For every set* Q *of* M *points and set* $\mathcal{F}$ *of* N *special 2-flats, both in* $\mathbb{R}^4$, *we have that*

$$I(Q, \mathcal{F}) \le \alpha_1 M^{2/3+\varepsilon} N^{2/3} + \alpha_2(M + N).$$

Proof The proof is a variant of the proof of Theorem 8.3. We prove the lemma by induction on $M + N$. For the induction basis, consider the case of $M + N \le 200$. In this case, the lemma holds by taking α_1 and α_2 to be sufficiently large. For the induction step, we consider M and N that satisfy $M + N > 200$.

Since two special 2-flats intersect in at most one point, the incidence graph of $Q \times \mathcal{F}$ contains no $K_{2,2}$. Lemma 8.1 implies $I(Q, \mathcal{F}) = O(M\sqrt{N} + N)$. If $M = O(\sqrt{N})$ then $I(Q, \mathcal{F}) = O(N)$, which completes the proof when α_2 is sufficiently large. We may thus assume that $M = \Omega(\sqrt{N})$, or equivalently

$$N = O(M^2). \tag{8.9}$$

We take r to be a sufficiently large constant, whose value depends on ε. Let f be an r-partitioning polynomial of Q. By the polynomial partitioning theorem, $\deg f = O(r)$ and $\mathbf{V}(f)$ partitions $\mathbb{R}^4$ into c cells, each containing at most M/r^4 points of Q. Warren's theorem implies that $c = O(r^4)$. We choose the constants in this proof so that

$$2^\varepsilon \ll r \ll \alpha_2 \ll \alpha_1.$$

We denote the cells of the partition as $C_1, \ldots, C_c$. For $1 \le j \le c$, we set $Q_j = Q \cap C_j$ and let $\mathcal{F}_j$ be the set of 2-flats of $\mathcal{F}$ that intersect C_j. Let $\mathcal{F}_0$ be the subset of 2-flats of $\mathcal{F}$ that are contained in $\mathbf{V}(f)$, and let $Q_0 = Q \cap \mathbf{V}(f)$. We note that

$$I(Q, \mathcal{F}) = \sum_{j=1}^{c} I(Q_j, \mathcal{F}_j) + I(Q_0, \mathcal{F}_0) + I(Q_0, \mathcal{F} \backslash \mathcal{F}_0).$$

We fix a 2-flat $\Pi \in \mathcal{F} \backslash \mathcal{F}_0$. By Theorem 4.15, $\Pi \backslash \mathbf{V}(f)$ consists of $O(r^2)$ connected components. Since every such connected component is in at most one cell of the partition, we get that Π intersects $O(r^2)$ cells.

Incidences in the cells: For $1 \le j \le c$, we let $M_j = |Q_j|$, $M' = \sum_{j=1}^{c} M_j$, and $N_j = |\mathcal{F}_j|$. Then $M_j \le M/r^2$ for every $1 \le j \le c$. By the induction hypothesis, we have that

$$\sum_{j=1}^{c} I(Q_j, \mathcal{F}_j) \leq \sum_{j=1}^{c} \left(\alpha_1 M_j^{2/3+\varepsilon} N_j^{2/3} + \alpha_2 (M_j + N_j) \right)$$

$$\leq \alpha_1 \left(\frac{M}{r^4} \right)^{2/3+\varepsilon} \sum_{j=1}^{c} N_j^{2/3} + \alpha_2 \left(M' + \sum_{j=1}^{c} N_j \right). \qquad (8.10)$$

Since every 2-flat intersects $O(r^2)$ cells, we get that $\sum_{j=1}^{c} N_j = O(Nr^2)$. Combining this with Hölder's inequality implies

$$\sum_{j=1}^{c} N_j^{2/3} \leq \left(\sum_{j=1}^{c} N_j \right)^{2/3} \cdot \left(\sum_{j=1}^{c} 1 \right)^{1/3} = O\left(\left(Nr^2 \right)^{2/3} \cdot r^{4/3} \right) = O(N^{2/3} r^{8/3}). \qquad (8.11)$$

By combining Inequalities (8.10) and (8.11), we obtain that

$$\sum_{j=1}^{c} I(Q_j, \mathcal{F}_j) \leq O\left(\frac{\alpha_1 M^{2/3+\varepsilon} N^{2/3}}{r^{4\varepsilon}} + \alpha_2 Nr^2 \right) + \alpha_2 M'.$$

We note that Equation (8.9) implies $N = N^{1/3} \cdot N^{2/3} = O(M^{2/3} N^{2/3})$. Taking α_1 to be sufficiently large with respect to α_2 and r leads to

$$\sum_{j=1}^{c} I(Q_j, \mathcal{F}_j) \leq O\left(\frac{\alpha_1 M^{2/3+\varepsilon} N^{2/3}}{r^{4\varepsilon}} \right) + \alpha_2 M'.$$

Finally, by taking r to be sufficiently large with respect to ε^{-1} and to the constant of the $O(\cdot)$-notation, we have that

$$\sum_{j=1}^{c} I(Q_j, \mathcal{F}_j) \leq \frac{\alpha_1}{3} M^{2/3+\varepsilon} N^{2/3} + \alpha_2 M'. \qquad (8.12)$$

Flats that intersect the partition: Consider a 2-flat $\Pi \in \mathcal{F}$ that is not contained in $\mathbf{V}(f)$. By Claim 4.4, the intersection $\mathbf{V}(f) \cap \Pi$ is a variety of dimension at most one. Since this variety is defined by f and two linear equations, it is of complexity $O(r)$. We set

$$\Gamma = \{ \Pi \cap \mathbf{V}(f) : \Pi \in \mathcal{F} \backslash \mathcal{F}_0 \}.$$

We also set $|Q_0| = M_0$, and note that $M_0 = M - M'$. Since the incidence graph of $Q \times \mathcal{F}$ contains no $K_{2,2}$, neither does the incidence graph of $Q_0 \times \Gamma$. Lemma 8.2 implies that

$$I(Q_0, \mathcal{F} \backslash \mathcal{F}_0) = I(Q_0, \Gamma) = O_r \left(M_0^{2/3} N^{2/3} + M_0 + N \right).$$

We recall that $N = O(M^{2/3}N^{2/3})$. By taking α_1 and α_2 to be sufficiently large with respect to r and to the constant of the $O(\cdot)$-notation, we get that

$$I(Q_0, \mathcal{F} \setminus \mathcal{F}_0) \le \frac{\alpha_1}{3} M^{2/3} N^{2/3} + \frac{\alpha_2}{2} M_0. \tag{8.13}$$

Flats that are contained in the partition: To handle incidences with 2-flats of $\mathcal{F}_0$, we show that two such flats cannot intersect at a regular point of $\mathbf{V}(f)$. In other words, every regular point of $\mathbf{V}(f)$ is incident to at most one flat of $\mathcal{F}$. Studying incidences with singular points of $\mathbf{V}(f)$ is easier, since these singular points form a variety of dimension at most two.

Consider a point $q \in \mathbf{V}(f)$ that is incident to two 2-flats $\Pi, \Pi' \in \mathcal{F}_0$. By performing a translation of $\mathbb{R}^4$ that takes q to the origin, we can think of Π and Π' as vector spaces. Since $\Pi \cap \Pi' = \{q\}$, these two vector spaces span $\mathbb{R}^4$. Since $\Pi = T_q \Pi$ and $\Pi' = T_q \Pi'$, the two tangent planes also span $\mathbb{R}^4$. If q is a regular point of $\mathbf{V}(f)$, then $T_q \Pi$ and $T_q \Pi'$ are both contained in $T_q \mathbf{V}(f)$. Since 2-flats that span $\mathbb{R}^4$ cannot be contained in the same hyperplane, we conclude that q is a singular point of $\mathbf{V}(f)$.

The above implies that every regular point of $\mathbf{V}(f)$ is incident to at most one 2-flat of $\mathcal{F}_0$. Thus, points of Q_0 that are regular points of $\mathbf{V}(f)$ form $O(M_0)$ incidences with $\mathcal{F}_0$. Let U_{sing} be the set singular points of $\mathbf{V}(f)$. It remains to consider incidences between $\mathcal{F}_0$ and point of $Q_0 \cap U_{\text{sing}}$.

Theorem 4.11 states that U_{sing} is a variety of dimension at most two and complexity $O_r(1)$. Every 2-flat that is contained in U_{sing} is also a component of U_{sing}. By combining this with Lemma 4.10, we get that U_{sing} contains $O_r(1)$ 2-flats of $\mathcal{F}_0$. Such 2-flats form $O_r(m_0)$ incidences with the points of Q_0. Every 2-flat of $\mathcal{F}_0$ that is not contained in U_{sing} intersects U_{sing} in a variety of dimension at most one and complexity $O_r(1)$. By Lemma 8.2, the number of incidences between points of $Q_0 \cap U_{\text{sing}}$ and 2-flats of $\mathcal{F}_0$ that are not contained in U_{sing} is

$$O_r \left(M^{2/3} N^{2/3} + M_0 + N \right).$$

By combining the singular and regular cases above, we get that

$$I(Q_0, \mathcal{F}_0) = O_r \left(M^{2/3} N^{2/3} + M_0 + N \right).$$

By recalling that $N = O(M^{2/3}N^{2/3})$ and taking α_1 and α_2 to be sufficiently large with respect to r and the constant of the $O(\cdot)$-notation, we have that

$$I(Q_0, \mathcal{F}_0) \le \frac{\alpha_1}{3} M^{2/3} N^{2/3} + \frac{\alpha_2}{2} M_0. \tag{8.14}$$

We complete the induction step by combining Equations (8.12)–(8.14). This in turn completes the proof of the lemma. $\qquad\square$

Lemma 8.5 implies that

$$I(\mathcal{P}, \mathcal{L}) = I(\mathcal{P}', \mathcal{L}') = O_\varepsilon \left(m^{2/3+\varepsilon} n^{2/3} + m + n \right).$$

This completes the proof of Theorem 8.4.   □

Solymosi and Tao (2012) prove an incidence result that is more general than the one stated for 2-flats in the proof of Lemma 8.5. The proof of this more general result is a simple generalization of the proof of Lemma 8.5.

Theorem 8.6 *Let* $\mathcal{P}$ *be a set of m points and let* $\mathcal{V}$ *be a set of n varieties of degree at most k, both in* $\mathbb{R}^d$, *such that:*

- *The dimension of every variety of* $\mathcal{V}$ *is at most* $d/2$.
- *The incidence graph of* $\mathcal{P} \times \mathcal{V}$ *contains no* $K_{s,t}$.
- *There is no incidence between a point* $p \in \mathcal{P}$ *and a variety* $U \in \mathcal{V}$ *where* p *is a singular point of* U.
- *If* $p \in \mathcal{P}$ *is a regular point of two varieties* $U, W \in \mathcal{V}$, *then the tangent spaces* $T_p U$ *and* $T_p W$ *intersect only in the origin.*

Then $\qquad I(\mathcal{P}, \mathcal{V}) = O_{\varepsilon,k,d,s,t} \left(m^{\frac{s}{2s-1}+\varepsilon} n^{\frac{2s-2}{2s-1}} + m + n \right).$

From lines to curves: The Szemerédi–Trotter theorem is a special case of Theorem 3.3. After extending Szemerédi–Trotter to $\mathbb{C}^2$, one may wish to extend Theorem 3.3 to $\mathbb{C}^2$ (with an extra ε in the exponent). At first glance, it might seem as if Theorem 8.6 is already such an extension. However, because of the additional restrictions in Theorem 8.6, we cannot apply this result to most types of curves in $\mathbb{C}^2$. We now discuss an issue that prevents us from easily extending the proof of Theorem 8.4.

Figure 8.2 Two circles with the same tangent line at an intersection point (in $\mathbb{R}^2$).

In the proof of Theorem 8.4 we relied on the following property: If two special 2-flats intersect at a point q, then the tangent planes at q also intersect at a single point. We used this to argue that special 2-flats that are contained in the partition intersect only at singular points of the partition. In an intersection point of two lines in $\mathbb{C}^2$, the two tangent lines indeed intersect at a single point. Hardly any other families of curves have this property. For example, two circles in $\mathbb{C}^2$ may have identical tangent lines at a point of intersection. See Figure 8.2.

To overcome the above difficulty, Sheffer et al. (2018) relies on basic differential geometry. This leads to the following result.

Theorem 8.7 *Let $\mathcal{P}$ be a set of m points and let Γ be a set of n irreducible curves of degree at most k, both in $\mathbb{C}^2$. If the incidence graph of $\mathcal{P} \times \Gamma$ contains no $K_{s,t}$, then for every $\varepsilon > 0$ we have that*

$$I(\mathcal{P}, \Gamma) = O_{s,t,k,\varepsilon}\left(m^{\frac{s}{2s-1}+\varepsilon} n^{\frac{2s-2}{2s-1}} + m + n\right).$$

The proof of Theorem 8.7 is beyond the scope of this book. We only mention that it relies on a variant of the Picard–Lindelöf theorem. We encounter this tool as Theorem 9.11 in the optional Section 9.3.

8.4 Exercises

In some of the exercises, we rely on the following result. This is a polynomial partitioning for varieties, rather than for points. That is, each cell is intersected by a bounded number of varieties.

Theorem 8.8 (Guth, 2015b) *Let $\mathcal{V}$ be a set of n varieties of dimension D in $\mathbb{R}^d$, and let $1 < r \le n$. Then there exists $f \in \mathbb{R}[x_1, \ldots, x_d]$ of degree at most r, such that every connected component of $\mathbb{R}^d \setminus \mathbf{V}(f)$ is intersected by $O(n/r^{d-D})$ varieties of $\mathcal{V}$.*

Use Theorem 8.8 only in exercises that explicitly refer to this theorem.

Exercise 8.1 Construct a set $\mathcal{P}$ of m points and a set H of n planes, both in $\mathbb{R}^3$, that satisfy the following. The incidence graph of $\mathcal{P} \times H$ contains no $K_{2,2}$ and $I(\mathcal{P}, H) = \Theta(m^{2/3}n^{2/3} + m + n)$.

Exercise 8.2 Let $\mathcal{P}$ be a set of m points and let H be a set of n planes, both in $\mathbb{R}^3$. Assume that the incidence graph of $\mathcal{P} \times H$ contains no $K_{2,2}$. Prove that $I(\mathcal{P}, H) = O(m^{4/5}n^{3/5} + m + n)$.

In your proof, you may apply the polynomial partitioning theorem (Theorem 4.14) and assume that no points are on the partition. As we saw in Exercise 3.1, such an assumption cannot be made. We allow this only to make a point: When ignoring incidences on the partition, it is not difficult to extend the proof of Theorem 3.3 to higher dimensions.

Exercise 8.3 We know how to prove the Szemerédi–Trotter theorem using polynomial partitioning. We also know how to prove a weaker variant by using constant-degree polynomial partitioning. You are now asked to prove the

Szemerédi–Trotter theorem by using Theorem 8.8. (Hint: Imitate the proof of Theorem 3.3.)

Exercise 8.4 Let $\mathcal{P}$ be a set of m points and let H be a set of n planes, both in $\mathbb{R}^3$. Assume that the incidence graph of $\mathcal{P} \times H$ contains no $K_{2,2}$. Prove that $I(\mathcal{P}, H) = O(m^{4/5+\varepsilon}n^{3/5} + m + n)$, for every $\varepsilon > 0$. (Hint: Use constant-degree polynomial partitioning. When studying incidences on the partition, use Lemma 8.1.)

Exercise 8.5 Let $\mathcal{P}$ be a set of m points and let H be a set of n planes, both in $\mathbb{R}^3$. Assume that the incidence graph of $\mathcal{P} \times H$ contains no $K_{2,2}$. By Exercise 8.4, we know that $I(\mathcal{P}, H) = O(m^{4/5+\varepsilon}n^{3/5} + m + n)$ for every $\varepsilon > 0$.

(a) Prove that $I(\mathcal{P}, H) = O(m^{3/5}n^{4/5+\varepsilon} + m + n)$.

(b) Prove that $I(\mathcal{P}, H) = O(m^{7/10+\varepsilon}n^{7/10+\varepsilon} + m + n)$.

(Hint: This exercise does not require any form of polynomial partitioning.)

Exercise 8.6 Joints were defined in Chapter 5. Use constant-degree polynomial partitioning to prove that a set of n lines in $\mathbb{R}^3$ spans $O(n^{3/2+\varepsilon})$ joints. Since there are no points in this problem, use Theorem 8.8 for the polynomial partitioning. (Hint: When you have an expression of the form $\sum n_j^{3/2+\varepsilon}$, take $n_j^{1+\varepsilon}$ out of the sum.)

Exercise 8.7 The goal of this exercise is to get more intuition about the transformation ϕ that is defined in Equation (8.1).

(a) Let $\tau\colon \mathbb{C}^2 \to \mathbb{C}^2$ be defined as $\tau(z_1, z_2) = (i \cdot z_1, i \cdot z_2)$. In other words, τ multiplies a point in the complex plane by i. Find the definition of $\phi' = \phi \circ \tau$.

(b) Let $\gamma \subset \mathbb{C}^2$ be a circle. Parameterize $\phi(\gamma) \subset \mathbb{R}^4$.

Exercise 8.8 Prove that there exist arbitrarily large point sets in $\mathbb{C}^2$ that span only one distance. (Hint: One option is to consider the line $y = ix$.)

Exercise 8.9 Consider an irreducible hypersurface $U \subset \mathbb{R}^d$ of degree $k > 1$, and let p be a regular point of U. Prove that that the number of $(d-2)$-flats that are contained in U and incident to p is at most k. (Hint: What trick from the current chapter might be useful here? Also, use Exercise 4.7.)

8.5 Open Problems

Constant-degree partitioning: The constant-degree polynomial partitioning technique allows us to obtain incidence bounds in dimension $d \geq 3$. One might say that the main issue with this technique is the extra ε that is added to the exponent of the incidence bound. Multiple works focus on removing this ε in specific cases (for example, see Basu and Sombra, 2016; Kaplan et al., 2012; Tóth, 2015; Zahl, 2013, 2015). This leads to the following question.

Open Problem 8.9 *Develop a technique for removing the extra ε from the exponent whenever constant-degree polynomial partitioning is used.*

Open Problem 8.9 demonstrates an issue with writing a book about a highly-active research front. It is plausible that a recent theorem of Walsh (2020) would allow us to solve the problem. We would not be surprised to find that a solution to this problem appeared shortly after the book is published.

Incidences in complex spaces: So far, only a a few incidence bounds have been extended from $\mathbb{R}^d$ to $\mathbb{C}^d$. Theorem 8.7 provides a bound for point-curve incidences in $\mathbb{C}^2$. However, this bound is weaker than Theorem 3.9, the current best point-curve bound in $\mathbb{R}^2$. Several works (Dvir and Gopi, 2015; Sheffer and Zahl, 2021; Zahl, 2016) study point-line incidences in $\mathbb{C}^d$. However, the main point-line incidence bounds in $\mathbb{R}^d$ have not been extended to $\mathbb{C}^d$ when $d \geq 4$. Elekes and Szabó (2012) derive a bound for arbitrary point-variety incidences in $\mathbb{C}^d$. Since Elekes and Szabó rely on older incidence tools, their bounds are weaker than the bounds that are obtained by using polynomial partitioning in $\mathbb{R}^d$ (see Chapter 11).

Open Problem 8.10 *For $d \geq 3$, extend the known point-variety incidence bounds in $\mathbb{R}^d$ to $\mathbb{C}^d$, possibly with an additional ε in the exponent.*

Point-plane incidences: In the exercises, we explored point-plane incidences in $\mathbb{R}^3$. Let $\mathcal{P}$ be a set of m points and let H be a set of n planes, both in $\mathbb{R}^3$. Assume that the incidence graph of $\mathcal{P} \times H$ contains no $K_{2,t}$ for some large constant t. Then, for every $\varepsilon > 0$ we have that

$$I(\mathcal{P}, H) = O(m^{4/5}n^{3/5} + m + n) \quad \text{and} \quad I(\mathcal{P}, H) = \Omega(m^{4/5}n^{3/5-\varepsilon} + m + n).$$

Thus, this incidence problem is solved up to subpolynomial factors. For more details, see for example Do and Sheffer (2021).

The problem of point-plane incidences in $\mathbb{R}^3$ is open when the incidence graph contains no $K_{3,t}$. In this case, the current best bounds are

$$I(\mathcal{P}, H) = O(m^{3/4}n^{3/4} + m + n) \quad \text{and} \quad I(\mathcal{P}, H) = \Omega(m^{7/10}n^{7/10} + m + n).$$

For more details, see for example Braß and Knauer (2003).

Open Problem 8.11 *Let* $\mathcal{P}$ *be a set of m points and let H be a set of n planes, both in* $\mathbb{R}^3$. *Assume that the incidence graph of* $\mathcal{P} \times H$ *contains no* $K_{3,t}$ *for some large constant t. Find the asymptotic value of* $I(\mathcal{P}, H)$.

Another interesting point-plane incidence problem in $\mathbb{R}^3$ is discussed in Section 14.7.

9

Lines in $\mathbb{R}^3$

In Chapter 7, we studied the ESGK framework. This was a reduction from the distinct distances problem to a problem about pairs of intersecting lines in $\mathbb{R}^3$. In the current chapter we further reduce the problem to a point-line incidence problem, and then solve this incidence problem. This completes the proof of the Guth–Katz distinct distances theorem.

The original proof of Guth and Katz is quite involved. In this book we study a simpler proof for a slightly weaker variant of Theorem 7.1. This simpler proof was introduced by Guth (2015a).

Theorem 9.1 *For every $\varepsilon > 0$, each set of n points in $\mathbb{R}^2$ determines $\Omega(n^{1-\varepsilon})$ distinct distances.*

Studying this weaker variant allows us to avoid several of the more technical parts of the original proof, such as properties of ruled surfaces and flat points.

9.1 From Intersecting Lines to Incidences

We first recall the situation that we reach after applying the ESGK framework. For points $a, b \in \mathbb{R}^2$, we define the line $\ell_{a,b} \subset \mathbb{R}^3$ as

$$\ell_{ab} = \left\{ \left(\frac{a_x + b_x}{2}, \frac{a_y + b_y}{2}, 0 \right) + t \left(\frac{b_y - a_y}{2}, \frac{a_x - b_x}{2}, 1 \right) \ : \ \text{for } t \in \mathbb{R} \right\}.$$
(9.1)

The line ℓ_{ab} is the set of parameterizations of the rotations of $\mathbb{R}^2$ that take a to b.

We get a set Q of n points in $\mathbb{R}^2$ and need to prove that Q spans $\Omega(n^{1-\varepsilon})$ distinct distances (we reserve the notation $\mathcal{P}$ for later). For this purpose, we define the set of lines

$$\mathcal{L} = \{ \ell_{ab} \ : \ a, b \in Q \}.$$
(9.2)

125

Consider points $a, b, p, q \in Q$ such that $(a, b) \neq (p, q)$. We saw in Section 7.2 that at most one rotation of $\mathbb{R}^2$ takes a to b and also p to q. Thus, the lines ℓ_{ab} and ℓ_{pq} intersect in at most one point. This implies that $\mathcal{L}$ is a set of n^2 distinct lines.

To prove that Q determines $\Omega(n^{1-\varepsilon})$ distinct distances, it suffices to prove that the number of pairs of intersecting lines in $\mathcal{L}^2$ is $O(n^{3+\varepsilon})$. Denote the number of such pairs as $X_{\mathcal{L}}$. To derive an upper bound for $X_{\mathcal{L}}$, we go over every point $u \in \mathbb{R}^3$ and check how many pairs of lines intersect at u. If u is incident to r lines of $\mathcal{L}$, then $2 \cdot \binom{r}{2}$ pairs of lines intersect at u. We recall that a point is r-*rich* if it is incident to at least r lines of $\mathcal{L}$ (this notation was introduced in Section 1.9). For a positive integer r, let $\mathcal{P}_r(\mathcal{L})$ be the set of r-rich points. We perform a dyadic decomposition of the intersection points, where for each j we consider points that are 2^j-rich but not 2^{j+1}-rich. This leads to

$$X_{\mathcal{L}} < \sum_{j=1}^{2\log n} |\mathcal{P}_{2^j}(\mathcal{L})| \cdot 2 \cdot \binom{2^{j+1}}{2} < \sum_{j=1}^{2\log n} |\mathcal{P}_{2^j}(\mathcal{L})| \cdot 2^{2j+2}. \qquad (9.3)$$

To obtain the desired bound for $X_{\mathcal{L}}$, we prove the following theorem.

Theorem 9.2 *Let Q be a set of n points in $\mathbb{R}^3$. Let $\mathcal{L}$ be the set of lines that is defined in Equation* (9.2). *Then for every $\varepsilon > 0$ and $2 \leq r \leq n^2$, we have that*

$$|\mathcal{P}_r(\mathcal{L})| = O\left(\frac{n^{3+\varepsilon}}{r^2}\right).$$

Let $0 < \varepsilon' < \varepsilon$. Applying Theorem 9.2 with ε' and combining the result with Inequality (9.3) gives

$$X_{\mathcal{L}} = O\left(\sum_{j=1}^{2\log n} \frac{n^{3+\varepsilon'}}{2^{2j}} \cdot 2^{2j}\right) = O(n^{3+\varepsilon'} \log n) = O(n^{3+\varepsilon}).$$

To prove Theorem 9.1, it remains to prove Theorem 9.2. That is, we reduced the problem of bounding the number of intersecting pairs of lines to a point-line incidence problem.

Theorem 9.2 is false for arbitrary sets of n^2 lines in $\mathbb{R}^3$. For example, consider a set of n^2 lines in the same plane, such that no two lines are parallel. Since every pair of lines intersect, the number of intersecting pairs is n^4. The same issue occurs when we take n^2 lines that are incident to the same point. To overcome this issue, we note that the set of lines that is defined in Equation (9.2) has additional properties.

Lemma 9.3 *Let Q be a set of n points in $\mathbb{R}^2$, and let $\mathcal{L}$ be the set of lines defined in Equation* (9.2). *Then*

(i) Every point of $\mathbb{R}^3$ is incident to at most n lines of $\mathcal{L}$.

(ii) Every plane in $\mathbb{R}^3$ contains at most n lines of $\mathcal{L}$.

Proof For a point $a \in Q$, we set

$$\mathcal{L}_a = \{\ell_{ab} \ : \ b \in Q\}.$$

(i) We consider two lines $\ell_{ab}, \ell_{ac} \in \mathcal{L}_a$. Recall that every point of ℓ_{ab} parameterizes a rotation of $\mathbb{R}^2$ that takes a to b, and similarly for ℓ_{ac}. Since a rotation cannot take a to both b and c, the lines ℓ_{ab} and ℓ_{ac} do not intersect.

We consider a point $p \in \mathbb{R}^3$. For every $a \in Q$, since the lines of $\mathcal{L}_a$ do not intersect, at most one line of $\mathcal{L}_a$ is incident to p. Since $\mathcal{L} = \bigcup_{a \in Q} \mathcal{L}_a$, we conclude that p is incident to at most n lines of $\mathcal{L}$.

(ii) Two lines in the same plane either intersect or are parallel. By part (i) of the current proof, two lines from the same $\mathcal{L}_a$ cannot intersect. By Equation (9.1), the direction of the line ℓ_{ab} is $\left(\frac{b_y - a_y}{2}, \frac{a_x - b_x}{2}, 1\right)$, so two lines from $\mathcal{L}_a$ cannot be parallel. Thus, two lines from the same $\mathcal{L}_a$ cannot be in the same plane. We conclude that every plane in $\mathbb{R}^3$ contains at most n lines of $\mathcal{L}$. □

When $r > n$, Lemma 9.3(i) implies that $\mathcal{P}_r(\mathcal{L}) = \emptyset$. It remains to prove Theorem 9.2 when $2 \le r \le n$. For that purpose, we prove the following two results. For a set of lines $\mathcal{L}$ and a two-dimensional variety S, both in $\mathbb{R}^3$, we define

$$\mathcal{L}_S = \{\ell \in \mathcal{L} \ : \ \ell \subset S\}.$$

Theorem 9.4 *For every $\varepsilon > 0$, there exist sufficiently large constants C and D that satisfy the following. Let $\mathcal{L}$ be a set of n lines in $\mathbb{R}^3$, let $2 \le r \le 2n^{1/2}$, and let $r' = \lceil 9r/10 \rceil$. Then there exists a set $\mathcal{V}$ of varieties in $\mathbb{R}^3$ such that*

- *Every variety of $\mathcal{V}$ is irreducible, of dimension two, and of degree at most D.*
- *Every variety of $\mathcal{V}$ contains at least $n^{1/2+\varepsilon}$ lines of $\mathcal{L}$.*
- $|\mathcal{V}| \le 2n^{1/2-\varepsilon}$.
- $|\mathcal{P}_r(\mathcal{L}) \backslash \cup_{S \in \mathcal{V}} \mathcal{P}_{r'}(\mathcal{L}_S)| \le Cn^{3/2+\varepsilon}/r^2.$

Lemma 9.5 *Let Q be a set of n points in $\mathbb{R}^2$, and let $\mathcal{L}$ be the set of lines defined in Equation (9.2). Then every irreducible two-dimensional variety $U \subset \mathbb{R}^3$ of degree D contains fewer than $2D^2n$ lines of $\mathcal{L}$.*

It is straightforward to prove Theorem 9.2 by using Theorem 9.4 and Lemma 9.5.

Proof of Theorem 9.2 We apply Theorem 9.4 with $\mathcal{L}$ and $\varepsilon/2$ to obtain a set $\mathcal{V}$ of two-dimensional varieties of degree at most D, each containing at least $n^{1+\varepsilon}$ lines of $\mathcal{L}$. (Theorem 9.4 has at least $n^{1/2+\varepsilon/2}$ lines in each variety. We obtain $n^{1+\varepsilon}$ since now there are n^2 lines, rather than n.) Lemma 9.5 implies that such varieties do not exist, so $\mathcal{V} = \emptyset$. Thus, Theorem 9.4 implies that

$$|\mathcal{P}_r(\mathcal{L})| = |\mathcal{P}_r(\mathcal{L})\setminus \cup_{S\in\mathcal{V}} \mathcal{P}_{r'}(\mathcal{L}_S)| \leq Cn^{3+\varepsilon}/r^2. \qquad \Box$$

We prove Theorem 9.4 in Section 9.2. We prove Lemma 9.5 in Section 9.3. The proof of Lemma 9.5 relies on analytic tools that are not used anywhere else in this book, so we mark Section 9.3 as optional.

Before concluding this section, we observe that Theorem 9.2 is tight up to subpolynomial factors.

Claim 9.6 *There exists a set $\mathcal{L}$ of n^2 lines in $\mathbb{R}^3$ such that $\mathcal{P}_3(\mathcal{L}) = \Theta(n^3)$, no point is 4-rich, and every plane in $\mathbb{R}^3$ contains $O(n)$ lines of $\mathcal{L}$.*

Proof We imitate the proof of Claim 5.2. Let H be a set of m generic planes in $\mathbb{R}^3$, for a parameter m that is set below. By generic planes, we mean that no two planes are parallel, no three intersect in a line, and no four intersect in a point. Set

$$\mathcal{L} = \{\Pi \cap \Pi' \ : \ \Pi, \Pi' \in H \text{ and } \Pi \neq \Pi'\}.$$

Since no three planes intersect in a line, $\mathcal{L}$ is a set of $\binom{m}{2}$ distinct lines. We may also assume that no two lines of $\mathcal{L}$ are parallel. We fix the value of m such that $|\mathcal{L}| = n^2$, and note that $m = \Theta(n)$. For any three distinct planes Π, Π', Π'', the three lines $\Pi \cap \Pi'$, $\Pi \cap \Pi''$, and $\Pi' \cap \Pi''$ intersect at a distinct point. Thus, $|\mathcal{P}_3(\mathcal{L})| = \binom{m}{3} = \Theta(n^3)$. $\qquad \Box$

9.2 Rich Points in $\mathbb{R}^3$

In this section, our goal is to prove Theorem 9.4. That is, we now study the rich points of a set of lines in $\mathbb{R}^3$. In Section 1.9 we briefly studied rich points of lines in $\mathbb{R}^2$. In particular, Lemma 1.15 states that the number of r-rich points in a set of n lines in $\mathbb{R}^2$ is $O\left(\frac{n^2}{r^3} + \frac{n}{r}\right)$. This bound easily extends to $\mathbb{R}^3$.

Claim 9.7 *Let $\mathcal{L}$ be a set of n lines in $\mathbb{R}^3$ and let $r \geq 2$. Then $|\mathcal{P}_r(\mathcal{L})| = O\left(\frac{n^2}{r^3} + \frac{n}{r}\right)$.*

Proof Let Π be a generic plane in $\mathbb{R}^3$, let $\pi: \mathbb{R}^3 \to \Pi$ be the projection on Π, let $\mathcal{P}' = \{\pi(p) \ : \ p \in \mathcal{P}_r(\mathcal{L})\}$, and let $\mathcal{L}' = \{\pi(\ell) \ : \ \ell \in \mathcal{L}\}$. Since Π is chosen generically, we may assume that every point of $\mathcal{P}_r(\mathcal{L})$ is projected to a distinct point in Π and that every line of $\mathcal{L}$ is projected to a distinct line in Π. In particular, $|\mathcal{L}'| = |\mathcal{L}|$ and $|\mathcal{P}'| = |\mathcal{P}_r(\mathcal{L})|$.

We apply Lemma 1.15 in Π, to obtain that $|\mathcal{P}_r(\mathcal{L})| = O\left(\frac{n^2}{r^3} + \frac{n}{r}\right)$. Since $\mathcal{P}' \subseteq \mathcal{P}_r(\mathcal{L}')$, we get that

$$|\mathcal{P}_r(\mathcal{L})| = |\mathcal{P}'| \le |\mathcal{P}_r(\mathcal{L}')| = O\left(\frac{n^2}{r^3} + \frac{n}{r}\right). \qquad \square$$

When $r > 2n^{1/2}$, Claim 9.7 implies that $|\mathcal{P}_r(\mathcal{L})| = O(n/r)$. We now examine an alternative proof of this case. This alternative proof is a warm-up for the proof of Lemma 9.10.

Lemma 9.8 *Let $\mathcal{L}$ be a set of n lines in $\mathbb{R}^3$ and let $r > 2n^{1/2}$. Then $|\mathcal{P}_r(\mathcal{L})| <$ $2n/r$.*

Proof We set $M = |\mathcal{P}_r(\mathcal{L})|$ and write $\mathcal{P}_r(\mathcal{L}) = \{p_1, \ldots, p_M\}$. We assume for contradiction that $M \ge r/2$. By definition, at least r lines of $\mathcal{L}$ are incident to p_1. At least $r - 1$ lines of $\mathcal{L}$ are incident to p_2 but not to p_1. At least $r - 2$ lines of $\mathcal{L}$ are incident to p_3 but to neither of p_1 and p_2. Continuing in this manner, we obtain that

$$n = |\mathcal{L}| \ge \sum_{j=0}^{r/2-1} (r - j) > (r/2) \cdot (r/2) = r^2/4. \qquad (9.4)$$

This contradicts the assumption $r > 2n^{1/2}$, so we get that $M < r/2$. This implies that the sum in Equation (9.4) should have only M terms:

$$n = |\mathcal{L}| \ge \sum_{j=0}^{M-1} (r - j) > (r/2) \cdot M.$$

The bound of the lemma is obtained by rearranging this inequality. $\qquad \square$

The following is a variant of Exercise 4.7(a).

Lemma 9.9 *Consider distinct irreducible two-dimensional varieties $U, W \subset$ $\mathbb{R}^3$ of degrees d_U and d_W. Then $U \cap W$ contains at most $d_U \cdot d_W$ lines.*

Proof Since U and W are distinct and irreducible, the variety $U \cap W$ is of dimension at most one. By Lemma 4.10, this variety has a finite number of components.

We consider a generic plane $\Pi \subset \mathbb{R}^3$, such that $\Pi \ne U$, $\Pi \ne W$, and Π does not contain any component of $U \cap W$. Since $U \cap W$ has a finite number of components, we get that $\Pi \cap U \cap W$ is finite. Let $f_U \in \mathbb{R}[x, y, z]$ satisfy $\deg f_U = d_U$ and $\mathbf{V}(f_U) = U$. Let $f_W \in \mathbb{R}[x, y, z]$ satisfy $\deg f_W = d_W$ and $\mathbf{V}(f_W) = W$. We define $f_{U,\Pi}$ and $f_{W,\Pi}$ as the restrictions of f_U and f_W to Π.

Then $\deg f_{U,\Pi} \leq d_U$ and $\deg f_{W,\Pi} \leq d_W$. By applying Bezout's theorem (Theorem 2.5) in Π, we obtain that $|\Pi \cap U \cap W| \leq d_U \cdot d_W$. Since Π is chosen generically, we may assume that every line that is contained in $U \cap W$ intersects Π at a distinct point. Since $|\Pi \cap U \cap W| \leq d_U \cdot d_W$, we conclude that $U \cap W$ contains at most $d_U \cdot d_W$ lines. $\square$

The following result is obtained by imitating the proof of Lemma 9.8. Exercise 9.2 is another instance of this proof technique.

Lemma 9.10   *Let* $\mathcal{L}$ *be a set of* n *lines in* $\mathbb{R}^3$. *Let* $\mathcal{V}$ *be a set of distinct irreducible varieties in* $\mathbb{R}^3$, *each of dimension two, of degree at most* D, *and containing at least* X *lines of* $\mathcal{L}$. *If* $X > 2Dn^{1/2}$ *then* $|\mathcal{V}| \leq 2n/X$.

Proof We set $M = |\mathcal{V}|$ and write $\mathcal{V} = \{S_1, \ldots, S_M\}$. We assume for contradiction that $M \geq X/2D^2$. By Lemma 9.9, for every distinct $S_j, S_k \in \mathcal{V}$, the intersection $S_j \cap S_k$ contains at most D^2 lines of $\mathcal{L}$.

By definition, at least X lines of $\mathcal{L}$ are contained in S_1. At least $X - D^2$ lines of $\mathcal{L}$ are contained in S_2 but not in S_1. At least $X - 2D^2$ lines of $\mathcal{L}$ are contained in S_3 but not in $S_1 \cup S_2$. By continuing in the same manner, we obtain that

$$n = |\mathcal{L}| \geq \sum_{j=0}^{X/2D^2-1} (X - j \cdot D^2) > (X/2D^2) \cdot (X/2) = X^2/4D^2. \qquad (9.5)$$

Since this contradicts the assumption $X > 2Dn^{1/2}$, we get that $M < X/2D^2$. This implies that the sum in Equation (9.5) should have only M terms:

$$n = |\mathcal{L}| \geq \sum_{j=0}^{M-1} (X - j \cdot D^2) > (X/2) \cdot M.$$

The bound of the lemma is obtained by rearranging this inequality. $\square$

We are now ready to prove Theorem 9.4, and first recall the statement of this theorem.

Theorem 9.4 *For every* $\varepsilon > 0$, *there exist sufficiently large constants* C *and* D *that satisfy the following. Let* $\mathcal{L}$ *be a set of* n *lines in* $\mathbb{R}^3$, *let* $2 \leq r \leq 2n^{1/2}$, *and let* $r' = \lceil 9r/10 \rceil$. *Then there exists a set* $\mathcal{V}$ *of varieties in* $\mathbb{R}^3$ *such that*

- *Every variety of* $\mathcal{V}$ *is irreducible, of dimension two, and of degree at most* D.
- *Every variety of* $\mathcal{V}$ *contains at least* $n^{1/2+\varepsilon}$ *lines of* $\mathcal{L}$.
- $|\mathcal{V}| \leq 2n^{1/2-\varepsilon}$.
- $|\mathcal{P}_r(\mathcal{L}) \setminus \bigcup_{S \in \mathcal{V}} \mathcal{P}_{r'}(\mathcal{L}_S)| \leq Cn^{3/2+\varepsilon}/r^2$.

Proof The proof technique is similar to the constant-degree polynomial partitioning technique from Chapter 8: We prove the theorem by induction on *n*, partition the space into a constant number of cells, and apply the induction hypothesis separately in each cell. Unlike the analysis in Chapter 8, we might not be able to apply the induction hypothesis in some cells. Specifically, we cannot apply the induction hypothesis in cells that do not satisfy the assumption $r \leq 2n^{1/2}$.

If a cell does not satisfy the assumption $r \leq 2n^{1/2}$, then we can apply Claim 9.7 in that cell. This leads to a reasonable bound when there are not too many *r*-rich points in the cell. It remains to handle cells that do not satisfy $r \leq 2n^{1/2}$ and contain many *r*-rich points. We do so by noting that such problematic cells contain a small fraction of the *r*-rich points.

Partitioning the space: As already stated, the proof is by induction on *n*. For the induction basis, consider the case of $n \leq C'$, where C' is a sufficiently large constant that is set below. Since $r \leq 2n^{1/2}$, this case holds by taking $C \geq (C')^3$ and $\mathcal{V} = \emptyset$.

We now consider the induction step, assume that $n > C'$, and set $m = |\mathcal{P}_r(\mathcal{L})|$. We apply the polynomial partitioning theorem (Theorem 3.1) with $\mathcal{P}_r(\mathcal{L})$ and a sufficiently large constant *s*. We obtain $f \in \mathbb{R}[x, y, z]$ of degree $O(s)$, such that each of connected component of $\mathbb{R}^3 \backslash \mathbf{V}(f)$ contains at most m/s^3 points of $\mathcal{P}_r(\mathcal{L})$. We denote the open cells of the partition as $C_1, \ldots, C_v$. Theorem 3.2 implies that $v = O(s^3)$. For each $1 \leq j \leq v$, we set $\mathcal{P}_j = \mathcal{P}_r(\mathcal{L}) \cap C_j$ and let $\mathcal{L}_j$ be the set of lines of $\mathcal{L}$ that intersect C_j.

We set $n_j = |\mathcal{L}_j|$. For a line $\ell \subset \mathbb{R}^3$, let Π be a generic plane that contains ℓ and let f_Π be the restriction of f to Π. Note that $\deg f_\Pi \leq \deg f$. We apply Bezout's theorem in Π with ℓ and f_Π, to obtain that either $\ell \subset \mathbf{V}(f)$ or $|\mathbf{V}(f) \cap \ell| \leq \deg f$. When traveling along ℓ, we get to a new cell of the partition only after crossing a point of $\mathbf{V}(f) \cap \ell$. This implies that every line of $\mathcal{L}$ intersects at most $\deg f + 1 = O(s)$ cells of $\mathbb{R}^3 \backslash \mathbf{V}(f)$. We conclude that $\sum_{j=1}^{v} n_j = O(ns)$.

For $0 < \alpha < 1$, we say that a cell C_j is *α-good* if $n_j \leq \alpha n/s^2$. The number of non-α-good cells is at most

$$\frac{\sum_{j=1}^{v} n_j}{\alpha n/s^2} = O\left(\frac{ns}{\alpha n/s^2}\right) = O(\alpha s^3).$$

Since each cell contains at most m/s^3 points of $\mathcal{P}_r(\mathcal{L})$, the number of points that are in non-α-good cells is $O(\alpha m)$. We take α to be a sufficiently small constant so that at most $m/10$ points of $\mathcal{P}_r(\mathcal{L})$ are contained in non-α-good cells.

Let $\mathcal{P}_B$ be the set of these points. For brevity, we say that a cell is *good* if it is α-good for the chosen value of α.

We choose the constants in this proof so that

$$\varepsilon^{-1}, \alpha \ll s \ll D \ll C, C'.$$

Rich points in good cells: Let G be the set of indices $1 \le j \le v$ for which C_j is a good cell. We can only apply the induction hypothesis in cells that satisfy $r \le 2n_j^{1/2}$. Let $G_r \subset G$ be the set of indices $j \in G$ that satisfy $r \le 2n_j^{1/2}$. For every $j \in G_r$, we apply the induction hypothesis on $\mathcal{L}_j$ to obtain a set of varieties $\mathcal{U}_j$ such that

$$|\mathcal{U}_j| \le 2n_j^{1/2-\varepsilon} \le 2(\alpha n/s^2)^{1/2-\varepsilon}.$$

Since $C_j \cap \mathcal{P}_r(\mathcal{L}) \subset \mathcal{P}_r(\mathcal{L}_j)$, the induction hypothesis leads to

$$\left| (C_j \cap \mathcal{P}_r(\mathcal{L})) \backslash \cup_{S \in \mathcal{U}_j} \mathcal{P}_{r'}(\mathcal{L}_S) \right| \le \left| \mathcal{P}_r(\mathcal{L}_j) \backslash \cup_{S \in \mathcal{U}_j} \mathcal{P}_{r'}(\mathcal{L}_S) \right|$$

$$\le \frac{Cn_j^{3/2+\varepsilon}}{r^2} \le \frac{C(\alpha n/s^2)^{3/2+\varepsilon}}{r^2}. \tag{9.6}$$

Every $j \in G \backslash G_r$ satisfies that $r > 2n_j^{1/2}$, so we can apply Lemma 9.8 with $\mathcal{L}_j$. By also recalling that $r \le 2n^{1/2}$, we obtain that

$$\left| C_j \cap \mathcal{P}_r(\mathcal{L}) \right| \le \left| \mathcal{P}_r(\mathcal{L}_j) \right| \le 2n_j/r < 2n/r \le 4n^{3/2}/r^2.$$

By setting $\mathcal{U}_j = \emptyset$ and taking C to be sufficiently large with respect to α and s, we also get the bound of Inequality (9.6) in this case. That is, Inequality (9.6) holds for every $j \in G$. By summing this bound over all good cells, we obtain that

$$\sum_{j \in G} \left| (C_j \cap \mathcal{P}_r(\mathcal{L})) \backslash \cup_{S \in \mathcal{U}_j} \mathcal{P}_{r'}(\mathcal{L}_S) \right| = O\left(s^3 \cdot \frac{C(\alpha n/s^2)^{3/2+\varepsilon}}{r^2} \right)$$

$$= O\left(\frac{C(\alpha n)^{3/2+\varepsilon}}{s^{2\varepsilon} r^2} \right).$$

By taking s to be sufficiently large with respect to ε and to the constant of the $O(\cdot)$-notation, we get that

$$\sum_{j \in G} \left| (C_j \cap \mathcal{P}_r(\mathcal{L})) \backslash \cup_{S \in \mathcal{U}_j} \mathcal{P}_{r'}(\mathcal{L}_S) \right| \le \frac{Cn^{3/2+\varepsilon}}{20r^2}. \tag{9.7}$$

Rich points on the partition: We denote the irreducible components of $\mathbf{V}(f)$ as $Z_1, Z_2, \ldots, Z_u$, and set $\mathcal{P}_j' = Z_j \cap \mathcal{P}_r(\mathcal{L})$. Some points of $\mathcal{P}_r(\mathcal{L})$ may appear in more than one $\mathcal{P}_j'$. Recall that all the components of $\mathbf{V}(f)$ are two-dimensional. We may thus refer to $\deg Z_j$.

We consider a component Z_j and a point $p \in \mathcal{P}'_j$ such that $p \notin \mathcal{P}_{r'}(\mathcal{L}_{Z_j})$. Since $r' = \lceil 9r/10 \rceil$, at least $r/10$ lines of $\mathcal{L}$ that are incident to p are not contained in Z_j. By Bezout's theorem, a line that is not contained in Z_j intersects Z_j in at most $\deg Z_j$ points. This implies that

$$\left| \mathcal{P}'_j \backslash \mathcal{P}_{r'}(\mathcal{L}_{Z_j}) \right| \leq \frac{n \cdot \deg Z_j}{r/10}.$$

Summing this over every Z_j, recalling that $r \leq 2n^{1/2}$, and taking C to be sufficiently large with respect to s, leads to

$$\left| (\mathcal{P}_r(\mathcal{L}) \cap \mathbf{V}(f)) \backslash \cup_{j=1}^u \mathcal{P}_{r'}(\mathcal{L}_{Z_j}) \right| \leq \sum_{j=1}^u \frac{10n \cdot \deg Z_j}{r} = \frac{10n \cdot \deg f}{r}$$

$$\leq \frac{20n^{3/2} \cdot \deg f}{r^2} \leq \frac{Cn^{3/2}}{20r^2}. \tag{9.8}$$

Let $\mathcal{V}'$ be a set of varieties, consisting of the component of $\mathbf{V}(f)$ and of the elements of $\mathcal{U}_j$ for every $j \in G$. By combining Inequalities (9.7) and (9.8), and recalling that $\mathcal{P}_B$ is the set of points in the non-good cells, we have that

$$\left| \mathcal{P}_r(\mathcal{L}) \backslash \cup_{S \in \mathcal{V}'} \mathcal{P}_{r'}(\mathcal{L}_S) \right| \leq |\mathcal{P}_B| + \frac{Cn^{3/2+\varepsilon}}{10r^2}. \tag{9.9}$$

By taking D that is larger than $\deg f = O(s)$, we get that every variety in $\mathcal{V}'$ is of degree at most D. The number of varieties in $\mathcal{V}'$ is at most

$$O(s) + v \cdot 2(\alpha n/s^2)^{1/2-\varepsilon} = O\left(s^3 \cdot (n^{1/2-\varepsilon}/s^{1-2\varepsilon}) \right) = O(s^{2+2\varepsilon} n^{1/2-\varepsilon}).$$

We note that Inequality (9.9) is similar to the bound of the theorem. To complete the proof, we still need to address three issues:

- We require an upper bound for $|\mathcal{P}_B|$.
- The set $\mathcal{V}'$ might contain more than $2n^{1/2-\varepsilon}$ varieties.
- Some varieties of $\mathcal{V}'$ might contain fewer than $n^{1/2+\varepsilon}$ lines of $\mathcal{L}$.

Rich points in non-good cells: We set $\mathcal{P}_1 = \mathcal{P}_r(\mathcal{L})$, $\mathcal{V}_1 = \mathcal{V}'$, and $\mathcal{P}_2 = \mathcal{P}_1 \backslash \cup_{S \in \mathcal{V}_1} \mathcal{P}_{r'}(\mathcal{L}_S)$. We then repeat the entire analysis for $\mathcal{P}_2$, and iteratively repeat this process. That is, we define

$$\mathcal{P}_j = \mathcal{P}_{j-1} \backslash \cup_{S \in \mathcal{V}_{j-1}} \mathcal{P}_{r'}(\mathcal{L}_S). \tag{9.10}$$

At each step, we repeat the above proof starting with the partitioning step. The set of lines remains unchanged, but the set of points becomes smaller at every step. That is, n and r remain unchanged while m decreases.

By adapting Inequality (9.9) to the iterative argument and recalling that $|\mathcal{P}_B| \leq |\mathcal{P}_1|/10$, we obtain that

$$|\mathcal{P}_j| \leq \frac{|\mathcal{P}_{j-1}|}{10} + \frac{Cn^{3/2+\varepsilon}}{10r^2}. \qquad (9.11)$$

Since every two lines intersect at most once, we have the trivial bound $|\mathcal{P}_1| \leq n^2$. If $|\mathcal{P}_j| \geq \frac{Cn^{3/2+\varepsilon}}{5r^2}$ then Inequality (9.11) implies that $|\mathcal{P}_{j+1}| < |\mathcal{P}_j|/2$. If $|\mathcal{P}_j| < \frac{Cn^{3/2+\varepsilon}}{5r^2}$ then $|\mathcal{P}_{j+1}| < \frac{Cn^{3/2+\varepsilon}}{5r^2}$. Thus, for $w = \frac{3}{2} \cdot \log n$ we have that

$$|\mathcal{P}_w| < \frac{Cn^{3/2+\varepsilon}}{5r^2}.$$

We set $\mathcal{V}^* = \bigcup_{j=1}^{w-1} \mathcal{V}_j$. Then

$$|\mathcal{P}_r(\mathcal{L}) \setminus \bigcup_{S \in \mathcal{V}^*} \mathcal{P}_{r'}(\mathcal{L}_S)| = |\mathcal{P}_w| < \frac{Cn^{3/2+\varepsilon}}{5r^2}. \qquad (9.12)$$

For every $1 \leq j \leq w - 1$, we have that $|\mathcal{V}_j| = O(s^{2+2\varepsilon} n^{1/2-\varepsilon})$. This implies that $|\mathcal{V}^*| = O(s^{2+2\varepsilon} n^{1/2-\varepsilon} \log n)$. Let $\mathcal{V}$ be the set of varieties of $\mathcal{V}^*$ that contain at least $n^{1/2+\varepsilon}$ lines of $\mathcal{L}$.

In the induction step we assume that $n > C'$. By taking C' to be sufficiently large with respect to D and ε, we get that $n^\varepsilon > 2D$. We may thus apply Lemma 9.10 with $\mathcal{L}$ and $X = n^{1/2+\varepsilon}$, obtaining that

$$|\mathcal{V}| \leq 2n/X = 2n^{1/2-\varepsilon}.$$

Completing the proof: It remains to show that $\mathcal{P}_r(\mathcal{L}) \setminus \bigcup_{S \in \mathcal{V}} \mathcal{P}_{r'}(\mathcal{L}_S)$ is not too large. Recalling Equation (9.12), it suffices to show that $\bigcup_{S \in \mathcal{V}^* \setminus \mathcal{V}} \mathcal{P}_{r'}(\mathcal{L}_S)$ is not too large. We perform a dyadic decomposition, defining $\mathcal{V}_j^*$ to be the set of varieties of $\mathcal{V}^*$ that contain at least 2^j lines of $\mathcal{L}$ and less than 2^{j+1} such lines. Note that $\mathcal{V}^* \setminus \mathcal{V} = \bigcup_{j=0}^{\log n^{1/2+\varepsilon}} \mathcal{V}_j^*$.

Consider a variety $S \in \mathcal{V}_j^*$. By Claim 9.7, the number of r'-rich points in a set of less than 2^{j+1} lines is

$$O\left(\frac{2^{2j}}{(r')^3} + \frac{2^j}{r'} \right) = O\left(\frac{2^{2j}}{r^3} + \frac{2^j}{r} \right). \qquad (9.13)$$

When $2^j > 2Dn^{1/2}$, applying Lemma 9.10 with $\mathcal{L}$ and $X = 2^j$ implies that $|\mathcal{V}_j^*| \leq 2n/2^j$. When $2^j \leq 2Dn^{1/2}$, we use the trivial bound $|\mathcal{V}_j^*| \leq |\mathcal{V}^*| = O(s^{2+2\varepsilon} n^{1/2-\varepsilon} \log n)$. Combining these two bounds with Equation (9.13) implies

$$\sum_{S \in \mathcal{V}^* \setminus \mathcal{V}} |\mathcal{P}_{r'}(\mathcal{L}_S)|$$

$$= \sum_{j=0}^{\log n^{1/2+\varepsilon}} \sum_{S \in \mathcal{V}_j^*} |\mathcal{P}_{r'}(\mathcal{L}_S)| = O\left(\sum_{j=0}^{\log n^{1/2+\varepsilon}} |\mathcal{V}_j^*| \cdot \left(\frac{2^{2j}}{r^3} + \frac{2^j}{r} \right) \right)$$

$$= O\left(\sum_{j=0}^{\log 2Dn^{1/2}} s^{2+2\varepsilon} n^{1/2-\varepsilon} \log n \cdot \left(\frac{2^{2j}}{r^3} + \frac{2^j}{r} \right) + \sum_{j=\log 2Dn^{1/2}}^{\log n^{1/2+\varepsilon}} \frac{2n}{2^j} \cdot \left(\frac{2^{2j}}{r^3} + \frac{2^j}{r} \right) \right)$$

$$= O\left(\left(\frac{s^{2+2\varepsilon} D^2 n^{3/2-\varepsilon} \log n}{r^3} + \frac{s^{2+2\varepsilon} Dn^{1-\varepsilon} \log n}{r} \right) + \left(\frac{n^{3/2+\varepsilon}}{r^3} + \frac{\log n}{r} \right) \right).$$

We recall that $r \le 2n^{1/2}$. By taking C to be sufficiently large with respect to the constant of the $O(\cdot)$-notation, to s, and to D, we obtain that

$$\sum_{S \in \mathcal{V}^* \setminus \mathcal{V}} |\mathcal{P}_{r'}(\mathcal{L}_S)| \le \frac{Cn^{3/2+\varepsilon}}{10r^2}.$$

Combining this with Equation (9.12) leads to

$$|\mathcal{P}_r(\mathcal{L}) \setminus \cup_{S \in \mathcal{V}} \mathcal{P}_{r'}(\mathcal{L}_S)| < \frac{Cn^{3/2+\varepsilon}}{r^2}.$$

This completes the induction step and the proof of the theorem. $\square$

9.3 (Optional) Lines in a Two-Dimensional Surface

In this section, we prove Lemma 9.5. That is, we study the structure of lines that are contained in a two-dimensional variety in $\mathbb{R}^3$. This section is optional because it relies on a few tools from basic differential geometry. These tools are not used in other parts of the book and may be safely skipped. Since this is not a differential geometry book, we are brief when presenting the relevant definitions and tools. Adding examples, intuition, and discussion would take us too far from our focus. One excellent source for additional information is Lee (2013).

We do not rigorously discuss the definition of a *d-dimensional smooth manifold*. We only state that an open subset S of a d-dimensional variety U is a d-dimensional smooth manifold if S does not contain any points of U_{sing}.[1]

---

[1] A rigorous definition: A set $M \subseteq \mathbb{R}^c$ is a d-dimensional smooth manifold if for every $p \in M$ there exists an open set of M that contains p and is homeomorphic to an open subset of a d-flat.

For example, every open subset of a plane in $\mathbb{R}^3$ is a two-dimensional smooth manifold. An open subset of the conical surface $\mathbf{V}(x^2 + y^2 - z^2)$ is a smooth manifold if and only if it does not contain the origin (see Figure 4.4).

A *vector field* of a smooth manifold $M \subset \mathbb{R}^c$ is a map $X : M \to TM$ that satisfies $X(p) = (p, v) \in TM$ for every p. That is, a vector field of M is an assignment of a tangent vector to each point of M. As with any function, a vector field X is said to be *smooth* if X is infinitely differentiable. We only discuss smooth vector fields, and omit the word "smooth" for brevity. We define an *arc* as a one-dimensional manifold. We can parameterize an arc γ with a smooth injective function $\alpha : [-c, c] \to \gamma$ (for some $c \in \mathbb{R}$). We refer to this arc both as γ and as α. The following is a variant of the Picard–Lindelöf theorem (for example, see Kelley and Peterson, 2010).

Theorem 9.11   *Let X be a vector field of a manifold M and let $p \in M$. Then for every sufficiently small $\varepsilon > 0$, there exists a unique arc $\alpha : [-\varepsilon, \varepsilon] \to M$ that contains p and whose tangent vectors are in X. In other words, there exists a unique arc α that solves the initial value problem*

$$\alpha(0) = p, \qquad \alpha'(t) = X(\alpha(t)) \quad \text{for all } t \in [-\varepsilon, \varepsilon]. \tag{9.14}$$

We are now ready to study the lines defined in Equation (9.1). For $a \in \mathbb{R}^2$, we define $\overline{\mathcal{L}}_a = \{\ell_{ab} : b \in \mathbb{R}^2\}$.

Lemma 9.12   *For a point $a \in \mathbb{R}^2$:*
(i) Every point in $\mathbb{R}^3$ is incident to exactly one line of $\overline{\mathcal{L}}_a$.
(ii) There exists a vector field $V_a : \mathbb{R}^3 \to \mathbb{R}^3 \backslash \{0\}$ with the following properties: For every $x \in \mathbb{R}^3$, the vector $V_a(x)$ has the direction of the unique line of $\overline{\mathcal{L}}_a$ incident to x. Each of the three coordinates of $V_a(x)$ is a polynomial of degree at most 1 in the coordinates of a and of degree at most 2 in the coordinates of x.

Proof (i) To distinguish between points in $\mathbb{R}^2$ and points in $\mathbb{R}^3$, we write $a = (a_x, a_y) \in \mathbb{R}^2$ but denote the coordinates of $x \in \mathbb{R}^3$ as x_1, x_2, x_3. For a point $b \in \mathbb{R}^2$, by inspecting Equation (9.1) we note that ℓ_{ab} can be defined by

$$2x_1 = a_x + b_x + x_3(b_y - a_y),$$
$$2x_2 = a_y + b_y + x_3(a_x - b_x).$$

We rewrite this as a system of equations in the coordinates of b:

$$\begin{pmatrix} 1 & x_3 \\ -x_3 & 1 \end{pmatrix} \begin{pmatrix} b_x \\ b_y \end{pmatrix} = (2x_1 - a_x + a_y x_3, 2x_2 - a_y - a_x x_3). \tag{9.15}$$

The determinant of the matrix on the left-hand side is $1 + x_3^2 > 0$. Since this determinant is nonzero, for any fixed $x \in \mathbb{R}^3$ there is a unique b that solves

Equation (9.15). That is, for every $x \in \mathbb{R}^3$ there is a unique line $\ell_{ab} \in \overline{\mathcal{L}}_a$ that is incident to x.

(ii) One can solve Equation (9.15) for b by using Cramer's rule. This leads to

$$b_x = \frac{2x_1 - a_x + a_y x_3 - x_3(2x_2 - a_y - a_x x_3)}{x_3^2 + 1},$$

$$b_y = \frac{2x_2 - a_y - a_x x_3 + x_3(2x_1 - a_x + a_y x_3)}{x_3^2 + 1}.$$

By Equation (9.1), the direction of $\ell_{a,b}$ is the direction of the vector $\left(\frac{b_y - a_y}{2}, \frac{a_x - b_x}{2}, 1\right)$, or equivalently $(b_y - a_y, a_x - b_x, 2)$. Plugging in the above values for b_x and b_y gives

$$\left(\frac{2x_2 - a_y - a_x x_3 + x_3(2x_1 - a_x + a_y x_3)}{x_3^2 + 1} - a_y, \right.$$
$$\left. a_x - \frac{2x_1 - a_x + a_y x_3 - x_3(2x_2 - a_y - a_x x_3)}{x_3^2 + 1}, 2 \right).$$

By multiplying each coordinate by $x_3^2 + 1$, we obtain the vector field

$$V_a(x) = \left(2x_2 - a_y - a_x x_3 + x_3(2x_1 - a_x + a_y x_3) - a_y \left(x_3^2 + 1 \right), \right.$$
$$\left. a_x \left(x_3^2 + 1 \right) - 2x_1 - a_x + a_y x_3 - x_3(2x_2 - a_y - a_x x_3), 2 \left(x_3^2 + 1 \right) \right).$$

The vector field $V_a(x)$ satisfies all the properties stated in the lemma. □

We recall that the gradient of a polynomial $f \in \mathbb{R}[x_1, x_2, x_3]$ is

$$\nabla f = \left(\frac{\partial f}{\partial x_1}, \frac{\partial f}{\partial x_2}, \frac{\partial f}{\partial x_3} \right).$$

Let $U \subset \mathbb{R}^3$ be a two-dimensional variety and let $f \in \mathbb{R}[x_1, x_2, x_3]$ satisfy $\mathbf{I}(U) = \langle f \rangle$. A point $p \in U$ is singular if $\nabla f(p) = (0, 0, 0)$. If p is a regular point of U then $\nabla f(p)$ is orthogonal to the tangent plane $T_p U$.

We are now ready to prove Lemma 9.5. We first recall the statement of this lemma.

Lemma 9.5 *Let Q be a set of n points in $\mathbb{R}^2$, and let $\mathcal{L}$ be the set of lines defined in Equation (9.2). Then every irreducible two-dimensional variety $U \subset \mathbb{R}^3$ of degree D contains fewer than $2D^2 n$ lines of $\mathcal{L}$.*

Proof Consider an irreducible two-dimensional variety $U \subset \mathbb{R}^3$ of degree D. The case where U is a plane is Lemma 9.3(ii). We may thus assume that $D > 1$. To prove the lemma, we show that at most one set $\overline{\mathcal{L}}_a$ has many of its lines contained in U.

By Lemma 4.5, there exists $f \in \mathbb{R}[x_1, x_2, x_3]$ of degree D such that $\langle f \rangle = \mathbf{I}(U)$. We consider a point $a \in Q$, a line $\ell \in \overline{\mathcal{L}}_a$ that is contained in U, and the vector field V_a from Lemma 9.12(ii). Let p be a regular point of U that is also incident to ℓ. Since $\ell \subset U$, this line is also contained in the tangent plane $T_p U$, which in turn implies that $V_a(p) \cdot \nabla f(p) = 0$. When p is a singular point we still have that $V_a(p) \cdot \nabla f(p) = 0$, since in this case $\nabla f(p) = 0_3$. We conclude that every line of $\overline{\mathcal{L}}_a$ that is contained in U is also contained in $\mathbf{V}(V_a \cdot \nabla f)$.

We assume that at least $2D^2$ lines of $\overline{\mathcal{L}}_a$ are contained in U, and set $W_a = \mathbf{V}(V_a \cdot \nabla f)$. By the previous paragraph, $W_a \cap U$ contains at least $2D^2$ lines. By Lemma 9.12(ii), each part of V_a is of degree at most 2 in the coordinates of p. Since every part of ∇f is of degree at most $D - 1$, we get that $\deg(V_a \cdot \nabla f) \le D + 1$. The number of lines that are contained in $W_a \cap U$ is at least

$$2D^2 > D(D + 1) \ge \deg f \cdot \deg(V_a \cdot \nabla f).$$

Thus, Exercise 4.7(a) implies that f and $V_a \cdot \nabla f$ have a common component.[2] Since U is irreducible, we get that $U \subseteq W_a$.

We next assume that there exist distinct points $a, b \in Q$ such that each of $\overline{\mathcal{L}}_a$ and $\overline{\mathcal{L}}_b$ has at least $2D^2$ lines in U. By Lemma 9.12(ii), the polynomial $V_a \cdot \nabla f$ is linear in the coordinates of a. Since both $V_a \cdot \nabla f$ and $V_b \cdot \nabla f$ vanish on U, so does $V_c \cdot \nabla f$ for any c in the linear space spanned by a and b.

By Theorem 4.11, a generic point of U is regular. Let p be a regular point of U, and let U_p be an open neighborhood of p in U that consists of regular points of U. Let c be in the span of a and b. By the preceding paragraph, $V_c \cdot \nabla f$ vanishes on U. This implies that the restriction of V_c to U_p is a vector field of U_p. We denote this restriction as V_c^p. By Theorem 9.11 with the manifold U_p and vector field V_c^p, there exists a unique arc $\alpha \colon [-\varepsilon, \varepsilon] \to U_p$ that solves Equation (9.14) (for a sufficiently small $\varepsilon > 0$). By Theorem 9.11 with manifold $\mathbb{R}^3$ and vector field V_c, there exists a unique arc $\beta \colon [-\varepsilon', \varepsilon'] \to U_p$ that solves Equation (9.14). By the definition of V_c, the arc β is a segment of a line of $\overline{\mathcal{L}}_c$. Since V_c^p is a restriction of V_c, the curves α and β coincide in a sufficiently small neighborhood of p. That is, in a sufficiently small neighborhood of p, the curve described by α is a segment of a line of $\overline{\mathcal{L}}_c$.

Let p and c be as in the preceding paragraph. Let $\ell \in \overline{\mathcal{L}}_c$ be the line that is contained in U and incident to p. Since the varieties U and ℓ have an infinite

---

[2] We cannot use Lemma 9.9 since W_a may not be an irreducible two-dimensional variety. Both results are proved in the same way.

intersection, we get that $\ell \subset U$. This holds for every c in the linear space that is spanned of a and b. Since no line in $\mathbb{R}^3$ is contained in more than one family $\overline{\mathcal{L}}_c$, we obtain infinitely many lines that are contained in U and incident to p. Since p is a regular point of U, these lines are contained in the plane $T_p U$. Applying Lemma 9.9 with U and $T_p U$ implies that these two varieties have a common component. Since U is irreducible and $T_p U$ is a plane, we get that $U = T_p U$. This contradicts the assumption $D > 1$, so at most one set $\overline{\mathcal{L}}_a$ has at least $2D^2$ lines in U.

One set $\mathcal{L}_a$ may have $n - 1$ lines of $\mathcal{L}$ that are contained in U. By the preceding paragraph, for every other $b \in Q$, at most $2D^2 - 1$ lines of $\mathcal{L}_b$ are in U. We conclude that the number of lines of $\mathcal{L}$ that are contained in U is at most

$$(n - 1) + (n - 1)(2D^2 - 1) < 2nD^2. \qquad \square$$

9.4 Exercises

Exercise 9.1

(i) Construct a set of n lines in $\mathbb{R}^2$ with $\Theta(n)$ points that are $n^{1/3}$-rich. (Hint: Start with Claim 1.3 when $m = n$.)

(ii) Construct a set of n lines in $\mathbb{R}^3$ such that no plane contains more than $n^{1/2}$ of the lines and the number of $n^{1/6}$-rich points is $\Theta(n)$.

Exercise 9.2 Consider a coloring of the edges of the complete graph K_n with the following properties: (i) No vertex is incident to two edges of the same color. (ii) There are no two colors c_1, c_2 and two vertices v, u such that both v and u are incident to an edge of color c_1 and to an edge of color c_2 (see Figure 9.1). Prove that the coloring consists of $\Theta(n^2)$ colors. (Hint: Consider the proofs of Lemmas 9.8 and 9.10.)

Figure 9.1 The forbidden configuration: Both v and u are incident to an edge of color c_1 and to an edge of color c_2.

Exercise 9.3

(a) We consider two triangles in $\mathbb{R}^2$ as distinct if they are not congruent or if they are congruent but with an opposite orientation. A point set *spans* a triangle

if the three vertices of the triangle are points of the set. Find a set of n points in $\mathbb{R}^2$ that span $\Theta(n^2)$ distinct triangles.
(b) Prove that every set of n points in $\mathbb{R}^2$ spans $\Omega(n^{2-\varepsilon})$ distinct triangles. (Hint: Use Theorem 9.2 without changing its proof. Change only the ESGK framework.)

Exercise 9.4   Guth and Katz proved Theorem 7.1, which is stronger than Theorem 9.2. Use this stronger bound to solve the following problem.

Let $\mathcal{P}$ be a set of n points in $\mathbb{R}^2$ that spans $O(n/\sqrt{\log n})$ distinct distances. Prove that there exists a rotation of $\mathbb{R}^2$ that takes $\Omega(2^{\log^{0.49} n})$ points of $\mathcal{P}$ to points of $\mathcal{P}$. (Hint: Take a look at the equations before and after the statement of Theorem 9.2.)

Exercise 9.5   We now consider a complex variant of Equation (9.1). That is, for points $a, b \in \mathbb{C}^2$, we have a line $\ell_{ab} \subset \mathbb{C}^3$. We also we replace $t \in \mathbb{R}$ with $t \in \mathbb{C}$.

Consider a point $a \in \mathbb{C}^2$ and a line $\ell = \mathbf{V}(y - ix - k)$ for some $k \in \mathbb{C}$. Prove that all lines ℓ_{ab} with $b \in \ell$ intersect at a single point. That is, the problem behaves rather differently over the complex field. (Hint: Look for a value of t that leads to several cancellations.)

Exercise 9.6   Let $\mathcal{L}$ be a set of n lines in $\mathbb{R}^3$. In Exercise 8.6, we used constant-degree polynomial partitioning to prove that $\mathcal{L}$ spans $O(n^{3/2+\varepsilon})$ joints. We did this by using the polynomial partitioning theorem for varieties (Theorem 8.8). You are now asked to prove the same result by using constant-degree polynomial partitioning for points (Theorem 3.1). Imitate the proof of Theorem 9.4: Partition the space according to the set of joints, define α-good cells, and handle the non-good cells by using an iterative process.

9.5 Open Problems

Theorem 9.4 provides an upper bound for the number of rich points in a set of n lines in $\mathbb{R}^3$, when every constant-degree two-dimensional variety contains $O(n^{1/2})$ of the lines. Claim 9.6 shows that our bound for this problem is tight up to subpolynomial factors. One main open problem asks what happens when we further restrict the lines.

Open Problem 9.13   *Let $\mathcal{L}$ be a set of n lines in $\mathbb{R}^3$ such that every constant-degree two-dimensional variety in $\mathbb{R}^3$ contains $O(1)$ lines of $\mathcal{L}$. For $r \geq 2$, what is the maximum number of r-rich points that $\mathcal{L}$ can have?*

Beyond being difficult to solve, this problem is also related to other main open problems, such as the unit distances problem. For $r = \Omega(\sqrt{n})$, the problem is solved by Claim 9.7 and Lemma 9.8. While the problem is open for every smaller r, the main case is $r = 2$.

Erdős (1946) originally asked the distinct distances problem in $\mathbb{R}^d$: What is the minimum number of distinct distances that can be determined by a set of n points in $\mathbb{R}^d$. When $d \geq 3$, the gap between the current best lower and upper bounds is polynomial in n. For $d \geq 3$, it is not difficult to show that an $n^{1/d} \times n^{1/d} \times \cdots \times n^{1/d}$ section of the integer lattice $\mathbb{Z}^d$ spans $\Theta(n^{2/d})$ distances. It is conjectured that no set of n points in $\mathbb{R}^d$ spans an asymptotically smaller number of distinct distances.

Open Problem 9.14 *For $d \geq 3$, what is the minimum number of distinct distances that can be determined by a set of n points in $\mathbb{R}^d$?*

The current best lower bound is a combination of a result of Solymosi and Vu (2008) with the planar bound of Guth and Katz (2015). For the full details, see for example Sheffer (2014).

It is conjectured that the approach of Guth and Katz in $\mathbb{R}^2$ could be extended to $\mathbb{R}^d$. A generalization of the ESGK framework can be found in Bardwell-Evans and Sheffer (2019). This generalization reduces the distinct distances problem in $\mathbb{R}^d$ to an incidence problem with $(d-1)$-flats in $\mathbb{R}^{2d-1}$. These flats are well behaved, similarly to the set lines from Equation (9.2). For example, every two flats intersect in at most one point and every constant-degree hypersurface contains a bounded number of flats.

We still do not know how to obtain good bounds for incidences with $(d-1)$-flats in $\mathbb{R}^{2d-1}$. For example, for the distinct distances problem in $\mathbb{R}^3$, we obtain an incidence problem with 2-flats in $\mathbb{R}^5$. In Chapter 8 we obtain good bounds for incidences with 2-flats in $\mathbb{R}^4$. The case of 2-flats in $\mathbb{R}^5$ is just beyond our reach.

10

Distinct Distances Variants

After the long and technical proof of the distinct distances theorem, we move to a lighter chapter. In this chapter we study two additional distinct distances problems. We then study a problem that does not involve distinct distances, but relies on a variant of Theorem 9.2.

10.1 Subsets with No Repeated Distances

A set $\mathcal{P} \subset \mathbb{R}^2$ *spans no repeated distances* if every distance is spanned by at most one pair of points from $\mathcal{P}^2$ (we ignore the distance 0). In other words, a set of n points spans no repeated distances if it determines $\binom{n}{2}$ distinct distances. For a finite set $\mathcal{P} \subset \mathbb{R}^2$, we denote by $\mathsf{subset}(\mathcal{P})$ the size of the largest subset $\mathcal{P}' \subset \mathcal{P}$ that spans no repeated distances. Figure 10.1 depicts a set of 25 points and a subset of 4 points that span no repeated distances.

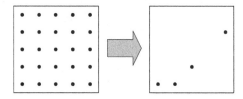

Figure 10.1 A set of 25 points and a subset of 4 points that span no repeated distances. No subset of five points has this property.

Let $\mathsf{subset}(n) = \min_{|\mathcal{P}|=n} \mathsf{subset}(\mathcal{P})$. That is, every set of n points contains $\mathsf{subset}(n)$ points that span no repeated distances. We are interested in the asymptotic value of $\mathsf{subset}(n)$.

If a subset $\mathcal{P}' \subset \mathcal{P}$ spans no repeated distances, then $\mathcal{P}'$ spans at least $\binom{|\mathcal{P}'|}{2}$ distinct distances. When $\mathcal{P}$ spans d distinct distances, we get that $|\mathcal{P}'| = O(\sqrt{d})$. Let $\mathcal{L}$ be a $\sqrt{n} \times \sqrt{n}$ section of the integer lattice $\mathbb{Z}^2$. By Theorem 1.9, the set $\mathcal{L}$ spans $\Theta(n/\sqrt{\log n})$ distinct distances. We conclude that

$$\mathsf{subset}(n) \leq \mathsf{subset}(\mathcal{L}) = O(\sqrt{n}/\log^{1/4} n).$$

This is the current best upper bound for $\mathsf{subset}(n)$. The following result is the current best lower bound for $\mathsf{subset}(n)$.

Theorem 10.1 (Charalambides, 2013) $\mathsf{subset}(n) = \Omega(n^{1/3}/\log^{1/3} n)$.

The proof of Theorem 10.1 combines Theorem 7.1 of Guth and Katz with a simple probabilistic argument. For this proof, we also require an upper bound on the maximum number of isosceles triangles that can be spanned by a set of n points in $\mathbb{R}^2$ (that is, isosceles triangles whose three vertices are in the point set). The current best bound, by Pach and Tardos (2002), is $O(n^{2.137})$. The following weaker bound suffices for proving Theorem 10.1.

Claim 10.2   *Let $\mathcal{P}$ be a set of n points in $\mathbb{R}^2$. Then $\mathcal{P}$ spans $O(n^{7/3})$ isosceles triangles.*

Proof We fix a point $p \in \mathcal{P}$ and consider the number of isosceles triangles where p is an endpoint of the base edge. Let $q \in \mathcal{P}\backslash\{p\}$ be the vertex of the triangle that is not an endpoint of the base edge. Then the third vertex of the isosceles triangle is on the circle centered at q and incident to p. See Figure 10.2(a). We denote this circle as C_q and set

$$C = \left\{ C_q \ : \ q \in \mathcal{P}\backslash\{p\} \right\}.$$

We note that $I(\mathcal{P}\backslash\{p\}, C)$ is the number of isosceles triangles in which p is an endpoint of the base edge.

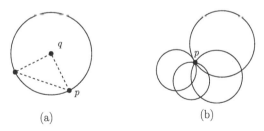

(a)   (b)

Figure 10.2 (a) An isosceles triangle where p is an endpoint of the base edge and q is not. The third vertex is on the circle centered at q and incident to p. (b) The incidence graph of circles that are incident to p contains no $K_{2,2}$ (when p is not a vertex of this graph).

No two circles of C are identical, since each circle has a distinct center. In general, the incidence graph of a set of points and a set of circles may contain $K_{2,2}$. However, since the circles of C are all incident to p, the incidence graph of $(\mathcal{P}\backslash\{p\}) \times C$ does not contain a $K_{2,2}$. See Figure 10.2(b). Theorem 3.3 implies that $I(\mathcal{P}\backslash\{p\}, C) = O(n^{4/3})$. That is, the number of isosceles triangles in which p is an endpoint of the base edge is $O(n^{4/3})$. By summing this bound over every $p \in \mathcal{P}$, we obtain that the number of isosceles triangles spanned by $\mathcal{P}$ is $O(n^{7/3})$. $\qquad\qquad\qquad\square$

We are now ready to derive an upper bound for subset(n).

Proof of Theorem 10.1 Consider a set $\mathcal{P}$ of n points in $\mathbb{R}^2$. Similarly to the ESGK framework (Section 7.2), we define the set

$$Q = \left\{ (a, b, c, d) \in \mathcal{P}^4 \ : \ |ab| = |cd| \right\}.$$

Unlike the ESGK framework, we only consider quadruples that consist of four distinct points. Theorem 7.1 implies that $|Q| = O(n^3 \log n)$. Let T be the set of isosceles triangles that are spanned by $\mathcal{P}$, including equilateral triangles. Claim 10.2 implies that $|T| = O(n^{7/3})$.

We consider a probability $0 < p < 1$ whose value is set below. Let $\mathcal{P}' \subseteq \mathcal{P}$ be a subset that is obtained by choosing every point of $\mathcal{P}$ with probability p. We have the expected size $\mathbb{E}[|\mathcal{P}'|] = pn$. Let $Q' \subseteq Q$ be the set of quadruples of Q that consist of four points from $\mathcal{P}'$. Every quadruple of Q is in Q' with a probability of p^4, so $\mathbb{E}[|Q'|] \leq \alpha p^4 n^3 \log n$ holds for a sufficiently large constant α. Let $T' \subseteq T$ be the set of isosceles triangles of T that have their three vertices in $\mathcal{P}'$. We have that $\mathbb{E}[|T'|] \leq \alpha p^3 n^{7/3}$, for a sufficiently large constant α. We fix a sufficiently large constant α that satisfies the above.

The set $\mathcal{P}'$ spans no repeated distances if and only if $|Q'| = |T'| = 0$. By linearity of expectation, we have that

$$\mathbb{E}\left[|\mathcal{P}'| - |Q'| - |T'|\right] \geq pn - \alpha p^4 n^3 \log n - \alpha p^3 n^{7/3}.$$

For sufficiently large n, setting $p = 1/(2\alpha n^2 \log n)^{1/3}$ leads to

$$\mathbb{E}[|\mathcal{P}'| - |Q'| - |T'|]$$
$$\geq \frac{n^{1/3}}{2^{1/3}\alpha^{1/3}\log^{1/3}n} - \frac{n^{1/3}}{2^{4/3}\alpha^{1/3}\log^{1/3}n} - \frac{n^{1/3}}{2\log n}$$
$$> \frac{n^{1/3}}{2^{1/3}\alpha^{1/3}\log^{1/3}n} - \frac{n^{1/3}}{2^{4/3}\alpha^{1/3}\log^{1/3}n} - \frac{n^{1/3}}{100\alpha^{1/3}\log^{1/3}n}$$
$$> \frac{n^{1/3}}{3\alpha^{1/3}\log^{1/3}n}.$$

Thus, there exists a subset $\mathcal{P}' \subset \mathcal{P}$ for which $|\mathcal{P}'| - |Q'| - |T'| > \frac{n^{1/3}}{3(\alpha_1 \log n)^{1/3}}$. We construct $\mathcal{P}'' \subseteq \mathcal{P}'$ by arbitrarily removing from $\mathcal{P}'$ a point from every tuple of Q' and T'. The subset $\mathcal{P}''$ does not span any repeated distances and contains $\Omega(n^{1/3}/\log^{1/3} n)$ points of $\mathcal{P}$. □

10.2 Point Sets with Few Distinct Distances

The structural distinct distances problem asks to characterize all sets of n points in $\mathbb{R}^2$ that determine a small number of distinct distances. We now briefly survey the embarrassing history of this problem.

- Erdős (1986) asked whether every set that spans $O(n/\sqrt{\log n})$ distances "has lattice structure." In Section 1.6, we saw that a $\sqrt{n} \times \sqrt{n}$ section of $\mathbb{Z}^2$ spans $O(n/\sqrt{\log n})$ distinct distances. The same holds for $\sqrt{n} \times \sqrt{n}$ sections of some other lattices (for example, see Sheffer, 2014).

- The lattice conjecture turned out to be frustratingly difficult, so Erdős suggested to first prove the following: For every point set that spans $O(n/\sqrt{\log n})$ distances, there exists a line that contains $\Omega(\sqrt{n})$ of the points. This would imply that the set can be covered by a small number of lines, which could be seen as a step towards the lattice conjecture.

- The line conjecture also turned out to be beyond our reach. Erdős then asked whether there exists a line that contains $\Omega(n^\varepsilon)$ points of the set, for any $\varepsilon > 0$. Even this weaker variant remains open after several decades.

The author of this book believes that the structural distinct distances problem might be the most difficult distinct distances variant. It is plausible that the tools required for solving this problem are not yet discovered.

We now present the current best bound for the above problem. To derive this bound, we require the following straightforward generalization of Theorem 3.3 (see also Problem 3.10).

Theorem 10.3 *Let $\mathcal{P}$ be a set of m points and let Γ be a set of n distinct irreducible curves of degree at most k, both in $\mathbb{R}^2$. If the incidence graph of $\mathcal{P} \times \Gamma$ contains no $K_{s,t}$, then*

$$I(\mathcal{P}, \Gamma) = O_{s,k}\left(m^{\frac{s}{2s-1}} n^{\frac{2s-2}{2s-1}} t^{\frac{1}{2s-1}} + tm + n\right).$$

Claim 10.4 *Let $\mathcal{P}$ be a set of n points in $\mathbb{R}^2$ that spans $O(n/\sqrt{\log n})$ distinct distances. Then there exists a line that contains $\Omega(\log n)$ points of $\mathcal{P}$.*

Proof Let D be the set of distances spanned by $\mathcal{P}$. For a distance $d \in D$ and a point $p \in \mathcal{P}$, let $C_{d,p}$ be the circle of radius d that is centered at p. We define the set of circles

$$C = \{C_{d,p} \: : \: d \in D, \, p \in \mathcal{P}\}.$$

Since $|\mathcal{P}| = n$ and $|D| = O(n/\sqrt{\log n})$, we have that $|C| = O(n^2/\sqrt{\log n})$. For any fixed $p \in \mathcal{P}$, the circles of C that are centered at p contain all the points of $\mathcal{P}\backslash\{p\}$. In other words, these circles form $n-1$ incidences with $\mathcal{P}$. By summing this over every p, we get that

$$I(\mathcal{P}, C) = n(n-1). \tag{10.1}$$

Let x denote the maximum number of points of $\mathcal{P}$ that are on a common line. For points $p, q \in \mathcal{P}$, the center of a circle that contains both p and q is on the perpendicular bisector of p and q. This implies that the incidence graph of $\mathcal{P} \times C$ contains no $K_{2,x+1}$. By Theorem 10.3,

$$I(\mathcal{P}, C) = O\left(n^{2/3} \left(\frac{n^2}{\sqrt{\log n}} \right)^{2/3} x^{1/3} + nx + \frac{n^2}{\sqrt{\log n}} \right)$$

$$= O\left(\frac{n^2 x^{1/3}}{\log^{1/3} n} + \frac{n^2}{\sqrt{\log n}} + nx \right).$$

Combining this bound with Equation (10.1) implies that $x = \Omega(\log n)$. $\quad\square$

10.3 Trapezoids Formed by Pairs of Intervals

The distinct distances proof of Guth and Katz reduces the question to a point-line incidence problem in $\mathbb{R}^3$ and solves this problem. In Section 10.1, we implicitly relied on the same reduction and on the same incidence bound. We now consider a different result that relies on a different reduction to lines in $\mathbb{R}^3$, and then applies the incidence bound of Guth and Katz.

We define an *interval* in $\mathbb{R}^2$ as a straight-line segment of a positive finite length. We denote by $(a, b; c, d)$ the interval whose endpoints are the points $(a, b), (c, d) \in \mathbb{R}^2$. Since an interval has a positive length, we have that $(a, b) \neq (c, d)$. For fixed $a, b, c, d \in \mathbb{R}$, the intervals $(a, b; c, d)$ and $(c, d; a, b)$ are identical. To have a unique notation for each interval, we only allow intervals $(a, b; c, d)$ where $a < c$ or $a = c$ and $b < d$.

Two intervals *form a trapezoid* if the four endpoints of the intervals are the vertices of a trapezoid. See Figure 10.3 for several examples. We consider the two intervals $(a, b; c, d)$ and $(a', b'; c', d')$. These two intervals form a trapezoid if at least one of the following cases holds.

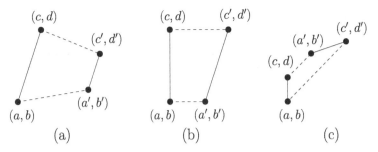

(a)   (b)   (c)

Figure 10.3 The intervals $(a, b; c, d)$ and $(a', b'; c', d')$ can form a trapezoid in three ways.

- The edge whose endpoints are (a, b) and (c, d) is parallel to the edge whose endpoints are (a', b') and (c', d'). See Figure 10.3(a). In this case, we have that

$$(b - d)(a' - c') = (a - c)(b' - d').\qquad(10.2)$$

- The edge whose endpoints are (a, b) and (a', b') is parallel to the edge whose endpoints are (c, d) and (c', d'). See Figure 10.3(b). In this case, we have that

$$(b - b')(c - c') = (a - a')(d - d').\qquad(10.3)$$

- The edge whose endpoints are (a, b) and (c', d') is parallel to the edge whose endpoints are (a', b') and (c, d). See Figure 10.3(c). In this case, we have that

$$(b - d')(a' - c) = (a - c')(b' - d).\qquad(10.4)$$

When the trapezoid is a parallelogram, more than one of Equations (10.2), (10.3), and (10.4) hold (see Figure 10.4(a)). When all four endpoints are collinear, all three equations hold (see Figure 10.4(b)). Another unusual case that satisfies one of the above equations happens when the two intervals share an endpoint (see Figure 10.4(c)). We count such degenerate cases as a valid trapezoids.

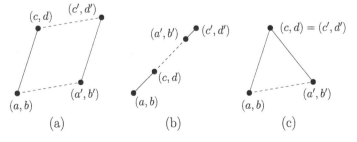

(a)   (b)   (c)

Figure 10.4 Additional cases that satisfy some of Equations (10.2), (10.3), and (10.4).

Let J be a set of n intervals. If J contains many parallel intervals, then many pairs of intervals form a trapezoid. But there are other interesting configurations of intervals that form many trapezoids. Di Benedetto et al. (2020) characterized all sets of intervals that form at least $cn^{3/2} \log n$ trapezoids, for a sufficiently large constant c. The following result is the first step of their analysis.

Theorem 10.5 *The following holds for every sufficiently large constant c. Let J be a set of n intervals in $\mathbb{R}^2$, such that at least $cn^{3/2} \log n$ pairs of intervals form a trapezoid. Then at least one of the following cases holds:*

- *There exist $\Omega(\sqrt{n} \log n)$ intervals in J with the same slope.*
- *There exist two parallel lines in $\mathbb{R}^2$, such that $\Omega(\sqrt{n})$ intervals from J have one endpoint on each line.*
- *There exist parallel lines ℓ_1 and ℓ_2, such that $\Omega(\sqrt{n})$ intervals $(a, b; c, d) \in J$ satisfy that $(a, c) \in \ell_1$ and $(b, d) \in \ell_2$.*
- *There exist two disjoint subsets $J_1, J_2 \subset J$ such that $|J_1| \cdot |J_2| = \Omega(n)$ and every pair of intervals from $J_1 \times J_2$ forms a trapezoid.*

The proof of Theorem 10.5 relies on point-line incidences in $\mathbb{R}^3$. In this book, a point-line incidence bound in $\mathbb{R}^3$ appeared in Theorem 9.2. As stated before, this theorem is a slightly weaker variant of the Guth–Katz result. We now state the original result of Guth and Katz (2015).

As discussed in Section 5.2, the *reguli* are a family of irreducible surfaces in $\mathbb{R}^3$. In particular, these are the hyperbolic paraboloids and the hyperboloids of one sheet. Let $S \subset \mathbb{R}^3$ be a regulus and let $\mathcal{L}$ be the set of lines that are contained in S. Then $\mathcal{L}$ can be partitioned into two disjoint families of lines $\mathcal{L}_1, \mathcal{L}_2 \subset \mathcal{L}$ with the following properties. Two lines from the same family do not intersect, while every pair of lines from $\mathcal{L}_1 \times \mathcal{L}_2$ intersect.[1] For more information, see Section 14.5.

Theorem 10.6 *Let $\mathcal{L}$ be a set of n lines in $\mathbb{R}^3$, such that every plane contains $O(\sqrt{n})$ lines of $\mathcal{L}$. Also, for every regulus $S \subset \mathbb{R}^3$, the lines of $\mathcal{L}$ that are contained in S form $O(n)$ intersecting pairs. Then for every $2 \leq r \leq \sqrt{n}$, we have that*

$$|\mathcal{P}_r(\mathcal{L})| = O\left(\frac{n^{3/2}}{r^2}\right).$$

The assumption of Theorem 10.6 allows for a regulus to contain $\Omega(\sqrt{n})$ lines from $\mathcal{L}$, if most of these lines belong to the same family. Indeed, in such a case we have a small number of pairs of intersecting lines in the regulus. We require the following corollary of Theorem 10.6.

---

[1] To be precise, it is possible that a few pairs from $\mathcal{L}_1 \times \mathcal{L}_2$ are parallel instead of intersecting. Also, in Theorem 10.6, the condition that involves reguli is required only for the case of $r = 2$.

Corollary 10.7 *The following holds for every sufficiently large constant c. Let $\mathcal{L}$ be a set of n lines in $\mathbb{R}^3$, such that at least $cn^{3/2}\log n$ pairs of lines from $\mathcal{L}^2$ intersect. Then at least one of the following holds:*

- *There exists a point that is incident to more than $\sqrt{n}$ lines of $\mathcal{L}$.*
- *There exists a plane that contains $\Omega(\sqrt{n})$ lines of $\mathcal{L}$.*
- *There exists a regulus that contains two disjoint sets of lines $\mathcal{L}_1, \mathcal{L}_2 \subset \mathcal{L}$, such that $|\mathcal{L}_1| \cdot |\mathcal{L}_2| = \Omega(n)$ and every pair of $\mathcal{L}_1 \times \mathcal{L}_2$ intersect.*

Proof Assume for contradiction that none of the three cases hold. Since no point is incident to more than $\sqrt{n}$ lines of $\mathcal{L}$, we have that $|\mathcal{P}_r(\mathcal{L})| = 0$ for every $r > \sqrt{n}$. By the assumption on planes and reguli, we may apply Theorem 10.6. That is, for every $2 \le r \le \sqrt{n}$, we have that $|\mathcal{P}_r(\mathcal{L})| = O(n^{3/2}/r^2)$. By a dyadic decomposition argument, the number of intersecting pairs of lines is at most

$$\sum_{j=0}^{\log\sqrt{n}} |\mathcal{P}_{2^j}(\mathcal{L})| \cdot \binom{2^{j+1}}{2} = O\left(\sum_{j=0}^{\log\sqrt{n}} \frac{n^{3/2}}{2^{2j}} \cdot 2^{2j}\right) = O\left(n^{3/2}\log n\right).$$

When c is sufficiently large, the above bound contradicts the assumption that at least $cn^{3/2}\log n$ pairs of lines intersect. Thus, at least one of the three cases holds. □

We are now ready to prove the trapezoids result.

Proof of Theorem 10.5 By the assumption of the theorem, at least $cn^{3/2}\log n$ pairs of intervals satisfy Equations (10.2), (10.3), or (10.4). By the pigeonhole principle, either at least $(cn^{3/2}\log n)/3$ pairs of intervals from J^2 satisfy Equation (10.2), at least $(cn^{3/2}\log n)/3$ pairs satisfy Equation (10.3), or at least at least $(cn^{3/2}\log n)/3$ pairs satisfy Equation (10.4).

We first consider the case where at least $(cn^{3/2}\log n)/3$ pairs of intervals from J^2 satisfy Equation (10.2). That is, at least $(cn^{3/2}\log n)/3$ pairs of intervals are parallel. If each interval of J is parallel to at most $(c\sqrt{n}\log n)/4$ other intervals of J, then the number of pairs of parallel intervals is at most $(cn^{3/2}\log n)/4$. This contradiction implies that there exist $\Omega(\sqrt{n}\log n)$ intervals with the same slope in J.

From trapezoids to line intersections in $\mathbb{R}^3$: Next, we consider the case where at least $(cn^{3/2}\log n)/3$ pairs of intervals from J^2 satisfy Equation (10.3). For an interval $j = (a, b; c, d) \in J$, we define the line

$$\ell_j = \left\{ \begin{pmatrix} b \\ d \\ 0 \end{pmatrix} + t \cdot \begin{pmatrix} a \\ c \\ 1 \end{pmatrix} : t \in \mathbb{R} \right\} \subset \mathbb{R}^3. \tag{10.5}$$

A line in $\mathbb{R}^3$ is *parallel to the xy-plane* if that line is contained in a plane that is parallel to the xy-plane. In other words, a line is parallel to the xy-plane if all its points have the same z-coordinate. We note that no line of the form ℓ_j is parallel to the xy-plane. Conversely, every line in $\mathbb{R}^3$ that is not parallel to the xy-plane can be defined as in Equation (10.5). Thus, there is a bijection between the intervals in $\mathbb{R}^2$ and the lines in $\mathbb{R}^3$ that are not parallel to the xy-plane. We consider the set of n distinct lines

$$\mathcal{L} = \{\ell_j \; : \; j \in J\}.$$

Consider intervals $j = (a, b; c, d)$ and $j' = (a', b'; c', d')$ such that $j \neq j'$. By inspecting Equation (10.5), we note that ℓ_j and $\ell_{j'}$ intersect when there exists a t that satisfies

$$b + at = b' + a't \qquad \text{and} \qquad d + ct = d' + c't.$$

By isolating t in both equations and comparing their values, we get that

$$(b - b')(c - c') = (a' - a)(d' - d).$$

We note that the above equation is equivalent to Equation (10.3). Since at least $(cn^{3/2} \log n)/3$ pairs of intervals from J satisfy Equation (10.3), at least $(cn^{3/2} \log n)/3$ pairs of lines from $\mathcal{L}$ intersect. We may thus apply Corollary 10.7 with $\mathcal{L}$. This corollary has three different outcomes, and we separately study each outcome.

We first assume that there exists a point $p \in \mathbb{R}^3$ that is incident to more than $\sqrt{n}$ lines of $\mathcal{L}$. We let $j = (a, b; c, d) \in J$ satisfy $p \in \ell_j$ and write $p = (p_x, p_y, p_z)$. Then Equation (10.5) implies that

$$p_x = b + ap_z \qquad \text{and} \qquad p_y = d + cp_z.$$

That is, the point $(a, b) \in \mathbb{R}^2$ is on the line defined by $y = p_x - xp_z$ and the point $(c, d) \in \mathbb{R}^2$ is on the line defined by $y = p_y - xp_z$. We conclude that, in this case, there exist two parallel lines in $\mathbb{R}^2$ such that $\Omega(\sqrt{n})$ intervals of J have one endpoint on each line.

We next assume that Corollary 10.7 implies that there exists a plane Π that contains $\Omega(\sqrt{n})$ lines of $\mathcal{L}$. We consider $A, B, C, D \in \mathbb{R}$ that satisfy $\Pi = \mathbf{V}(Ax + By + Cz + D)$. An interval $j = (a, b; c, d) \in J$ satisfies $\ell_j \subset \Pi$ if and only if

$$0 = A(b + ta) + B(d + tc) + Ct + D = t(Aa + Bc + C) + Ab + Bd + D,$$

for every $t \in \mathbb{R}$. This in turn implies that $\ell_j \subset \Pi$ if and only if

$$Aa + Bc + C = 0 \quad \text{and} \quad Ab + Bd + D = 0.$$

The above condition asks for the point (a, c) to be on the line defined by $Ax + By + C = 0$ and for the point (b, d) to be on the line defined by $Ax + By + D = 0$. Thus, in this case, there exist two parallel lines $\ell_1, \ell_2 \subset \mathbb{R}^2$ such that $\Omega(\sqrt{n})$ intervals $(a, b; c, d) \in J$ satisfy $(a, c) \in \ell_1$ and $(b, d) \in \ell_2$.

We move to consider the case where many pairs of lines of $\mathcal{L}$ in a regulus intersect. That is, there exists a regulus $S \subset \mathbb{R}^3$ that contains two disjoint sets of lines $\mathcal{L}_1, \mathcal{L}_2 \subset \mathcal{L}$, such that $|\mathcal{L}_1| \cdot |\mathcal{L}_2| = \Omega(n)$ and every pair of $\mathcal{L}_1 \times \mathcal{L}_2$ intersect. Let J_1 and J_2 be the disjoint sets of intervals from J that correspond to $\mathcal{L}_1$ and $\mathcal{L}_2$. By definition, we have that $|J_1| \cdot |J_2| = \Omega(n)$ and that every pair of intervals from $J_1 \times J_2$ forms a trapezoid. This completes the case where at least $(cn^{3/2} \log n)/3$ pairs of intervals from J satisfy Equation (10.3).

A more involved case: Finally, we consider the case where at least $(cn^{3/2} \log n)/3$ pairs of intervals from J satisfy Equation (10.4). Consider an interval $j = (a, b; c, d)$. Similarly to Equation (10.5), we define the line

$$\ell'_j = \left\{ \begin{pmatrix} d \\ b \\ 0 \end{pmatrix} + t \cdot \begin{pmatrix} c \\ a \\ 1 \end{pmatrix} \; : \; t \in \mathbb{R} \right\} \subset \mathbb{R}^3. \tag{10.6}$$

Intuitively, ℓ'_j is the line that corresponds to the interval $(c, d; a, b)$. (Because of the above restrictions, $(c, d; a, b)$ is not a valid way to define an interval. However, we can still define a line according to it.)

We consider two intervals $j = (a, b; c, d)$ and $j' = (a', b'; c', d')$ such that $j \neq j'$. The lines ℓ_j and $\ell'_{j'}$ intersect when there exists a t that satisfies

$$b + ta = d' + tc' \quad \text{and} \quad d + tc = b' + ta'.$$

By repeating the analysis of the previous case, we obtain that ℓ_j and $\ell'_{j'}$ intersect if and only if Equation (10.4) holds.

We define the set of lines

$$\mathcal{L} = \{\ell_j \; : \; j \in J\} \cup \{\ell'_j \; : \; j \in J\} \subset \mathbb{R}^3.$$

Since at least $(cn^{3/2} \log n)/3$ pairs of intervals from J satisfy Equation (10.4), at least $(cn^{3/2} \log n)/3$ pairs of lines from $\mathcal{L}$ intersect. We may thus apply Corollary 10.7 with $\mathcal{L}$. The remainder of the proof is very similar to the analysis of the previous case, so we do not repeat it here. □

For a more advanced characterization of intervals that form many trapezoids, see Di Benedetto et al. (2020).

10.4 Exercises

Exercise 10.1   Let $A \subset \mathbb{R}$ be a set of n real numbers. Prove that there exists a subset $A' \subset A$ such that $|A'| = \Omega(n^{1/3})$ and no *difference* repeats more than once in A'. That is, there do not exist $a, b, c, d \in A'$ such that $(a, b) \neq (c, d)$ and $a - b = c - d > 0$. (Hint: Define a set of quadruples and derive an upper bound on its size.)

Exercise 10.2   Let $\mathcal{P} \subset \mathbb{R}^2$ be a set of n points, such that no four points are on the same circle. Prove that there exists a subset $\mathcal{P}' \subset \mathcal{P}$ such that $|A'| = \Omega(n^{1/5})$ and every three points from $\mathcal{P}'$ determine a circle with a distinct radius. That is, no two triples of points from $\mathcal{P}'$ determine circles with the same radius. (Hint: Double count the number of 6-tuples.)

Exercise 10.3   Let $\mathcal{P}$ be a set of n points in $\mathbb{R}^2$ such that no four distinct points $a, b, c, d \in \mathcal{P}$ are the vertices of an isosceles trapezoid (see Figure 10.5). Prove that $\mathcal{P}$ determines $\Omega(n)$ distinct distances. You might like to use the following approach.

Let x be the number of distinct distances that are spanned by $\mathcal{P}$. Consider the set of isosceles triangles that are spanned by $\mathcal{P}$:

$$T = \left\{ (a, b, c) \in \mathcal{P}^3 \ : \ |ab| = |ac| \text{ and } b \neq c \right\}. \tag{10.7}$$

Prove the claim by double counting $|T|$. Use the Cauchy–Schwarz inequality to find a lower bound for the number of triangles that involve a specific $a \in \mathcal{P}$. The points of $\mathcal{P} \setminus \{a\}$ are on at most x circles centered at a. Two points $b, c \in \mathcal{P}$ form a triple with a if they are on the same circle. To derive an upper bound for $|T|$, find a connection to perpendicular bisectors of pairs of points of $\mathcal{P}$.

Figure 10.5 In an isosceles trapezoid, the two legs have the same length and the base angles are identical. Equivalently, an isosceles trapezoid has a symmetry line that bisects a pair of parallel sides.

Exercise 10.4   Let $\mathcal{P}$ be a set of n points in $\mathbb{R}^2$, such that no three points of $\mathcal{P}$ are collinear. Prove that there exists a point $p \in \mathcal{P}$ such that the number of distinct distances between p and $\mathcal{P} \setminus \{p\}$ is at least $(n - 1)/3$.

You might like to imitate the proof of Problem 10.3 by defining T as in Equation (10.7) and double counting $|T|$. In this case, x is the maximum number of distinct distances between any point $p \in \mathcal{P}$ and $\mathcal{P} \setminus \{p\}$.

Exercise 10.5 A quadrilateral is *orthodiagonal* if the two diagonals of the quadrilateral are orthogonal. See Figure 10.6. Two intervals *form an orthodiagonal quadrilateral* if the four endpoints of the intervals are the vertices of an orthodiagonal quadrilateral.

Repeat the first part of the proof of Theorem 10.5 for orthodiagonal quadrilaterals. That is, reduce the problem to line intersections in $\mathbb{R}^3$ and apply Corollary 10.7. There is no need to check what each of the cases of the corollary implies in $\mathbb{R}^2$. (Hint: Just like with the case of trapezoids, some line intersections may correspond to cases that are not orthodiagonal.)

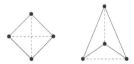

Figure 10.6 Two orthodiagonal quadrilaterals.

Exercise 10.6 We define a *pseudo-trapezoid* to be a quadrilateral with a side that has twice the slope of the opposite side. See Figure 10.7. Two intervals *form a pseudo-trapezoid* if the four endpoints of the intervals are the vertices of a pseudo-trapezoid. Prove an analog of Theorem 10.5 for pseudo-trapezoids.

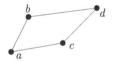

Figure 10.7 A pseudo-trapezoid: The edge ab has slope $1/2$ while the edge cd has slope 1.

10.5 Open Problems

In Section 10.1 we discussed subset(n): the largest number satisfying that every set of n points contains subset(n) points that span no repeated distances. The current best bounds for this expression are

$$\text{subset}(n) = O(\sqrt{n}/\log^{1/4} n) \qquad \text{and} \qquad \text{subset}(n) = \Omega(n^{1/3}/\log^{1/3} n).$$

Open Problem 10.8 *Determine the asymptotic value of* subset(n).

In Section 10.2 we discuss characterizing the sets of n points in $\mathbb{R}^2$ that span $O(n/\sqrt{\log n})$ distinct distances. As mentioned in that section, this topic includes many conjectures and hardly any results. We first repeat a conjecture of Erdős.

Conjecture 10.9 *Let $\mathcal{P}$ be a set of n points in $\mathbb{R}^2$, such that $\mathcal{P}$ spans $O(n/\sqrt{\log n})$ distinct distances. Then there exists a line that contains $\Omega(\sqrt{n})$ points of $\mathcal{P}$. (Weaker variant: Then there exists a line that contains $\Omega(n^\varepsilon)$ points of $\mathcal{P}$, for some $\varepsilon > 0$.)*

The current best bound for Conjecture 10.9 is that there exists a line that contains $\Omega(\log n)$ points of $\mathcal{P}$. One may also ask whether, for every set $\mathcal{P}$ that spans few distinct distances, no curve contains many points of $\mathcal{P}$.

Open Problem 10.10 (Sheffer et al., 2016) *Prove or disprove: Let $\mathcal{P}$ be a set of n points in $\mathbb{R}^2$ such that $\mathcal{P}$ spans $O(n/\sqrt{\log n})$ distinct distances. Then, for every $\varepsilon > 0$, each constant-degree curve contains $O(n^{1/2+\varepsilon})$ points of $\mathcal{P}$.*

It is easier to make progress on Open Problem 10.10 than on Conjecture 10.9. Let $\mathcal{P}$ be a set of n points in $\mathbb{R}^2$, such that $\mathcal{P}$ spans $O(n/\sqrt{\log n})$ distinct distances. The current best bounds are:

- (Raz et al., 2015) Every line contains $O(n^{43/52})$ points of $\mathcal{P}$ (up to polylogarithmic factors).
- (Sheffer et al., 2016) Every circle contains $O(n^{7/8})$ points of $\mathcal{P}$.
- (Pach and De Zeeuw, 2017) Every constant-degree curve that is not a line or a circle contains $O(n^{3/4})$ points of $\mathcal{P}$.

11

Incidences in $\mathbb{R}^d$

In this chapter, we study general incidence bounds in $\mathbb{R}^d$. As a warm-up, we first derive an incidence bound for curves in $\mathbb{R}^3$. The main result of this chapter is a general point-variety incidence bound in $\mathbb{R}^d$. This result relies on another polynomial partitioning theorem, for the case where the points are on a constant-degree variety. The proof of this partitioning theorem relies on Hilbert polynomials.

11.1 Warmup: Incidences with Curves in $\mathbb{R}^3$

Chapter 8 contains two incidence bounds in $\mathbb{R}^d$: Theorem 8.6 and Lemma 8.2. Both of these results hold for curves in $\mathbb{R}^3$. For incidences between m points and n lines in $\mathbb{R}^3$, both results lead to the Szemerédi–Trotter bound $O(m^{2/3}n^{2/3} + m + n)$, possibly with an extra ε in the exponent. To see that this bound is tight, we take a point-line configuration in $\mathbb{R}^2$ with $\Theta(m^{2/3}n^{2/3})$ incidences (such as Claim 1.3) and place it in an arbitrary plane in $\mathbb{R}^3$.

The above bounds are tight in the degenerate case where the point-line configuration is mostly two-dimensional. When every plane in $\mathbb{R}^3$ contains $O(\sqrt{n})$ lines, the maximum number of point line incidences decreases to $\Theta(m^{1/2}n^{3/4} + m + n)$. Similarly, for point-circle incidences in $\mathbb{R}^3$, we obtain a better bound when every plane and sphere contains a bounded number of circles. Intuitively, the number of incidences decreases when the configuration is truly three-dimensional.

We now study a general point-curve incidence bound in $\mathbb{R}^3$ for the case where every surface contains a bounded number of curves. This result is presented as a warm-up before getting to a more involved technique for incidences in $\mathbb{R}^d$. A reader who does not wish to see yet another use of constant-degree polynomial partitioning may safely skip this section.

155

Theorem 11.1   *For any $\varepsilon > 0$ there exists a constant c_ε that satisfies the following. Let $\mathcal{P}$ be a set of m points and let Γ be a set of n irreducible curves of complexity at most k, both in $\mathbb{R}^3$. Assume that the incidence graph of $\mathcal{P} \times \Gamma$ contains no $K_{s,t}$. If every two-dimensional variety in $\mathbb{R}^3$ of degree at most c_ε contains $O(q)$ curves of Γ, then*

$$I(\mathcal{P}, \Gamma) = O_{s,t,k,\varepsilon}\left(m^{\frac{s}{3s-2}+\varepsilon} n^{\frac{3s-3}{3s-2}} + m^{\frac{s}{2s-1}+\varepsilon} n^{\frac{3s-3}{4s-2}} q^{\frac{s-1}{4s-2}} + m + n \right).$$

As an example, consider the case where Γ is a set of lines, $m = n$, and every surface of degree at most c_ε contains at most $\sqrt{n}$ lines of Γ. Applying Theorem 11.1 with $s = t = 2$ and $q = \sqrt{n}$ gives $I(\mathcal{P}, \Gamma) = O_\varepsilon(n^{5/4+\varepsilon})$. Theorem 8.6 and Lemma 8.2 lead to the weaker bound $I(\mathcal{P}, \Gamma) = O(n^{4/3})$.

Proof of Theorem 11.1   Let α_1 and α_2 be sufficiently large constants that depend on s, t, k, and ε. We prove by induction on $m + n$ that

$$I(\mathcal{P}, \Gamma) \leq \alpha_1 \left(m^{\frac{s}{3s-2}+\varepsilon} n^{\frac{3s-3}{3s-2}} + m^{\frac{s}{2s-1}+\varepsilon} n^{\frac{3s-3}{4s-2}} q^{\frac{s-1}{4s-2}} \right) + \alpha_2(m + n).$$

For the induction base, the case where $m + n \leq 100$. By taking $\alpha_2 \geq 100$, we obtain that

$$I(\mathcal{P}, \Gamma) \leq mn < \alpha_2(m + n).$$

We move to consider the induction step, where we assume that $m + n > 100$. The hidden constants in the $O(\cdot)$-notations throughout the proof may also depend on s, t, k, and ε. For brevity we write $O(\cdot)$ instead of $O_{s,t,k,\varepsilon}(\cdot)$. Since the incidence graph contains no $K_{s,t}$, Lemma 8.1 implies that $I(\mathcal{P}, \Gamma) = O(mn^{1-1/s} + n)$. When $m = O(n^{1/s})$ this implies $I(\mathcal{P}, \Gamma) = O(n)$. This completes the proof when α_2 is larger than the constant hidden by the $O(\cdot)$-notation. We may thus assume that

$$n = O(m^s). \tag{11.1}$$

Partitioning the space: Let f be an r-partitioning polynomial of $\mathcal{P}$, for a sufficiently large constant r. The asymptotic relations between the constants in this proof are

$$2^{1/\varepsilon}, k, s, t \ll r \ll \alpha_2 \ll \alpha_1.$$

By the polynomial partitioning theorem (Theorem 3.1), we have $\deg f = O(r)$. By Warren's theorem (Theorem 3.2), the number of cells in $\mathbb{R}^3 \backslash \mathbf{V}(f)$ is $c = O(r^3)$. We denote the cells of the partition as $C_1, \ldots, C_c$. For each $1 \leq j \leq c$, we set $\mathcal{P}_j = P \cap C_j$ and let Γ_j be the set of curves of Γ that intersect C_j. We also set $m_j = |\mathcal{P}_j|$, $m' = \sum_{j=1}^{c} m_j$, and $n_j = |\Gamma_j|$. By definition, $m_j \leq m/r^3$ for every $1 \leq j \leq c$.

We consider a curve $\gamma \in \Gamma$. Applying Theorem 4.15 with $U = \gamma$ and $W = \mathbf{V}(f)$ implies that γ intersects $O(r)$ cells of $\mathbb{R}^3 \backslash \mathbf{V}(f)$. By summing this over every $\gamma \in \Gamma$, we obtain that $\sum_{j=1}^{c} n_j = O(nr)$. Then, Hölder's inequality (Theorem A.3) leads to

$$\sum_{j=1}^{c} n_j^{\frac{3s-3}{3s-2}} \leq \left(\sum_{j=1}^{c} n_j \right)^{\frac{3s-3}{3s-2}} \left(\sum_{j=1}^{c} 1 \right)^{\frac{1}{3s-2}} = O\left((nr)^{\frac{3s-3}{3s-2}} r^{\frac{3}{3s-2}} \right) = O\left(n^{\frac{3s-3}{3s-2}} r^{\frac{3s}{3s-2}} \right),$$

$$\sum_{j=1}^{c} n_j^{\frac{3s-3}{4s-2}} \leq \left(\sum_{j=1}^{c} n_j \right)^{\frac{3s-3}{4s-2}} \left(\sum_{j=1}^{c} 1 \right)^{\frac{s+1}{4s-2}} = O\left((nr)^{\frac{3s-3}{4s-2}} r^{\frac{3s+3}{4s-2}} \right) = O\left(n^{\frac{3s-3}{4s-2}} r^{\frac{3s}{2s-1}} \right).$$

Applying the induction hypothesis separately in each cell gives

$$\sum_{j=1}^{c} I(\mathcal{P}_j, \Gamma_j)$$

$$\leq \sum_{j=1}^{c} \left(\alpha_1 \left(m_j^{\frac{s}{3s-2}+\varepsilon} n_j^{\frac{3s-3}{3s-2}} + m_j^{\frac{s}{2s-1}+\varepsilon} n_j^{\frac{3s-3}{4s-2}} q^{\frac{s-1}{4s-2}} \right) + \alpha_2 (m_j + n_j) \right)$$

$$\leq \alpha_1 \left(\frac{m^{\frac{s}{3s-2}+\varepsilon}}{r^{\frac{3s}{3s-2}+3\varepsilon}} \sum_{j=1}^{c} n_j^{\frac{3s-3}{3s-2}} + \frac{m^{\frac{s}{2s-1}+\varepsilon} q^{\frac{s-1}{4s-2}}}{r^{\frac{3s}{2s-1}+3\varepsilon}} \sum_{j=1}^{c} n_j^{\frac{3s-3}{4s-2}} \right) + \sum_{j=1}^{c} \alpha_2 (m_j + n_j)$$

$$= \alpha_1 \cdot O\left(\frac{m^{\frac{s}{3s-2}+\varepsilon} n^{\frac{3s-3}{3s-2}}}{r^{3\varepsilon}} + \frac{m^{\frac{s}{2s-1}+\varepsilon} n^{\frac{3s-3}{4s-2}} q^{\frac{s-1}{4s-2}}}{r^{3\varepsilon}} \right) + \alpha_2 \left(m' + O(nr) \right).$$

By Equation (11.1), we have that $n = O\left(m^{\frac{s}{3s-2}} n^{\frac{3s-3}{3s-2}} \right)$. When α_1 is sufficiently large with respect to r and α_2, we get that

$$\sum_{j=1}^{c} I(\mathcal{P}_j, \Gamma_j) = \alpha_1 \cdot O\left(\frac{m^{\frac{s}{3s-2}+\varepsilon} n^{\frac{3s-3}{3s-2}}}{r^{3\varepsilon}} + \frac{m^{\frac{s}{2s-1}+\varepsilon} n^{\frac{3s-3}{4s-2}} q^{\frac{s-1}{4s-2}}}{r^{3\varepsilon}} \right) + \alpha_2 m'.$$

When r is sufficiently large with respect to ε and to the constant hidden in the $O(\cdot)$-notation, we have that

$$\sum_{j=1}^{c} I(\mathcal{P}_j, \Gamma_j) \leq \frac{\alpha_1}{2} \left(m^{\frac{s}{3s-2}+\varepsilon} n^{\frac{3s-3}{3s-2}} + m^{\frac{s}{2s-1}+\varepsilon} n^{\frac{3s-3}{4s-2}} q^{\frac{s-1}{4s-2}} \right) + \alpha_2 m'. \quad (11.2)$$

Incidences on the partition: It remains to study incidences with points that lie on $\mathbf{V}(f)$. We set $\mathcal{P}_0 = \mathcal{P} \cap \mathbf{V}(f)$ and $m_0 = |\mathcal{P}_0| = m - m'$. Let Γ_0 denote the set of curves that are contained in $\mathbf{V}(f)$. We also set $\Gamma' = \Gamma \backslash \Gamma_0$ and $n_0 = |\Gamma_0|$. By Theorem 4.15, every curve of Γ' intersects $\mathbf{V}(f)$ in $O(r)$ points. This implies that every such curve is incident to $O(r)$ points of $\mathcal{P}_0$. Summing this over all curves of Γ' and taking α_2 to be sufficiently large gives

$$I(\mathcal{P}_0, \Gamma') = O(nr) \le \frac{\alpha_2}{2}n. \qquad (11.3)$$

It remains to derive an upper bound for $I(\mathcal{P}_0, \Gamma_0)$. We set c_ε to be larger than the maximum possible value of deg f, which is $\Theta(r)$. By the assumption of the theorem, we get that $|\Gamma_0| \le q$. By applying Lemma 8.2 with $\mathcal{P}_0$ and Γ_0, we obtain that

$$I(\mathcal{P}_0, \Gamma_0) = O\left(m_0^{\frac{s}{2s-1}} q^{\frac{2s-2}{2s-1}} + m_0 + q \right).$$

Since $q \le n$ and $m_0 \le m$, we have that

$$m_0^{\frac{s}{2s-1}} q^{\frac{2s-2}{2s-1}} \le m^{\frac{s}{2s-1}} n^{\frac{3s-3}{4s-2}} q^{\frac{s-1}{4s-2}}.$$

Combining this with a sufficiently large choice of α_1 and α_2 implies that

$$I(\mathcal{P}_0, \Gamma_0) = O\left(m^{\frac{s}{2s-1}} n^{\frac{3s-3}{4s-2}} q^{\frac{s-1}{4s-2}} + n + m_0 \right)$$

$$\le \frac{\alpha_1}{2} m^{\frac{s}{2s-1}} n^{\frac{3s-3}{4s-2}} q^{\frac{s-1}{4s-2}} + \frac{\alpha_2}{2}(n + m_0). \qquad (11.4)$$

By combining Equations (11.2)–(11.4), we complete the induction step and thus the proof of the theorem. □

It is not difficult to extend the proof of Theorem 11.1 to incidences with curves in $\mathbb{R}^d$, for any d. As d grows, the incidence bound becomes smaller while the restrictions over the set of curves become more involved (see Sharir et al., 2016).

We recall Lemma 1.16: The Szemerédi–Trotter theorem is equivalent to an upper bound for the number of r-rich lines that are spanned by n points in $\mathbb{R}^2$. That is, each result can be easily derived from the other by using only basic arguments. This is no longer the case in dimension $d \ge 3$. For example, Theorem 9.2 provides an upper bound for the number of r-rich lines in $\mathbb{R}^3$. Theorem 9.2 immediately implies Theorem 11.1 for the case of lines. However, as far as we know, Theorem 11.1 cannot easily lead to Theorem 9.2 for small values of r. In general, bounds for r-rich varieties in dimension $d \ge 3$ are significantly more difficult to prove with the current machinery.

11.2 Hilbert Polynomials

As already stated, the main goal of this chapter is to derive a general point-variety incidence bound in $\mathbb{R}^d$. We prove this bound in Section 11.3. In the current section, we study tools from algebraic geometry that are required for that proof.

For an integer $m \geq 0$, we denote by $\mathbb{R}[x_1, \ldots, x_d]_{\leq m}$ the set of polynomials of degree at most m in $\mathbb{R}[x_1, \ldots, x_d]$. Similarly, for an ideal $J \subset \mathbb{R}[x_1, \ldots, x_d]$, we write $J_{\leq m} = J \cap \mathbb{R}[x_1, \ldots, x_d]_{\leq m}$. In other words, $J_{\leq m}$ is the set of polynomials in J with degree at most m. We note that $\mathbb{R}[x_1, \ldots, x_d]_{\leq m}$ is not a ring and that $J_{\leq m}$ is not an ideal.

As shown in the proof of Theorem 3.6, there are $\binom{d+m}{m}$ monomials in the variables $x_1, \ldots, x_d$ of degree at most m (ignoring the coefficient of the monomial). Thus, we can consider $\mathbb{R}[x_1, \ldots, x_d]_{\leq m}$ as a vector space of dimension $\binom{d+m}{m}$. Specifically, $\mathbb{R}[x_1, \ldots, x_d]_{\leq m}$ is isomorphic to $\mathbb{R}^{\binom{d+m}{m}}$.[1] Similarly, $J_{\leq m}$ is a vector space of a finite dimension and a subspace of $\mathbb{R}[x_1, \ldots, x_d]_{\leq m}$.

An ideal that is generated by monomials is called *a monomial ideal*. For example, $M = \langle x^2y, xy^2 \rangle \subset \mathbb{R}[x, y]$ is a monomial ideal. We note that every term of every polynomial in M is a multiple of x^2y or of xy^2. While all of the following definitions and results hold for arbitrary ideals, we use monomial ideals as examples, since their behavior is more intuitive. We think of the set of monomials in $\mathbb{R}[x, y]$ as a lattice, as shown in Figure 11.1(a). Figure 11.1(b) shows the structure of M, and Figure 11.1(c) shows $M_{\leq 4}$.

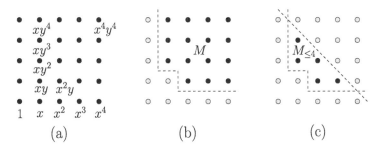

(a)   (b)   (c)

Figure 11.1 (a) Every lattice point corresponds to a monomial of $\mathbb{R}[x, y]$. (b) The ideal $M = \langle x^2y, xy^2 \rangle$. (c) The vector space $M_{\leq 4}$.

Every polynomial from $\mathbb{R}[x, y]_{\leq 3}$ can be written as

$$c_1 + c_2x + c_3y + c_4x^2 + c_5xy + c_6y^2 + c_7x^3 + c_8x^2y + c_9xy^2 + c_{10}y^3,$$

---

[1] Since constant multiples of a polynomial define the same variety, we can consider such multiples as equivalent. Under this equivalence relation, $\mathbb{R}[x_1, \ldots, x_d]_{\leq m}$ behaves like the projective space $\mathbb{R}\mathbf{P}^{\binom{d+m}{m}-1}$. We do not rely on projective properties in this chapter, so we use the affine $\mathbb{R}^{\binom{d+m}{m}}$.

where $c_1, \ldots, c_{10} \in \mathbb{R}$. We say that this polynomial corresponds to the point $(c_1, \ldots, c_{10}) \in \mathbb{R}^{10}$. We write the coordinates of $\mathbb{R}^{10}$ as $x_1, \ldots, x_{10}$. For the ideal $M = \langle x^2 y, xy^2 \rangle \subset \mathbb{R}[x, y]$, we get that $M_{\leq 3}$ corresponds to the 2-flat

$$\mathbf{V}(x_1, x_2, x_3, x_4, x_5, x_6, x_7, x_{10}) \subset \mathbb{R}^{10}. \tag{11.5}$$

For any ideal $J \subset \mathbb{R}[x_1, \ldots, x_d]$, the quotient $\mathbb{R}[x_1, \ldots, x_d]_{\leq m}/J_{\leq m}$ is also a vector space.[2] The *Hilbert function* of J is

$$h_J(m) = \dim \left(\mathbb{R}[x_1, \ldots, x_d]_{\leq m}/J_{\leq m} \right).$$

We return to the example $M = \langle x^2 y, xy^2 \rangle \subset \mathbb{R}[x, y]$. We can think of $\mathbb{R}[x, y]_{\leq m}/M_{\leq m}$ as the set of polynomials in $\mathbb{R}[x, y]_{\leq m}$ with no monomials that are multiples of $x^2 y$ and xy^2. When $m = 3$, two polynomials are equivalent in $\mathbb{R}[x, y]_{\leq 3}/M_{\leq 3}$ if their difference is in the 2-flat from Equation (11.5). We can think of $\mathbb{R}[x, y]_{\leq 3}/M_{\leq 3}$ as the set of polynomials of the form $c_1 + c_2 x + c_3 y + c_4 x^2 + c_5 xy + c_6 y^2 + c_7 x^3 + c_{10} y^3$. This implies that

$$h_M(3) = \dim(\mathbb{R}[x, y]_{\leq 3}/M_{\leq 3}) = 8.$$

Similarly to the above, we have that

$$h_M(0) = \dim(\mathbb{R}[x, y]_{\leq 0}/M_{\leq 0}) = \dim(\mathbb{R}[x, y]_{\leq 0}) = 1,$$
$$h_M(1) = \dim(\mathbb{R}[x, y]_{\leq 1}/M_{\leq 1}) = \dim(\mathbb{R}[x, y]_{\leq 1}) = 3.$$

When $m \geq 2$, we note that $\mathbb{R}[x, y]_{\leq m}/M_{\leq m}$ is spanned by xy and the monomials of degree at most m that do not include both x and y. This implies that

$$h_M(m) = \dim \left(\mathbb{R}[x, y]_{\leq m}/M_{\leq m} \right) = \begin{cases} 1, & m = 0, \\ 3, & m = 1, \\ 2 + 2m, & m \geq 2. \end{cases}$$

For every ideal $J \subset \mathbb{R}[x_1, \ldots, x_d]$, there exists an integer $m_J \geq 0$ and a polynomial $H_J \in \mathbb{R}[m]$ that satisfy the following: For every $m > m_J$, we have that $h_J(m) = H_J(m)$. That is, for sufficiently large m the Hilbert function behaves like a polynomial. The polynomial H_J is called the *Hilbert polynomial of* J, and m_J is called the *regularity* of J. In the above example, $H_M = 2 + 2m$ and $m_M = 1$.

Let $U \subset \mathbb{R}^d$ be a variety. We can obtain properties of U by studying the Hilbert polynomial $H_{\mathbf{I}(U)}(m)$. For example, the dimension of U is equal to the degree of $H_{\mathbf{I}(U)}(m)$. As with previous uses of ideals of varieties, we cannot use an arbitrary ideal $J \subset \mathbb{R}[x_1, \ldots, x_d]$ that satisfies $U = \mathbf{V}(J)$. To get the correct

---

[2] Reminder: Consider a vector space A and let B be a subspace of A. In the *quotient* A/B, two elements $a, a' \in A$ are equivalent when there exists $b \in B$ such that $a + b = a'$.

dimension of U, we must use the Hilbert polynomial of the ideal $\mathbf{I}(U)$. The following lemma collects several properties of Hilbert polynomials.

Lemma 11.2 *Let $U \subset \mathbb{R}^d$ be a variety of complexity k and dimension d'. Let $J = \mathbf{I}(U)$. Then*

- $\deg H_J(m) = d'$.
- $m_J = O_{d,k}(1)$.
- *The coefficient of the leading term $m^{d'}$ of $H_J(m)$ is positive (Exercise 11.4).*

Let $U \subset \mathbb{R}^d$ be a variety of complexity k and dimension d'. By combining the properties of Lemma 11.2, we get that

$$h_{\mathbf{I}(U)}(m) = \Theta_{d,k}(m^{d'}), \tag{11.6}$$

for every $m > m_J$.

A more detailed introduction to Hilbert polynomials, including proofs for most of the above claims, can be found in Cox et al. (2013, Chapter 9). For the bound on the regularity m_J, see Giusti (1984, Theorem B) and Fox et al. (2017).

11.3 A General Point-Variety Incidence Bound

We are now ready to state our general incidence bound in $\mathbb{R}^d$. This result was derived in Fox et al. (2017).

Theorem 11.3 *Let $\mathcal{P}$ be a set of m points and let $\mathcal{V}$ be a set of n varieties of complexity at most k, both in $\mathbb{R}^d$. Assume that the incidences graph of $\mathcal{P} \times \mathcal{V}$ contains no $K_{s,t}$. Then for every $\varepsilon > 0$, we have that*

$$I(\mathcal{P}, \mathcal{V}) = O_{k,s,t,d,\varepsilon}\left(m^{\frac{(d-1)s}{ds-1}+\varepsilon}n^{\frac{d(s-1)}{ds-1}} + m + n\right).$$

Note that Theorem 11.3 holds for varieties of any dimension. It is known to be tight for hypersurfaces when $s = 2$, up to the extra ε in the exponent (see Do and Sheffer, 2021). An improved bound for varieties that are not hypersurfaces can be found in Exercise 11.5. This bound is a corollary of Theorem 11.3 and is also tight when $s = 2$, up to the extra ε. At first look, this tightness statement may seem to contradict the stronger bounds of Theorems 8.6 and 11.1. However, these stronger bounds rely on additional assumptions about the varieties.

Proof strategy: We now discuss a high-level nonrigorous strategy for proving Theorem 11.3. The proof begins with the constant-degree polynomial partitioning technique: We prove the incidence bound by induction, use a constant-degree partitioning, and apply the induction hypothesis separately in each cell. As before, bounding the number of incidences in the cells is straightforward. It then remains to bound the number of incidences with points that are on the partition. How did we address this issue before?

- In Theorems 9.2 and 11.1, we avoided the issue by assuming that every variety with constant complexity contains a bounded number of curves.
- In Lemma 8.5 and Theorem 8.6, we relied on an additional assumption about the tangent spaces of the varieties.

Theorem 11.3 does not include additional assumptions that we can rely on.

We let $U \subset \mathbb{R}^d$ be the partition and set $\mathcal{P}_0 = \mathcal{P} \cap U$. We note that U is a constant-degree hypersurface. To bound the number of incidences with $\mathcal{P}_0$, we take a *second constant-degree partitioning polynomial* $f_2 \in \mathbb{R}[x_1, \ldots, x_d]$. The polynomial f_2 partitions U rather than $\mathbb{R}^d$. That is, the cells of the second partition are the connected components of $U \backslash \mathbf{V}(f_2)$. As before, we want each cell to contain a bounded number of points from $\mathcal{P}_0$.

After applying the second polynomial partitioning, it is straightforward to bound the number of incidences in the new cells. Unfortunately, there might be points of $\mathcal{P}_0$ that are on the second partition. Thus, it remains to study the number of incidences with points on $U_2 = U \cap \mathbf{V}(f_2)$. To handle these incidences, we take a *third* partition that divides U_2 into cells. By assuming that U and $\mathbf{V}(f_2)$ do not have common components, the dimension of U_2 is at most $d - 2$. Similarly, after the third partition, it remains to study points on a variety of dimension at most $d - 3$. We continue to apply more partitioning steps until the remaining points are on a one-dimensional variety. It is not difficult to bound the number of incidences with points on such a variety.

The full proof: The above proof strategy requires a polynomial partitioning theorem for points on varieties. We now prove such a theorem by adapting the proof of the original polynomial partitioning theorem from Section 3.3. The original proof starts with the polynomial ham sandwich theorem (Theorem 3.6). We now derive a variant of this result. We recall that $f \colon \mathbb{R}^d \to \mathbb{R}$ *bisects* a finite set $\mathcal{P} \subset \mathbb{R}^d$ if at most $|\mathcal{P}|/2$ points $p \in \mathcal{P}$ satisfy $f(p) > 0$ and at most $|\mathcal{P}|/2$ points satisfy $f(p) < 0$.

Lemma 11.4 *Let $U \subset \mathbb{R}^d$ be an irreducible variety of dimension $d' \geq 1$ and complexity k. Let $\mathcal{P}_1, \ldots, \mathcal{P}_t \subset U$ be finite sets. Then there exists $f \in \mathbb{R}[x_1, \ldots, x_d]$ of degree $O_{d,k}(t^{1/d'})$ such that $f \notin \mathbf{I}(U)$ and f simultaneously bisects all the sets $\mathcal{P}_j$.*

Proof This proof is a variant of the proof of Theorem 3.6, combined with Hilbert polynomials. We set $J = \mathbf{I}(U)$. Let d_m be the dimension of the vector space $\mathbb{R}[x_1, \ldots, x_d]_{\leq m}/J_{\leq m}$. By Equation (11.6), for every $m > m_J$ we have that $d_m = \Theta_{d,k}(m^{d'})$. We fix the value of m to be the minimum integer that satisfies $d_m \geq t$. If $m > m_J$ then

$$t = \Theta_{d,k}(m^{d'}), \qquad \text{or equivalently} \qquad m = \Theta_{d,k}(t^{1/d'}).$$

Lemma 11.2 implies that $m_J = O_{d,k}(1)$. Thus, when $m \leq m_J = O_{d,k}(1)$, we get that $m = O_{d,k}(t^{1/d'})$ by taking the constant of the $O(\cdot)$-notation to be sufficiently large.

Let $\phi_1, \ldots, \phi_{d_m}$ be a basis for the vector space $\mathbb{R}[x_1, \ldots, x_d]_{\leq m}/J_{\leq m}$. We consider the map $\phi : \mathbb{R}^d \to \mathbb{R}^{d_m}$ defined by

$$\phi(x_1, \ldots, x_d) = (\phi_1(x_1, \ldots, x_d), \ldots, \phi_{d_m}(x_1, \ldots, x_d)).$$

For every $1 \leq j \leq t$, let $\mathcal{P}'_j = \phi(\mathcal{P}_j) \subset \mathbb{R}^{d_m}$. For any points $p_1, p_2 \in U$, the quotient $\mathbb{R}[x_1, \ldots, x_d]_{\leq m}/J_{\leq m}$ contains a polynomial that vanishes on p_1 but not on p_2. For example, a generic linear polynomial that vanishes on p_1 is not in $J_{\leq m}$ and does not vanish on p_2. This implies that ϕ is injective on $U = \mathbf{V}(J)$, which in turn implies that $|\mathcal{P}'_j| = |\mathcal{P}_j|$. Since $d_m \geq t$, the ham sandwich theorem (Theorem 3.5) states that there exists a hyperplane $\Pi \subset \mathbb{R}^{d_m}$ that simultaneously bisects each $\mathcal{P}'_j$. We write the coordinates of $\mathbb{R}^{d_m}$ as $y_1, \ldots, y_{d_m}$. There exist $a_1, \ldots, a_{d_m} \in \mathbb{R}$ that satisfy $\Pi = \mathbf{V}(a_1 y_1 + \cdots + a_{d_m} y_{d_m})$. That is, for each $1 \leq j \leq t$, we have

$$\left| \left\{ y \in \mathcal{P}'_j \ : \ a_1 y_1 + \cdots + a_{d_m} y_{d_m} > 0 \right\} \right| \leq |\mathcal{P}'_j|/2,$$

and

$$\left| \left\{ y \in \mathcal{P}'_j \ : \ a_1 y_1 + \cdots + a_{d_m} y_{d_m} < 0 \right\} \right| \leq |\mathcal{P}'_j|/2.$$

We set $f = a_1 \phi_1 + \cdots + a_{d_m} \phi_{d_m} \in \mathbb{R}[x_1, \ldots, x_d]$. By the definition of the basis elements ϕ_j, the polynomial f is of degree at most m and is not in J. For a point $p \in \mathbb{R}^d$, we note that

$$f(p) > 0 \quad \text{if and only if} \quad a_1 \phi_1(p) + \cdots + a_{d_m} \phi_{d_m}(p) > 0.$$

Thus, for each $1 \leq j \leq d_m$, we have that

$$\left| \left\{ p \in \mathcal{P}_j \ : \ f(p) > 0 \right\} \right| \leq |\mathcal{P}_j|/2 \quad \text{and} \quad \left| \left\{ p \in \mathcal{P}_j \ : \ f(p) < 0 \right\} \right| \leq |\mathcal{P}_j|/2.$$

In other words, f simultaneously bisects every $\mathcal{P}_j$. $\qquad\square$

The main difference between the proofs of Theorem 3.6 and Lemma 11.4 is that the Veronese map is replaced with a basis for $\mathbb{R}[x_1, \ldots, x_d]_{\leq m}/J_{\leq m}$.

The reason for this change is that the only polynomial in $\mathbb{R}[x_1, \ldots, x_d]_{\leq m}/J_{\leq m}$ that vanishes on U is 0. This guarantees that our partitioning polynomial would not vanish on U.

The proof of Lemma 11.4 is a generalization of the proof of Theorem 3.6. Indeed, Lemma 11.4 implies Theorem 3.6 when setting $U = \mathbb{R}^d$. We then have that $J = \mathbf{I}(U) = \{0\}$, which leads to $\mathbb{R}[x_1, \ldots, x_d]_{\leq m}/J_{\leq m} = \mathbb{R}[x_1, \ldots, x_d]_{\leq m}$. By taking the basis of $\mathbb{R}[x_1, \ldots, x_d]_{\leq m}$ that consists of all monomials of degree at most m, we obtain that ϕ is a Veronese map. Then, the two proofs are identical.

We are now ready to state our polynomial partitioning theorem for points on varieties.

Theorem 11.5   *Let* $U \subset \mathbb{R}^d$ *be an irreducible variety of dimension* d' *and complexity* k. *Let* $\mathcal{P} \subset U$ *be a set of* m *points. Then there exists an* r-*partitioning polynomial* f *of* $\mathcal{P}$ *such that* $f \notin \mathbf{I}(U)$ *and* $\deg f = O_{d,k}(r^{d/d'})$.

Proof   This proof is almost identical to the proof of Theorem 3.1. Let $c_{d,k}$ be the hidden constant in the $O(\cdot)$-notation of the bound on $\deg f$ in the statement of Lemma 11.4. That is, applying Lemma 11.4 on t sets of points on U leads to a polynomial of degree at most $c_{d,k} t^{1/d'}$. Let $J = \mathbf{I}(U)$.

We show that there exists a sequence of polynomials $f_0, f_1, f_2, \ldots$ that satisfy the following:

(i)   The degree of f_j is smaller than $c_{d,k} 2^{(j+1)/d'}/(2^{1/d'} - 1)$.

(ii) $f_j \notin J$.

(iii)   Every connected component of $\mathbb{R}^d \backslash \mathbf{V}(f_j)$ contains at most $m/2^j$ points of $\mathcal{P}$.

To complete the proof, we then set $f = f_s$ with the minimum integer s that satisfies $2^s \geq r^d$. By property (i), we have that $\deg f = O(r^{d/d'})$. By property (iii), every cell of $\mathbb{R}^d \backslash \mathbf{V}(f)$ contains fewer than m/r^d points of $\mathcal{P}$.

We prove the existence of f_j by induction on j. For the induction basis we may take $f_0 = 1$. For the induction step, we assume that f_j exists and prove that f_{j+1} also exists. We have that $\deg f_j \leq c_{d,k} 2^{(j+1)/d'}/(2^{1/d'} - 1)$, that $f_j \notin J$, and that every connected component of $\mathbb{R}^d \backslash \mathbf{V}(f_j)$ contains at most $m/2^j$ points of $\mathcal{P}$. Let t be the number of connected components of $\mathbb{R}^d \backslash \mathbf{V}(f_j)$ that contain more than $m/2^{j+1}$ points of $\mathcal{P}$. We denote the subsets of $\mathcal{P}$ in these connected components as $\mathcal{P}_1, \ldots, \mathcal{P}_t$. Since $|\mathcal{P}| = m$, we get that $t < 2^{j+1}$. By Lemma 11.4, there is a polynomial g_j of degree smaller than $c_d 2^{(j+1)/d'}$ that simultaneously bisects every $\mathcal{P}_j$.

We set $f_{j+1} = f_j \cdot g_j$. Every connected component of $\mathbb{R}^d \backslash \mathbf{V}(f_j \cdot g_j)$ contains at most $m/2^{j+1}$ points of $\mathcal{P}$. We also have that

$$\deg f_{j+1} = \deg f_j + \deg g_j < \frac{c_d 2^{(j+1)/d'}}{2^{1/d'} - 1} + c_d 2^{(j+1)/d'}$$

$$= c_d 2^{(j+1)/d} \cdot \left(\frac{1}{2^{1/d'} - 1} + 1 \right)$$

$$= \frac{c_d 2^{(j+2)/d'}}{2^{1/d'} - 1}.$$

Since $f_j \notin J$ and $g_j \notin J$, we get that $f_{j+1} \notin J$. This completes the induction step and thus the proof of the theorem. $\square$

We are now ready to prove a general incidence bound in $\mathbb{R}^d$. Theorem 11.3 is obtained from the following theorem by setting $U = \mathbb{R}^d$.

Theorem 11.6 *Let $U \subset \mathbb{R}^d$ be an irreducible variety of dimension d' and complexity at most k. Let $\mathcal{P} \subset U$ be a set of m points and let $\mathcal{V}$ be a set of n varieties of complexity at most k in $\mathbb{R}^d$, such that the incidence graph of $\mathcal{P} \times \mathcal{V}$ contains no $K_{s,t}$. In addition, no variety of $\mathcal{V}$ contains U. Then for every $\varepsilon > 0$, we have that*

$$I(\mathcal{P}, \mathcal{V}) = O_{k,s,t,d,\varepsilon} \left(m^{\frac{(d'-1)s}{d's-1} + \varepsilon} n^{\frac{d'(s-1)}{d's-1}} + m + n \right).$$

Theorem 11.3 does not require that the varieties of $\mathcal{V}$ do not contain U. Since the incidence graph contains no $K_{s,t}$, at most $t - 1$ such varieties contain U. These varieties contribute at most $(t - 1)m$ incidences, which does not change the bound.

Proof of Theorem 11.6 Let $\alpha_{1,d',k}$ and $\alpha_{2,d',k}$ be sufficiently large constants that depend on s, t, d, k, ε, and d'. The subscripts of $\alpha_{1,d',k}$ and $\alpha_{2,d',k}$ explicitly mention only d' and k, to make some of the following computations clearer. The hidden constants in the $O(\cdot)$-notations throughout the proof may also depend on s, t, d, k, and ε. For brevity we write $O(\cdot)$ instead of $O_{s,t,d,k,\varepsilon}(\cdot)$. To prove the theorem, we prove by induction on d' that

$$I(\mathcal{P}, \mathcal{V}) \le \alpha_{1,d',k} m^{\frac{(d'-1)s}{d's-1} + \varepsilon} n^{\frac{d'(s-1)}{d's-1}} + \alpha_{2,d',k}(m + n). \tag{11.7}$$

For the induction basis, we consider the case of $d' = 1$. Since no variety of $\mathcal{V}$ contains U, every such variety has a finite intersection with U. By the Milnor–Thom theorem (Theorem 4.9), every such variety intersects U in $O(1)$ points and contributes $O(1)$ incidences. We conclude that $I(\mathcal{P}, \mathcal{V}) = O(m+n)$, so the claim holds when $\alpha_{2,1,k}$ is sufficiently large.

For the induction step, consider $d' > 1$ and assume that the theorem holds for all smaller values of d'. We prove the induction step with a second induction on $m + n$. That is, we prove Inequality (11.7) for a fixed $d' > 1$ by induction on the size of $\mathcal{P}$ and $\mathcal{V}$. For the induction basis, when $m + n \le 100$, we have that $I(\mathcal{P}, \mathcal{V}) \le mn < 10^4$. We then obtain Inequality (11.7) by taking $\alpha_{2,d',k} \ge 10^4$.

We move to the induction step of the induction on $m + n$. Since the incidence graph contains no $K_{s,t}$, Theorem 8.1 implies $I(\mathcal{P}, \mathcal{V}) = O(mn^{1-1/s})$. When $m = O(n^{1/s})$, we get that $I(\mathcal{P}, \mathcal{V}) = O(n)$. We may thus assume that

$$n = O(m^s). \tag{11.8}$$

This in turn implies that

$$n = n^{\frac{d'-1}{d's-1}} n^{\frac{d'(s-1)}{d's-1}} = O\left(m^{\frac{(d'-1)s}{d's-1}} n^{\frac{d'(s-1)}{d's-1}} \right). \tag{11.9}$$

Partitioning U: Let r be a sufficiently large constant. We apply Theorem 11.5 on U to obtain an r-partitioning polynomial $f \notin \mathbf{I}(U)$ for $\mathcal{P}$. By that theorem $\deg f = O(r^{d/d'})$. The asymptotic relations between the constants in this proof are

$$d, k, s, t, 2^{1/\varepsilon} \ll r \ll \alpha_{2,d',k} \ll \alpha_{1,d',k}.$$

We recall that the cells of the partition are the connected components of $U \backslash \mathbf{V}(f)$. By Theorem 4.15, the number of cells is $c = O((r^{d/d'})^{d'}) = O(r^d)$. We denote the cells of the partition as $C_1, \ldots, C_c$. For each $1 \le j \le c$, let $\mathcal{V}_j$ be the set of varieties of $\mathcal{V}$ that intersect C_j and let $\mathcal{P}_j = C_j \cap \mathcal{P}$. We also set $m_j = |\mathcal{P}_j|$, $m' = \sum_{j=1}^{c} m_j$, and $n_j = |\mathcal{V}_j|$. Note that $m_j \le m/r^d$ for every $1 \le j \le c$. For each $W \in \mathcal{V}$, since $\dim(W \cap U) \le d' - 1$, Theorem 4.15 implies that W intersects $O(r^{d(d'-1)/d'})$ cells. This implies that $\sum_{j=1}^{c} n_j = O(nr^{d(d'-1)/d'})$. By Hölder's inequality, we have that

$$\sum_{j=1}^{c} n_j^{\frac{d'(s-1)}{d's-1}} \le \left(\sum_{j=1}^{c} n_j \right)^{\frac{d'(s-1)}{d's-1}} \left(\sum_{j=1}^{c} 1 \right)^{\frac{d'-1}{d's-1}} = O\left(\left(nr^{d(d'-1)/d'} \right)^{\frac{d'(s-1)}{d's-1}} r^{\frac{d(d'-1)}{d's-1}} \right)$$

$$= O\left(n^{\frac{d'(s-1)}{d's-1}} r^{\frac{ds(d'-1)}{d's-1}} \right).$$

Applying the hypothesis of the second induction separately in each cell leads to

$$\sum_{j=1}^{c} I(\mathcal{P}_j, \mathcal{V}_j) \le \sum_{j=1}^{c} \left(\alpha_{1,d',k} m_j^{\frac{(d'-1)s}{d's-1}+\varepsilon} n_j^{\frac{d'(s-1)}{d's-1}} + \alpha_{2,d',k}(m_j + n_j) \right)$$

$$\le \alpha_{1,d',k} \frac{m^{\frac{(d'-1)s}{d's-1}+\varepsilon}}{r^{\frac{ds(d'-1)}{d's-1}+d\varepsilon}} \sum_{j=1}^{c} n_j^{\frac{d'(s-1)}{d's-1}} + \sum_{j=1}^{c} \alpha_{2,d',k}(m_j + n_j)$$

$$= O\left(\alpha_{1,d',k} \frac{m^{\frac{(d'-1)s}{d's-1}+\varepsilon} n^{\frac{d'(s-1)}{d's-1}}}{r^{d\varepsilon}} \right) + \alpha_{2,d',k} \left(m' + O\left(nr^{d(d'-1)/d'} \right) \right).$$

By applying Equation (11.9) and taking $\alpha_{1,d',k}$ to be sufficiently large with respect to r and to $\alpha_{2,d',k}$, we obtain that

$$\sum_{j=1}^{c} I(\mathcal{P}_j, \mathcal{V}_j) = O\left(\alpha_{1,d',k} \frac{m^{\frac{(d'-1)s}{d's-1}+\varepsilon} n^{\frac{d'(s-1)}{d's-1}}}{r^{d\varepsilon}}\right) + \alpha_{2,d',k} m'.$$

When r is sufficiently large with respect to ε and to the constant hidden in the $O(\cdot)$-notation, we have that

$$\sum_{j=1}^{c} I(\mathcal{P}_j, \mathcal{V}_j) \leq \frac{\alpha_{1,d',k}}{2} m^{\frac{(d'-1)s}{d's-1}+\varepsilon} n^{\frac{d'(s-1)}{d's-1}} + \alpha_{2,d',k} m'. \tag{11.10}$$

Incidences on the partition: We set $U_0 = U \cap \mathbf{V}(f)$, $\mathcal{P}_0 = \mathcal{P} \cap U_0$, and $m_0 = |\mathcal{P}_0| = m - m'$. Let $\mathcal{V}_0$ denote the set of varieties of $\mathcal{V}$ that are contained in $\mathbf{V}(f)$. Since U is an irreducible variety and $U \not\subseteq \mathbf{I}(f)$, we get that U_0 is a variety of dimension at most $d' - 1$ and of complexity $O(r^{d/d'})$. We denote this complexity as k_0.

It remains to bound incidences with the points of $\mathcal{P}_0$. We partition these incidences into two types:

- Let I_1 be the number of incidences $(p, W) \in \mathcal{P}_0 \times \mathcal{V}$ such that there exists an irreducible component of U_0 contains p and is contained in W.
- Let I_2 be the number of incidences $(p, W) \in \mathcal{P}_0 \times \mathcal{V}$ such that no irreducible component of U_0 contains p and is contained in W.

We note that $I(\mathcal{P}_0, \mathcal{V}) = I_1 + I_2$.

We first derive an upper bound for I_1. By Lemma 4.10, the number of irreducible components of U_0 is $O_r(1)$. The complexity of each such component is at most $k_0 = O(r^{d/d'})$.

We consider an irreducible component A of U_0. If A contains at most $s - 1$ points of $\mathcal{P}_0$, then these points contribute at most $(s - 1)n$ incidences to I_1. If A contains at least s points of $\mathcal{P}_0$ then at most $t - 1$ varieties of $\mathcal{V}$ contain A. In this case, the points in A contribute at most $(t - 1)m_0$ incidences to I_1. By summing these bounds over every irreducible component of U_0, choosing sufficiently large $\alpha_{1,d',k}$ and $\alpha_{2,d',k}$, and recalling Equation (11.9), we obtain that

$$I_1 = O_r(n + m_0) < \frac{\alpha_{1,d',k}}{4} m^{\frac{s(d'-1)}{d's-1}} n^{\frac{d'(s-1)}{d's-1}} + \frac{\alpha_{2,d',k}}{2} m_0. \tag{11.11}$$

We next derive an upper bound for I_2. Since U_0 is of dimension at most $d' - 1$, we can apply the first induction hypothesis with each irreducible component A of U_0. Specifically, we apply the hypothesis in A with the point set $\mathcal{P}_0 \cap A$ and the set of varieties of $\mathcal{V}$ that do not contain A. Since U_0 has $O_r(1)$ irreducible components and each has complexity at most k_0, we get that

$$I_2 = O_r\left(\alpha_{1,d'-1,k_0} m_0^{\frac{s(d'-2)}{(d'-1)s-1}+\varepsilon} n^{\frac{(d'-1)(s-1)}{(d'-1)s-1}} + \alpha_{2,d'-1,k_0}(m_0+n)\right). \qquad (11.12)$$

By Equation (11.8) we have that

$$m_0^{\frac{s(d'-2)}{(d'-1)s-1}+\varepsilon} n^{\frac{(d'-1)(s-1)}{(d'-1)s-1}} \le m^{\frac{s(d'-2)}{(d'-1)s-1}+\varepsilon} n^{\frac{d'(s-1)}{d's-1}} n^{\frac{s-1}{(d's-s-1)(d's-1)}}$$

$$= O\left(m^{\frac{s(d'-2)}{(d'-1)s-1}+\varepsilon} n^{\frac{d'(s-1)}{d's-1}} m^{\frac{s(s-1)}{(d's-s-1)(d's-1)}}\right)$$

$$= O\left(m^{\frac{s(d'-1)}{d's-1}+\varepsilon} n^{\frac{d'(s-1)}{d's-1}}\right).$$

Combining the above with Equations (11.9) and (11.12), and taking $\alpha_{1,d',k}$ and $\alpha_{2,d',k}$ to be sufficiently large with respect to $\alpha_{1,d'-1,k_0}$ and $\alpha_{2,d'-1,k_0}$, leads to

$$|I_2| \le \frac{\alpha_{1,d',k}}{4} m^{\frac{s(d'-1)}{d's-1}+\varepsilon} n^{\frac{d'(s-1)}{d's-1}} + \frac{\alpha_{2,d',k}}{2} m_0. \qquad (11.13)$$

Combining Inequalities (11.10), (11.11), and (11.13) completes the induction step and thus the proof of the theorem. $\qquad \square$

11.4 Exercises

Exercise 11.1   Let $\ell \subset \mathbb{R}^2$ be a line.
(a) Find all the cases in which $\mathbf{I}(\ell)$ is a monomial ideal.
(b) Find the dimension of the vector space $\mathbf{I}(\ell)_{\le m}$ (as a function of m).

Exercise 11.2
(a) Find the Hilbert polynomial and the regularity of the ideal $\langle x^2 \rangle \subset \mathbb{R}[x, y]$.

(b) Find the Hilbert polynomial and the regularity of the ideal $\langle x^3, x^2 y \rangle \subset \mathbb{R}[x, y]$.

Exercise 11.3   Find the Hilbert polynomial and the regularity of the ideal $\mathbf{I}(z) \subset \mathbb{R}[x, y, z]$. In other words, find the Hilbert polynomial and the regularity of the plane $\mathbf{V}(z) \subset \mathbb{R}^3$.

Exercise 11.4   Let $U \subset \mathbb{R}^d$ be a variety. Prove that the coefficient of the leading term $m^{\dim U}$ of $H_{\mathbf{I}(U)}(m)$ is positive.

Exercise 11.5 Let $\mathcal{P}$ be a set of m points and let $\mathcal{V}$ be a set of n varieties of complexity at most k and dimension at most $d' < d$, both in $\mathbb{R}^d$. The incidence graph of $\mathcal{P} \times \mathcal{V}$ contains no $K_{s,t}$. Prove that for every $\varepsilon > 0$, we have that

$$I(\mathcal{P}, \mathcal{V}) = O_{k,s,t,d,\varepsilon} \left(m^{\frac{d's}{d's+s-1}+\varepsilon} n^{\frac{(d'+1)(s-1)}{d's+s-1}} + m + n \right).$$

(Hint: This is a simple corollary of Theorem 11.3. It does not require understanding the proof of that theorem.)

Exercise 11.6 Explain why the proof of Lemma 11.4 is no longer valid when $\mathcal{P}_1, \ldots, \mathcal{P}_t$ are unrestricted point sets in $\mathbb{R}^d$. That is, when these point sets are no longer required to be on U.

Exercise 11.7 Change the proof of Theorem 11.6 so that it would show the dependency of the bound in t. In particular, prove that

$$I(\mathcal{P}, \mathcal{V}) = O_{k,s,d,\varepsilon} \left(m^{\frac{(d'-1)s}{d's-1}+\varepsilon} n^{\frac{d'(s-1)}{d's-1}} t^{\frac{d'-1}{d's-1}} + mt + n \right).$$

Exercise 11.8 (Kaplan et al., 2012) In this exercise we provide an alternative proof for Theorem 11.5, for the special case where $d = 3$ and $d' = 2$. This alternative proof is more elementary, since it does not rely on Hilbert polynomials. It also provides a bound with an explicit dependency on $\deg U$.

(a) Prove a variant of Lemma 11.4 for the case where $U \subset \mathbb{R}^3$ is an irreducible surface of degree k. If $t < (k/3)^3$, then we require that $\deg f = O(t^{1/3})$. If $t \geq (k/3)^3$, then we require that $\deg f = O((t/k)^{1/2})$. (Hint: Find a basis for a vector space of polynomials that are not divisible by f, as follows. Let $x^a y^b z^c$ be a degree k term of f. The basis should consist of monomials that are not divisible by $x^a y^b z^c$.)

(b) Prove a variant of Theorem 11.5 for the case where $U \subset \mathbb{R}^3$ is an irreducible surface of degree k. We require that $\deg f = O(r^{3/2}/k^{1/2} + k)$. (Hint: In the proof of Theorem 11.5, replace Lemma 11.4 with part (a) of the current exercise.)

Exercise 11.9 Let $\mathcal{P}$ be a set of m points and let H be a set of n planes, both in $\mathbb{R}^3$. Assume that the incidence graph of $\mathcal{P} \times H$ contains no $K_{2,2}$. In Exercise 8.4, we proved that $I(\mathcal{P}, H) = O(m^{4/5+\varepsilon} n^{3/5} + m + n)$, for every $\varepsilon > 0$. You are now asked to provide a different proof, which also removes the ε from the exponent of m. To do that, use a nonconstant-degree polynomial partitioning. To handle incidences on the partition, rely on the result of Exercise 11.8(b). There is no need to understand the proof of that exercise. (Warning: The answer to this exercise is quite long.)

Exercise 11.10 Let $U \subset \mathbb{R}^3$ be an irreducible variety of degree k. Let $\mathcal{P} \subset U$ be a set of m points. Prove that there exists a one-dimensional variety of complexity $O(m^{1/2}/k^{1/2})$ that contains $\mathcal{P}$. (Hint: Combine the proof of Lemma 5.4 with the basis from Exercise 11.8(a).)

11.5 Open Problems

The first 11 chapters of this book contain a wide variety of incidence results in $\mathbb{R}^d$. It might seem as if these are unrelated: Each result relies on different assumptions, applies different tools, and consists of a different bound. For a while, these were indeed treated as a collection of unrelated results. However, one might say that a general incidence theory is beginning to emerge in $\mathbb{R}^d$.

Let $\mathcal{P}$ be a set of m points and let $\mathcal{V}$ be a set of n varieties, both in $\mathbb{R}^d$. Assume that the incidence graph of $\mathcal{P} \times \mathcal{V}$ contains no $K_{s,t}$. We saw several bounds that fit this scenario:

- Theorem 11.3 is known to be tight in some cases where the varieties are $(d-1)$-dimensional. In these cases, the theorem leads to the bound

$$I(\mathcal{P}, \mathcal{V}) = O\left(m^{\frac{(d-1)s}{ds-1}+\varepsilon} n^{\frac{ds-d}{ds-1}} + m + n \right).$$

- Theorem 8.6 is known to be tight in some cases where the varieties are $(d/2)$-dimensional. In these cases, the theorem leads to the bound

$$I(\mathcal{P}, \mathcal{V}) = O(m^{2/3+\varepsilon} n^{2/3} + m + n).$$

- It is straightforward to extend Theorem 11.1 to incidences with curves in $\mathbb{R}^d$ (see Sharir et al., 2016). This more general result assumes that every constant-degree variety contains a bounded number of the curves. It leads to the bound

$$I(\mathcal{P}, \mathcal{V}) = O\left(m^{\frac{s}{ds-d+1}+\varepsilon} n^{\frac{ds-d}{ds-d+1}} + m + n \right).$$

At first, the three above results may seem unrelated. However, they all lead to the bound

$$I(\mathcal{P}, \mathcal{V}) = O\left(m^{\frac{sd'}{ds-d+d'}+\varepsilon} n^{\frac{ds-d}{ds-d+d'}} + m + n \right), \tag{11.14}$$

where d' is the dimension of the varieties. It is known that Equation (11.14) holds for every d' under certain assumptions (see Do and Sheffer, 2021). This bound is also known to be tight for every d' in some specific cases.

The above could be seen as a first step towards a unified incidence theory in $\mathbb{R}^d$. The author of this book believes that such a unified theory exists and that uncovering it could lead to significant progress. This belief leads to the following vague open problem.

Open Problem 11.7 *Continue to expose the unified incidence theory in $\mathbb{R}^d$, or prove that one does not exist.*

The bound of Theorem 11.3 is known to be tight up to an ε in the exponent when $s = 2$, for many types of hypersurfaces (see Do and Sheffer, 2021; Sheffer, 2016). This bound is not tight when the varieties are not hypersurfaces (see Exercise 11.5). It is not tight in $\mathbb{R}^2$ when $s \geq 3$ (see Section 3.6) and in some cases in $\mathbb{R}^d$ when s is larger (see Section 14.2). When $s \geq 3$, $d \geq 3$, and the varieties are hypersurfaces, we do not know whether the bound of Theorem 11.3 is close to being tight.

Open Problem 11.8 *Determine when the bound of Theorem 11.3 is tight up to an ε when $s \geq 3$, $d \geq 3$, and the varieties are hypersurfaces. That is, either find constructions that provide matching lower bounds or derive improved upper bounds.*

12

Incidence Applications in $\mathbb{R}^d$

The older I get, the more I believe that at the bottom of most deep mathematical problems there is a combinatorial problem.
Attributed to Israel Gelfand.

In Chapter 11 we derived general point-variety incidence bounds in $\mathbb{R}^d$. We now study two applications of these bounds. These applications do not require reading any part of Chapter 11, except for the statement of Theorem 11.3.

The first application comes from discrete geometry, and is another distinct distances problem. The second application comes from a discrete Fourier restriction problem in harmonic analysis. For simplicity, we only discuss the combinatorial aspect of that problem.

12.1 Distinct Distances with Local Properties

Erdős (1985) posed the following family of distinct distances problems. Let n be asymptotically large and let k and ℓ be fixed constants. We consider sets of n points in $\mathbb{R}^2$ such that every k points span at least ℓ distinct distances. For example, when $k = \ell = 3$, we consider sets where every three points span three distinct distances. In other words, sets that do not span isosceles and equilateral triangles. Let $\phi(n, k, \ell)$ denote the minimum number of distinct distances that can be spanned by such a point set. Intuitively, we study how a local property of small subsets affects a global property of the entire set.

As a first example, we consider $\phi(n, 3, 3)$. This is the minimum number of distinct distances that are spanned by n points that do not form isosceles and equilateral triangles. By definition, the sets cannot form isosceles triangles with three collinear vertices either. Let $\mathcal{P}$ be such a point set and let $p \in \mathcal{P}$. Since there are no isosceles triangles, p cannot have the same distance from two points of $\mathcal{P} \setminus \{p\}$. This implies that $\phi(n, 3, 3) \geq n - 1$.

• •   • •   • •   • •
(1, 0) (2, 0)   (4, 0) (5, 0)   (10, 0) (11, 0) (12, 0) (13, 0)

Figure 12.1 Eight points that span few distinct distances, while every three points span three distinct distances.

Behrend (1946) constructed a set A of positive integers $a_1 < a_2 < \cdots < a_n$ such that

• the set A contains no 3-term arithmetic progression, and
• $a_n < n2^{O\left(\sqrt{\log n}\right)}$.

We define the point set $\mathcal{P} = \{(a_1, 0), (a_2, 0), \ldots, (a_n, 0)\}$. See Figure 12.1 for an example. Since A does not contain 3-term arithmetic progressions, the set $\mathcal{P}$ does not span isosceles and equilateral triangles. The distances that are spanned by $\mathcal{P}$ are positive integers of size smaller than $n2^{O\left(\sqrt{\log n}\right)}$. This implies that

$$\phi(n, 3, 3) \leq D(\mathcal{P}) < n2^{O\left(\sqrt{\log n}\right)}.$$

Closing the small gap between $\phi(n, 3, 3) = \Omega(n)$ and $\phi(n, 3, 3) < n2^{O\left(\sqrt{\log n}\right)}$ is an open problem.

We move to a more general case, considering bounds that hold for every sufficiently large k. As a first trivial bound, we note that $\phi\left(n, k, \binom{k}{2}\right) = \Theta(n^2)$ for every $k \geq 4$. Indeed, k points span at most $\binom{k}{2}$ distinct distances, so no distance repetitions are allowed in this case. It is not difficult to prove that $\phi\left(n, k, \binom{k}{2} - \lfloor k/2 \rfloor + 2\right) = \Omega(n^2)$, for every $k \geq 4$ (Exercise 12.2). That is, when decreasing the value of ℓ by about $k/2$, the global number of distinct distances is still asymptotically maximal.

We obtain a more difficult problem when ℓ is smaller than the maximal $\binom{k}{2}$ by about k. That is, when considering the value of $\phi\left(n, k, \binom{k}{2} - k + c\right)$, for some small constant c. We begin to study this case with the following warm-up claim.

Claim 12.1 *For every $k \geq 4$, we have that $\phi\left(n, k, \binom{k}{2} - k + 3\right) = \Omega(n)$.*

Proof Let $\mathcal{P} \subset \mathbb{R}^2$ be a set of n points, such that every k points of $\mathcal{P}$ span at least $\binom{k}{2} - k + 3$ distinct distances. We arbitrarily fix a point $p \in \mathcal{P}$. Assume for contradiction that there exists a distance δ such that p is at distance δ from at least $k - 1$ points of $\mathcal{P} \backslash \{p\}$. Let $\mathcal{P}' \subset \mathcal{P}$ consist of p and of $k - 1$ points at a distance of δ from p. Then $\mathcal{P}'$ is a set of k points that determine at most $\binom{k}{2} - k + 2$ distinct distances. This contradicts the assumption on $\mathcal{P}$, so p can have the same distance from at most $k - 2$ points of $\mathcal{P}$. Since k is a constant, we get that

$$\phi\left(n, k, \binom{k}{2} - k + 3\right) \geq \frac{n}{k-2} = \Omega(n).$$ □

We now derive a stronger bound by relying on incidences in $\mathbb{R}^4$. The following theorem is a variant of a result of Fox et al. (2018).

Theorem 12.2 *For any* $k \geq 6$ *and* $\varepsilon > 0$,

$$\phi\left(n, k, \binom{k}{2} - k + 5\right) = \Omega_{k,\varepsilon}\left(n^{8/7-\varepsilon}\right).$$

Proof Let $\mathcal{P}$ be a set of n points in $\mathbb{R}^2$ such that every k points of $\mathcal{P}$ span at least $\binom{k}{2} - k + 5$ distinct distances. We begin, as in Section 7.2, by setting

$$Q = \left\{(a, p, b, q) \in \mathcal{P}^4 \ : \ |ap| = |bq|\right\}.$$

(Recall that $|ap|$ denotes the distance between the points a and p.) The quadruples are ordered, so (a, p, b, q) and (p, a, b, q) are distinct quadruples of Q.

We denote the number of distinct distances that are spanned by $\mathcal{P}$ as x. We denote these distances as $\delta_1, \ldots, \delta_x$. For $1 \leq j \leq x$, let

$$E_j = \left\{(a, b) \in \mathcal{P}^2 \ : \ |ab| = \delta_j\right\}.$$

Since every ordered pair of $\mathcal{P}^2$ appears in exactly one E_j, we have that $\sum_{j=1}^{x} |E_j| = n^2$. The number of quadruples in Q that satisfy $|ap| = |bq| = \delta_j$ is $|E_j|^2$. Combining this with the Cauchy–Schwarz inequality (Theorem A.1) leads to

$$|Q| = \sum_{j=1}^{x} |E_j|^2 \geq \frac{\left(\sum_{j=1}^{x} |E_j|\right)^2}{x} = \frac{n^4}{x}. \tag{12.1}$$

To complete the proof, it remains to derive an upper bound for $|Q|$. We reduce this problem to an incidence problem in $\mathbb{R}^4$, as follows. For $a, b \in \mathbb{R}^2$, we define the point $v_{a,b} = (a_x, a_y, b_x, b_y) \in \mathbb{R}^4$. We consider the point set

$$\mathcal{P}_4 = \{v_{a,b} \ : \ a, b \in \mathcal{P}\} \subset \mathbb{R}^4.$$

For $p, q \in \mathbb{R}$, we define the variety

$$S_{p,q} = \mathbf{V}\left((x_1 - p_x)^2 + (x_2 - p_y)^2 - (x_3 - q_x)^2 - (x_4 - q_y)^2\right) \subset \mathbb{R}^4.$$

We also consider the set of varieties

$$\mathcal{V} = \{S_{p,q} \ : \ p, q \in \mathcal{P}\}.$$

We note that a point $v_{a,b}$ is incident to a variety $S_{p,q}$ if and only if

$$(a_x - p_x)^2 + (a_y - p_y)^2 = (b_x - q_x)^2 + (b_y - q_y)^2.$$

That is, $v_{a,b}$ is incident to $S_{p,q}$ if and only if $|ap| = |pq|$. This leads to a bijection between the incidences of $\mathcal{P}_4 \times \mathcal{V}$ and the quadruples of Q. Thus, to derive an upper bound on $|Q|$, it suffices to obtain an upper bound for $I(\mathcal{P}_4, \mathcal{V})$.

Assume that there is a copy of $K_{2,k-4}$ in the incidence graph of $\mathcal{P}_4 \times \mathcal{V}$. Then there exist points $a_1, b_1, a_2, b_2, p_1, q_1, \ldots, p_{(k-4)/2}, q_{(k-4)/2} \in \mathcal{P}$ such that $|a_j p_\ell| = |b_j q_\ell|$ for every $1 \le j \le 2$ and $1 \le \ell \le (k-4)/2$. These k points determine at most $\binom{k}{2} - k + 4$ distinct distances, which contradicts the assumption on $\mathcal{P}$. We conclude that the incidence graph of $\mathcal{P}_4 \times \mathcal{V}$ contains no $K_{2,k-4}$.

We note that $|\mathcal{P}_4| = n^2$ and that $|\mathcal{V}| = n^2$. By applying Theorem 11.3 on $\mathcal{P}_4$ and $\mathcal{V}$ with $d = 4$, $s = 2$, and $t = k - 4$, we obtain that

$$I(\mathcal{P}_4, \mathcal{V}) = O_{k,\varepsilon}\left((n^2)^{6/7+\varepsilon}(n^2)^{4/7} + n^2\right) = O_{k,\varepsilon}\left(n^{20/7+\varepsilon}\right).$$

Combining this with Equation (12.1) gives $x = \Omega_{k,\varepsilon}(n^{8/7-\varepsilon})$, which completes the proof. $\square$

12.2 Additive Energy on a Hypersphere

We now study an application of incidences from a harmonic analysis work of Bourgain and Demeter (2015). This is a discrete Fourier restriction problem. Since harmonic analysis is outside the scope of this book, we do not explain what restriction problems are. We only present the following discrete restriction problem in $\mathbb{R}^4$. For additional information, see for example Demeter (2020).

Given a set $\mathcal{P} \subset \mathbb{R}^d$, the *additive energy* of $\mathcal{P}$ is

$$E(\mathcal{P}) = \left|\left\{(a, b, p, q) \in \mathcal{P}^4 \ : \ a + b = p + q\right\}\right|.$$

The additive energy of a set $\mathcal{P}$ is a main object in additive combinatorics. It provides information about the additive structure of $\mathcal{P}$. For additional information, see Tao and Vu (2006). We study additive energy in more detail in Section 13.4.

We denote the coordinates of $\mathbb{R}^4$ as (x_1, x_2, x_3, x_4). We denote the coordinates of a point $p \in \mathbb{R}^4$ as $p = (p_1, p_2, p_3, p_4)$. For positive integers n and d, we consider the hypersphere

$$S_n^{(d)} = \mathbf{V}\left(x_1^2 + x_2^2 + \cdots + x_d^2 - n^2\right) \subset \mathbb{R}^d.$$

In other words, S_n is the hypersphere in R^d of radius n that is centered at the origin. The result of Bourgain and Demeter is about additive properties of sets of points on $S_n^{(4)}$. In particular, they study points that have only integer coordinates.

Theorem 12.3   *Let* $\mathcal{P} \subset S_n^{(4)} \cap \mathbb{Z}^4$. *Then for every* $\varepsilon > 0$ *we have that*

$$E(\mathcal{P}) = O_\varepsilon \left(|\mathcal{P}|^{7/3} n^\varepsilon \right).$$

To prove Theorem 12.3, we need to know how large $S_n^{(4)} \cap \mathbb{Z}^4$ can be. This question has a number-theoretic formulation: In how many ways can we write n^2 as a sum of four squares? The following theorem collects the known results in every dimension (see for example Grosswald, 1985; for part (d) see Sheffer, 2016).

Theorem 12.4   *The following hold for an absolute constant c and for every* $\varepsilon > 0$.

(a) $\left| S_n^{(2)} \cap \mathbb{Z}^2 \right| < n^{c/\log\log n}$.

(b) $\left| S_n^{(3)} \cap \mathbb{Z}^3 \right| = O(n^{1+\varepsilon})$.

(c) For every $d \geq 4$, we have that $\left| S_n^{(d)} \cap \mathbb{Z}^d \right| = O(n^{d-2})$.

(d) Consider hyperplanes $\Pi, \Pi' \subset \mathbb{R}^4$ that are defined by linear equations with integer coefficients of size $O(n)$. Set $\gamma = S_n^{(4)} \cap \Pi \cap \Pi' \subset \mathbb{R}^4$. Then $|\gamma \cap \mathbb{Z}^4| < n^{c/\log\log n}$.

We also need a variant of Theorem 11.3 that shows the dependency of the bound in t. To prove this variant, we only need to make minor changes to the proof of the original theorem. (Exercise 11.7 asks to prove this variant.)

Theorem 12.5   *Let $\mathcal{P}$ be a set of m points and let $\mathcal{V}$ be a set of n varieties of complexity at most k, both in $\mathbb{R}^d$. Assume that the incidences graph of $\mathcal{P} \times \mathcal{V}$ contains no $K_{s,t}$. Then for every* $\varepsilon > 0$, *we have that*

$$I(\mathcal{P}, \mathcal{V}) = O_{k,s,d,\varepsilon} \left(m^{\frac{(d-1)s}{ds-1}+\varepsilon} n^{\frac{d(s-1)}{ds-1}} t^{\frac{d-1}{ds-1}} + mt + n \right).$$

We are now ready to prove the result of Bourgain and Demeter.

Proof of Theorem 12.3   We set $m = |\mathcal{P}|$. For $v \in \mathbb{Z}^4$, we set

$$m_v = \left| \left\{ (p, q) \in \mathcal{P}^2 \ : \ v = p + q \right\} \right|.$$

In this above definition, the pairs (p, q) and (q, p) are considered as distinct, so they contribute 2 to m_v. Given a fixed $v \in \mathbb{Z}^4$, for every $p \in \mathcal{P}$ there is at most one $q \in \mathcal{P}$ such that $p + q = v$. This implies that $m_v \leq m$.

We note that the number of quadruples (a, b, p, q) that satisfy $a + b = p + q = v$ is m_v^2. For an integer $j \geq 0$, let $k_j = |\{v \in \mathbb{Z}^4 \ : \ m_v \geq 2^j\}|$. In other words, k_j

is the number of points in $\mathbb{Z}^4$ that can be written as a sum of two points of $\mathcal{P}$ in at least 2^j distinct ways. A dyadic decomposition argument leads to

$$E(\mathcal{P}) = \sum_{v \in \mathbb{Z}^4} m_v^2 = \sum_{j=0}^{\log m} \sum_{\substack{v \in \mathbb{Z}^4 \\ 2^j \le m_v < 2^{j+1}}} m_v^2 < \sum_{j=0}^{\log m} 2^{2j+2} k_j. \tag{12.2}$$

To derive an upper bound on $E(\mathcal{P})$, it remains to derive an upper bound on k_j. There are m^2 ordered pairs $(p, q) \in \mathcal{P}^2$ and each contributes to m_v for exactly one $v \in \mathbb{Z}^4$, so $\sum_{v \in \mathbb{Z}^4} m_v \le m^2$. Since every v that contributes to k_j has at least 2^j corresponding pairs in $\mathcal{P}^2$, we get that

$$k_j \le \frac{m^2}{2^j}. \tag{12.3}$$

To obtain the required upper bound for $E(\mathcal{P})$, we can plug Inequality (12.3) into Inequality (12.2) when $2^j = O(m^{1/3} n^\varepsilon)$. To handle larger values of j, we need to study the geometry of the problem.

From energy to hyperplanes: We denote the origin of $\mathbb{R}^4$ as o. The number of pairs $(p, q) \in \mathcal{P}$ that satisfy $p + q = o$ is at most m. Such pairs contribute at most m^2 to $E(\mathcal{P})$. We may thus assume that $v \ne o$ for the remainder of the proof.

When $v = p + p$, we have that $|ov| = 2n$. In this case, no other pair $(p', q') \in \mathcal{P}^2$ satisfies $p' + q' = v$. Such values of v contribute m to $E(\mathcal{P})$. We may thus ignore such values of v for the remainder of the proof.

Consider $p, q \in \mathcal{P}$ and let $v = p + q$. Note that v is in the two-dimensional plane spanned by p, q, and o. Since $p, q \in S_n^{(4)}$ we have that $|op| = |oq| = n$. This implies that the quadrilateral $vpqo$ is a rhombus of side length n. Figure 12.2 depicts two such cases.

Let S_v be the hypersphere of radius n in $\mathbb{R}^4$ that is centered at v. By the previous paragraph, points $p, q \in \mathcal{P}$ that satisfy $p + q = v$ are incident to S_v. Thus, every two points $p, q \in \mathcal{P}$ that satisfy $p + q = v$ are contained in the two-dimensional sphere $S_n^{(4)} \cap S_v$. We note that the sphere $S_n^{(4)} \cap S_v$ is centered

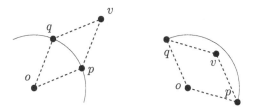

Figure 12.2 Two examples of a rhombus $vpqo$.

at $v/2$. Let Π_v be the unique hyperplane in $\mathbb{R}^4$ that contains the sphere $S_n^{(4)} \cap S_v$. By the above, if two points $p, q \in \mathcal{P}$ satisfy $p + q = v$ then p and q are incident to Π_v. This implies that

$$m_v \leq |\mathcal{P} \cap \Pi_v|. \tag{12.4}$$

The sphere $S_n^{(4)} \cap S_v$ is the set of points that are at a distance of n from both o and v. This implies that Π_v is the *perpendicular bisector* of o and v (the set of points in $\mathbb{R}^4$ that are at the same distance from o and from v). This means that no hyperplane Π_v is incident to o. Let H_j be the set of hyperplanes in $\mathbb{R}^4$ that contain at least 2^j points of $\mathcal{P}$ and are not incident to o. By Inequality (12.4), we get that $k_j \leq |H_j|$. Thus, we have reduced the problem to a point-hyperplane incidence problem in $\mathbb{R}^4$.

The incidence problem: We fix an integer $j \geq 0$ and consider $\Pi_v, \Pi_u \in H_j$, where $v \neq u$. The intersection $\Pi_v \cap \Pi_u \cap S_n^{(4)}$ is either a circle or a set of at most two points. Moreover, the absolute value of any coordinate of v is at most $2n$, since otherwise $m_v = 0$ (and symmetrically for u). We may thus apply Theorem 12.4(d), to obtain that $\Pi_v \cap \Pi_u \cap S_n^{(4)}$ contains $O(n^{c/\log\log n})$ points of $\mathbb{Z}^4$. Instead of the bound $O(n^{c/\log\log n})$, we use the weaker bound $O(n^{\varepsilon'})$, where $\varepsilon' = \varepsilon/100$. We conclude that the incidence graph of $\mathcal{P} \times H_j$ contains no copy of $K_{t,2}$ where $t = O(n^{\varepsilon'})$.

To replace the property "no $K_{t,2}$" with "no $K_{2,t}$," we move to a dual space, as follows. For a point $p \in \mathbb{R}^4$, we define the *dual hyperplane*

$$p^* = \mathbf{V}(p_1 x_1 + p_2 x_2 + p_3 x_3 + p_4 x_4 - 1) \subset \mathbb{R}^4.$$

We consider a hyperplane $\Pi \in H_j$ and recall that Π is not incident to the origin. Thus, we can write $\Pi = \mathbf{V}(a_1 x_1 + a_2 x_2 + a_3 x_3 + a_4 x_4 - 1)$ for some $a_1, a_2, a_3, a_4 \in \mathbb{R}$. We define the *dual point* of Π as $\Pi^* = (a_1, a_2, a_3, a_4) \in \mathbb{R}^4$.

We consider the set of planes

$$\mathcal{P}^* = \{p^* \ : \ p \in \mathcal{P}\},$$

and the set of points

$$H_j^* = \left\{\Pi^* \ : \ \Pi \in H_j\right\}.$$

Consider $p \in \mathcal{P}$ and $\Pi \in H_j$. We have that $p \in \Pi$ if and only if $\Pi^* \in p^*$, since both are equivalent to $a_1 p_1 + a_2 p_2 + a_3 p_3 + a_4 p_4 = 1$. This implies that $I(\mathcal{P}, H_j) = I(H_j^*, \mathcal{P}^*)$. In addition, since the incidence graph of $\mathcal{P} \times H_j$ contains no $K_{t,2}$, the incidence graph of $H_j^* \times \mathcal{P}^*$ contains no $K_{2,t}$. Applying Theorem 12.5 on H_j^* and $\mathcal{P}^*$ implies that

$$I(H_j^*, \mathcal{P}^*) = O_\varepsilon\left(|H_j|^{6/7+\varepsilon'} m^{4/7} n^{\varepsilon'} + |H_j| n^{\varepsilon'} + m\right).$$

Since each hyperplane of H_j is incident to at least 2^j points of $\mathcal{P}$, we have that

$$I(H_j^*, \mathcal{P}^*) = I(\mathcal{P}, H_j) \geq 2^j |H_j|.$$

Combining the two above bounds for $I(H_j^*, \mathcal{P}^*)$ leads to

$$2^j |H_j| = O_\varepsilon \left(|H_j|^{6/7 + \varepsilon'} m^{4/7} n^{\varepsilon'} + |H_j| n^{\varepsilon'} + m \right). \tag{12.5}$$

Recall that we already addressed the case where $2^j = O(m^{1/3} n^\varepsilon)$. We may thus assume that $2^j = \Omega(m^{1/3} n^\varepsilon)$. In this case, the right-hand side of Equation (12.5) cannot be dominated by the term $|H_j| n^{\varepsilon'}$. Removing the term $|H_j| n^{\varepsilon'}$ from this equation and rearranging gives

$$k_j \leq |H_j| = O_\varepsilon \left(\frac{m^{\frac{4}{1-7\varepsilon'}} n^{\frac{7\varepsilon'}{1-7\varepsilon'}}}{2^{\frac{7j}{1-7\varepsilon'}}} + \frac{m}{2^j} \right) = O_\varepsilon \left(\frac{m^{\frac{4}{1-7\varepsilon'}} n^{\frac{7\varepsilon'}{1-7\varepsilon'}}}{2^{7j}} + \frac{m}{2^j} \right).$$

The statement of the theorem is trivial when $\varepsilon \geq 1$. We may thus assume that $\varepsilon < 1$, which implies that $\varepsilon' < 1/100$. For such ε' we have that $\frac{1}{1-7\varepsilon'} < 1 + 10\varepsilon'$ and that $\frac{7\varepsilon'}{1-7\varepsilon'} < 8\varepsilon'$. This leads to

$$k_j = O_\varepsilon \left(\frac{m^{4+40\varepsilon'} n^{8\varepsilon'}}{2^{7j}} + \frac{m}{2^j} \right). \tag{12.6}$$

Completing the proof: We use Equations (12.3) and (12.6) to bound k_j in Inequality (12.2). Let $b = m^{1/3} n^\varepsilon$. When $2^j < b$, we apply Inequality (12.2). For larger values of j, we apply Equation (12.6). This gives

$$E(\mathcal{P}) = O_\varepsilon \left(\sum_{j=1}^{\log b} 2^j m^2 + \sum_{j=1+\log b}^{\log m} \left(\frac{m^{4+40\varepsilon'} n^{8\varepsilon'}}{2^{5j}} + 2^j m \right) \right)$$

$$= O_\varepsilon \left(m^{7/3} n^\varepsilon + \frac{m^{7/3 + 40\varepsilon'} n^{8\varepsilon'}}{n^{5\varepsilon}} + m^2 \right) = O_\varepsilon \left(m^{7/3} n^\varepsilon \right).$$

In the last transition, we used Theorem 12.4(c), which states that $m = O(n^2)$. □

12.3 Exercises

Exercise 12.1 Prove that $\phi(n, 5, 4) = \Omega(n)$.

Exercise 12.2 Prove that $\phi \left(n, k, \binom{k}{2} - \lfloor k/2 \rfloor + 2 \right) = \Theta(n^2)$ for every $k \geq 4$.

Exercise 12.3 Prove that $\phi \left(n, k, \binom{k}{2} - \lfloor 2k/3 \rfloor + 3 \right) = \Omega(n^{3/2})$ for every $k \geq 6$. Instructions:

- Consider a set $\mathcal{P}$ that satisfies the local restriction.
- Show that there exists a distance δ that is spanned by $\Omega(n^{1/2})$ pairs of points.
- Let $\mathcal{P}'$ be the set of points of $\mathcal{P}$ that are at distance δ from at least one point of $\mathcal{P}$. How large is $|\mathcal{P}'|$? How many times can a distance δ' repeat in $\mathcal{P}'$?

Exercise 12.4

(a) Let S be a sphere in $\mathbb{R}^3$ that is centered at the origin, and let $\mathcal{P} \subset S$ be a set of n points. Unlike Theorem 12.3, the points of $\mathcal{P}$ may have coordinates that are not integers. Prove that $E(\mathcal{P}) = O(|\mathcal{P}|^{9/4})$. (Hint: Use Lemma 8.2.)

(b) Prove that $E(\mathcal{P}) = O(|\mathcal{P}|^{20/9+\varepsilon})$, for every $\varepsilon > 0$. (Hint: Replace Lemma 8.2 with Theorem 3.9.)

Exercise 12.5 Let $\mathcal{P}$ be a set of m points and let Γ be a set of n circles, both in $\mathbb{R}^2$. Theorem 3.3 provides an upper bound for $I(\mathcal{P}, \Gamma)$. In this problem, we study an alternative proof of the same bound, up to an extra ε in the exponent.

Consider a circle $\gamma = \mathbf{V}((x-a)^2 + (y-b)^2 - r^2) \subset \mathbb{R}^2$ and a point $p \in \mathbb{R}^2$. Let the *dual* of γ be the point $\gamma^* = (a, b, r^2) \in \mathbb{R}^3$. Let the *dual* of p be the set p^* of points in $\mathbb{R}^3$ that are dual to a circle that is incident to p. We set $\Gamma^* = \{\gamma^* : \gamma \in \Gamma\}$ and $\mathcal{P}^* = \{p^* : p \in \mathcal{P}\}$. Derive an upper bound for $I(\mathcal{P}, \Gamma)$ by obtaining an upper bound for $I(\Gamma^*, \mathcal{P}^*)$.

Exercise 12.6 We prove a weaker variant of Theorem 12.4(b). Specifically, we use incidences to prove that $\|S_n^{(3)} \cap \mathbb{Z}^3\| = O(n^{6/5+\varepsilon})$, for every $\varepsilon > 0$. Instructions:

- Adapt the proof of Claim 3.7 up to the part where an incidence bound is applied.
- Assume, without proof, that every two spheres intersect in $O(n^\varepsilon)$ points of $\mathbb{Z}^3$ (this is a simpler variant of Theorem 12.4(d)).
- Use duality between points and spheres of the same radii in $\mathbb{R}^3$. Apply an incidence bound in the dual space.

12.4 Open Problems

Both of the problems that we study in this chapter are still wide open.

Distinct distances with local properties: Section 12.1 deals with the asymptotic value of $\phi(n, k, \ell)$ for various values of k and ℓ. The following open problem considers two main open cases.

Open Problem 12.6
(a) Find the asymptotic size of $\phi(n, 4, 5)$.
(b) Find the asymptotic size of $\phi(n, 5, 9)$.

No nontrivial bound is known for either part of Open Problem 12.6. In both cases, we only have the trivial upper bound $O(n^2)$ and the trivial lower bound $\Omega(n)$.

In Section 12.1, we focus on the value of $\phi\left(n, k, \binom{k}{2} - k + c\right)$, for some small constant c. Theorem 12.2 consists of the current best lower bound $\phi\left(n, k, \binom{k}{2} - k + 5\right) = \Omega_{k,\varepsilon}(n^{8/7-\varepsilon})$. The current best upper bound is the trivial $\phi\left(n, k, \binom{k}{2} - k + 5\right) = O(n^2)$.

Open Problem 12.7 *Find the asymptotic size of $\phi\left(n, k, \binom{k}{2} - k + c\right)$ where c is a constant and k is sufficiently large.*

More information about this family of distinct distances problems can be found in Sheffer (2014).

Additive energy on a hypersphere: In Section 12.2, we studied $E(\mathcal{P})$ where $\mathcal{P}$ is on a hypersphere of radius n. The trivial lower bound $E(\mathcal{P}) = \Omega(|\mathcal{P}|^2)$ is obtained from quadruples $(a, b, p, q) \in \mathcal{P}^4$ where $a = p$ and $b = q$. There is a large gap between this lower bound and the bound of Theorem 12.3: $E(\mathcal{P}) = O(|\mathcal{P}|^{7/3}n^{\varepsilon})$.

Open Problem 12.8 *Let $\mathcal{P} \subset S_n^{(4)} \cap \mathbb{Z}^4$. Find the maximum asymptotic size that $E(\mathcal{P})$ can have.*

Bourgain and Demeter (2015) showed that, when replacing the hypersphere with a truncated paraboloid $\mathbf{V}(x_1^2 + x_2^2 + x_3^2 - x_4)$, there exists $\mathcal{P}$ that satisfies $E(\mathcal{P}) = \Omega(|\mathcal{P}|^{7/3})$. People also studied the following variant of the problem in $\mathbb{R}^3$.

Open Problem 12.9 *Let S be a sphere in $\mathbb{R}^3$ that is centered at the origin. Let $\mathcal{P} \subset S$ be a finite set (not necessarily in $\mathbb{Z}^3$). Find the maximum size that $E(\mathcal{P})$ can have.*

Exercise 12.4(b) is the current best bound for the three-dimensional problem. When replacing the sphere with a paraboloid, Demeter derived the bound $E(\mathcal{P}) = O(|\mathcal{P}|^{2+\varepsilon})$, for any $\varepsilon > 0$. For more details about this family of problems, see Demeter (2014).

13

Incidences in Spaces Over Finite Fields

In Chapter 6 we saw several applications of polynomial methods in finite fields. In this chapter, we continue our study of finite fields, focusing on point-line incidences in finite planes. Much less is known about incidences over finite fields, since many incidence problems become more difficult in this case. Unlike the case of $\mathbb{R}^d$, there is no one main technique that leads to most of the current best bounds. Instead, each bound that we derive in this chapter requires a rather different set of tools.

13.1 First Incidence Bounds in $\mathbb{F}_q^2$

As discussed in Chapter 8, the Szemerédi–Trotter theorem can be generalized to $\mathbb{C}^2$. Other works generalize this theorem for semi-algebraic sets (Fox et al., 2017), for definable curves and points in o-minimal structures (Basu and Raz, 2018), and more (Chernikov et al., 2020). However, the situation is more involved in $\mathbb{F}_q^2$. For example, let $\mathcal{P}$ be the set of all q^2 points of $\mathbb{F}_q^2$, and let $\mathcal{L}$ be the set of all lines in $\mathbb{F}_q^2$. Each line in $\mathbb{F}_q^2$ is defined by an equation of the form $y \equiv ax + b$ or $x \equiv b$, where $a, b \in \mathbb{F}_q$. This implies that there are $q + 1$ line slopes, each with q lines. Thus, we have that $|\mathcal{L}| = q^2 + q$ and that each point of $\mathcal{P}$ is incident to $q + 1$ lines of $\mathcal{L}$. This leads to

$$I(\mathcal{P}, \mathcal{L}) = \Theta(q^3) = \Theta\left(|\mathcal{P}|^{3/4}|\mathcal{L}|^{3/4}\right). \tag{13.1}$$

The above example shows that, in $\mathbb{F}_q^2$, we do not have the Szemerédi–Trotter bound $I(\mathcal{P}, \mathcal{L}) = O(|\mathcal{P}|^{2/3}|\mathcal{L}|^{2/3} + |\mathcal{P}| + |\mathcal{L}|)$. We now show that Equation (13.1) is asymptotically tight. Throughout this chapter, we use the $\mathbf{V}(\cdot)$ notation to define varieties in $\mathbb{F}_q^d$.

Lemma 13.1 *Let $\mathcal{P}$ be a set of m points and let $\mathcal{L}$ be a set of n lines, both in $\mathbb{F}_q^2$. Then $I(\mathcal{P}, \mathcal{L}) = O(m^{3/4}n^{3/4} + m + n)$.*

Proof Let $\mathcal{L}_x$ be the set of lines of $\mathcal{L}$ of the form $\mathbf{V}(x - b)$. That is, $\mathcal{L}_x$ is the set of lines from $\mathcal{L}$ that are vertical. Since each point of $\mathbb{F}_q^2$ is incident to at most one line of $\mathcal{L}_x$, we have that $I(\mathcal{P}, \mathcal{L}_x) \leq m$. We may thus remove the lines of $\mathcal{L}_x$ from $\mathcal{L}$, and assume that no line of $\mathcal{L}$ is parallel to the x-axis.

Two distinct lines in $\mathbb{F}_q^2$ intersect in at most one point. Indeed, a point on two lines satisfies two equations of the form $y \equiv ax + b$, and such a system has at most one solution. Thus, the incidence graph of $\mathcal{P} \times \mathcal{L}$ contains no $K_{2,2}$. Applying Lemma 8.1 leads to

$$I(\mathcal{P}, \mathcal{L}) = O(m\sqrt{n} + n). \tag{13.2}$$

We recall the point-line duality from Section 1.10: The *dual of a point* $p = (a, b) \in \mathbb{F}_q^2$ is the line $p^* = \mathbf{V}(y - ax + b)$. Similarly, the *dual of a line* $\ell = \mathbf{V}(y - cx + d)$ is the point $\ell^* = (c, d) \in \mathbb{F}_q^2$. A point p is incident to a line ℓ if and only if the point ℓ^* is incident to the line p^*, since both are equivalent to $d \equiv ac - b$. Consider the point set $\mathcal{L}^* = \{\ell^* : \ell \in \mathcal{L}\}$ and the set of lines $\mathcal{P}^* = \{p^* : p \in \mathcal{P}\}$. By the observation above, we have $I(\mathcal{P}, \mathcal{L}) = I(\mathcal{L}^*, \mathcal{P}^*)$. Thus, by applying Lemma 8.1 on $\mathcal{L}^*$ and $\mathcal{P}^*$, we get that

$$I(\mathcal{P}, \mathcal{L}) = I(\mathcal{L}^*, \mathcal{P}^*) = O(n\sqrt{m} + m). \tag{13.3}$$

By multiplying (13.2) and (13.3), we have that

$$I(\mathcal{P}, \mathcal{L}) \cdot I(\mathcal{P}, \mathcal{L}) = O\left((m\sqrt{n} + n)(n\sqrt{m} + m)\right)$$
$$= O\left(m^{3/2}n^{3/2} + m^2 n^{1/2} + m^{1/2}n^2 + mn\right). \tag{13.4}$$

If $m = \Omega(n^2)$, then Equation (13.3) implies that $I(\mathcal{P}, \mathcal{L}) = m$. We may thus assume that $m = O(n^2)$, which in turn implies that $m^2 n^{1/2} = O(m^{3/2}n^{3/2})$. A symmetric argument leads to $m^{1/2}n^2 = O(m^{3/2}n^{3/2})$. We also note that $mn = O(m^2 + n^2)$. Combining these observations with Equation (13.4) gives that

$$I(\mathcal{P}, \mathcal{L})^2 = O\left(m^{3/2}n^{3/2} + m^2 + n^2\right).$$

By taking the square root of both sides, we obtain the assertion of the lemma.   □

The example that leads to Equation (13.1) shows that the bound of Lemma 13.1 is tight. This is an extreme case, since it includes all of the points and lines in $\mathbb{F}_q^2$. As the sets of points and lines become smaller, better incidence bounds can be obtained. The following result studies the other extreme case, where m and n are tiny compared to p. (Recall that we think of p as a prime and of q as a prime power p^r.)

Theorem 13.2 (Grosu, 2014) *Let p be a prime number. Let $\mathcal{P}$ be a set of n points and let $\mathcal{L}$ be a set of n lines, both in $\mathbb{F}_p^2$. If $n = O(\log\log\log p)$ then $I(\mathcal{P}, \mathcal{L}) = O(n^{4/3})$.*

To show that the bound of Theorem 13.2 is also tight, we note that Elekes's construction from Claim 1.3 also holds in $\mathbb{F}_p^2$. Indeed, in that construction the points have integer coordinates and the lines are defined by equations with integer coordinates. Thus, the construction exists in $\mathbb{F}_p^2$, when p is sufficiently large with respect to m and n.

To recap, we know the maximum number of point-line incidences in $\mathbb{F}_p^2$ when m and n are very large or very small. However, for most ranges of m and n, this problem is an open problem. In Sections 13.3 and 13.5 we study the current best bounds for this problem.

13.2 A Brief Introduction to the Projective Plane

In Section 13.3, we derive an incidence bound by relying on the projective plane. In the current section, we first explain what this plane is. Intuitively, a projective plane is obtained by taking a plane, such as $\mathbb{R}^2$, $\mathbb{C}^2$, or $\mathbb{F}_q^2$, and adding one additional line "at infinity." The resulting plane has nice properties, such as that every two lines intersect in *exactly* one point. We only consider the case of $\mathbb{F}_q$, but the projective plane over $\mathbb{R}$ is defined in the same way. When reading the explanations below, one might get additional geometric intuition by also thinking about the case of $\mathbb{R}$. For a more detailed introduction of projective spaces, see for example Cox et al. (2013, Chapter 8).

A formal definition: We now study a rigorous definition of the *projective plane* $\mathbb{PF}_q^2$. Recall that 0_d is the origin of a d-dimensional space. We consider points $u, v \in \mathbb{F}_q^3 \backslash \{0_3\}$ to be equivalent if there exists $c \in \mathbb{F}_q$ such that $u \equiv cv$. For example, the points $(0, 1, 1)$ and $(0, 2, 2)$ are equivalent in $\mathbb{F}_5^3$. An equivalence class is the set of points on a line that is incident to 0_3, without the point 0_3. The points of the projective plane $\mathbb{PF}_q^2$ are these equivalence classes. That is, a projective point is an equivalence class of points in $\mathbb{F}_q^3 \backslash \{0\}$. For example, one point of $\mathbb{PF}_5^2$ corresponds to the line in $\mathbb{F}_5^3$ that is defined by $x = 0$ and $y = z$. These points and line correspond to the equivalence class $\{(0, 1, 1), (0, 2, 2), (0, 3, 3), (0, 4, 4)\}$.

A nonprojective space is called an *affine* space. To distinguish between points in affine and projective spaces, we use the notation $[x : y : z]$ for the coordinates of a point in $\mathbb{PF}_q^2$. In the projective plane, every point has multiple equivalent names. For example, $[0 : 1 : 1], [0 : 2 : 2]$, and $[0 : 3 : 3]$ are all the same point in $\mathbb{PF}_5^2$.

We note that the q^2 points of the form $[a : b : 1] \in \mathbb{PF}_q^2$ are all distinct. The q points of the form $[1 : b : 0]$ are also distinct and not equivalent to any of the preceding q^2 points, and so is $[0 : 1 : 0]$. Every other point of the form $[x : y : z]$ is equivalent to one of the $q^2 + q + 1$ points above. Thus, the projective plane $\mathbb{PF}_q^2$ consists of $q^2 + q + 1$ points.

When working in $\mathbb{PF}_q^2$, we cannot define a variety as before. For example, the object that is defined by $x + y - 3 \equiv 0$ contains the point $[2 : 1 : 1]$ but not the point $[4 : 2 : 2]$. This is impossible, since these are different names for the same point. To overcome this issue, in $\mathbb{PF}_q^2$ we define varieties with homogeneous polynomials.[1] For a homogeneous polynomial $f \in \mathbb{F}_q[x, y, z]$ of degree k, a point $u \in \mathbb{F}_q^3$, and $c \in \mathbb{F}_q \backslash \{0\}$, we have that $f(c \cdot u) \equiv c^k f(u)$. Thus, $f(c \cdot u) \equiv 0$ if and only if $f(u) \equiv 0$. We conclude that, when using only homogeneous polynomials, varieties are well defined in $\mathbb{PF}_q^2$.

We move from the affine plane $\mathbb{F}_q^2$ to the corresponding projective plane $\mathbb{PF}_q^2$ by taking a point $(p_x, p_y) \in \mathbb{F}_q^2$ to the point $[p_x : p_y : 1] \in \mathbb{PF}_q^2$. To move a variety U from $\mathbb{F}_q^2$ to $\mathbb{PF}_q^2$, we *homogenize* the polynomials that define U, as follows. Given a polynomial $f \in \mathbb{F}_q[x, y]$ of degree k, we multiply a monomial of f of degree j by z^{k-j}. This leads to a homogeneous polynomial f^* of degree k in $\mathbb{F}_q[x, y, z]$. For example, $f(x, y) = x^3 + y^3 + xy - 3$ leads to $f^*(x, y, z) = x^3 + y^3 + xyz - 3z^3$. We note that f vanishes on $(p_x, p_y) \in \mathbb{F}_q^2$ if and only if f^* vanishes on $[p_x : p_y : 1]$. Thus, the transition from $\mathbb{F}_q^2$ to $\mathbb{PF}_q^2$ maintains point-variety incidences.

We recall that the projective plane $\mathbb{PF}_q^2$ contains additional points that do not correspond to points of $\mathbb{F}_q^2$. In particular, these are the points with a zero z-coordinate. When moving varieties from $\mathbb{F}_q^2$ to $\mathbb{PF}_q^2$, we may create incidences with those additional points. For example, consider the line in $\mathbb{F}_q^2$ that is defined by $f(x, y) = x - y + 1$. The homogenized $f^*(x, y, z) = x - y + z$ also vanishes on the point $[1 : 1 : 0]$, which does not correspond to a point of $\mathbb{F}_q^2$.

More intuition: We now consider a more geometric interpretation of $\mathbb{PF}_q^2$. Let Π be the plane in $\mathbb{F}_q^3$ that is defined by $z \equiv 1$. We can think of the move from $\mathbb{F}_q^2$ to $\mathbb{PF}_q^2$ as placing $\mathbb{F}_q^2$ in Π. Then, a point $u \in \mathbb{PF}_q^2$ becomes the line that is incident to 0_3 and to u. These lines cover all the points of $\mathbb{F}_q^3$ that have a nonzero z-coordinate. The points that have a zero z-coordinate are on lines that pass through 0_3 and are parallel to Π. These lines correspond to points of $\mathbb{PF}_q^2$ that are not in $\mathbb{F}_q^2$. In the plane $\mathbb{PF}_q^2$, we can think of these additional points

---

[1] Recall that a polynomial is *homogeneous* if all of its monomials have the same degree.

as lying on the line that is defined by $z \equiv 0$. We think of this line as being at infinity, for reasons that are explained below.

Let ℓ be a line in $\mathbb{F}_q^2$ that is defined by $ax + by + c \equiv 0$. By homogenizing this equation, we obtain that the corresponding projective line $\ell^* \subset \mathbb{PF}_q^2$ is defined by $ax + by + cz \equiv 0$. As explained above, every point on ℓ in $\mathbb{F}_q^2$ corresponds to a point on ℓ^* in $\mathbb{PF}_q^2$. However, ℓ^* also contains the extra point $[1 : -b^{-1}a : 0]$ on the projective line at infinity. If $b \equiv 0$ then the extra point is $[1 : 0 : 0]$. The extra point is defined by the slope of ℓ^*, so parallel projective lines intersect at a point at infinity. For example, every projective line with slope 1 can be defined as $y \equiv x + zc$ for some $c \in \mathbb{F}_q$, so all of these lines intersect at $[1 : 1 : 0]$.

The above implies that every two lines from $\mathbb{PF}_q^2$ intersect in exactly one point. Nonparallel lines intersect at a point that is also in the affine plane. Parallel lines intersect at a point at infinity. Why should parallel lines intersect at infinity? For an intuitive reason, recall that in paintings and photographs parallel lines appear to meet at infinity (see Figure 13.1).

Figure 13.1 Parallel lines meet at infinity. (Picture by Gesundheit/E+/Getty Images.)

Projective transformations: In affine planes, we work with transformations that take lines to lines, such as translations, rotations, and reflections. Similarly, *projective transformations* are bijections between $\mathbb{PF}_q^2$ and itself that take lines to lines. We can define such a transformation from $[x : y : z]$ to $[x' : y' : z']$ as

$$\begin{pmatrix} x' \\ y' \\ z' \end{pmatrix} = \begin{pmatrix} a_{1,1} & a_{1,2} & a_{1,3} \\ a_{1,2} & a_{2,2} & a_{2,3} \\ a_{1,3} & a_{2,3} & a_{3,3} \end{pmatrix} \begin{pmatrix} x \\ y \\ z \end{pmatrix}, \tag{13.5}$$

where $a_{j,k} \in \mathbb{F}_q$ and the 3×3 matrix is invertible.

In the above geometric interpretation of $\mathbb{PF}_q^2$, the points of the affine plane $\mathbb{F}_q^2$ form the plane that is defined by $z \equiv 1$. From this perspective, a projective transformation changes the plane that corresponds to $\mathbb{F}_q^2$.

We conclude this section with a useful observation. Consider a point $u \in \mathbb{PF}_q^2$ with a nonzero z-coordinate and let $\mathcal{L}$ be a set of projective lines that are incident to u. Let τ be a projective transformation that takes u to the line at infinity (that is, to a point with a zero z-coordinate). Since τ takes the intersection point of the lines of $\mathcal{L}$ to infinity, all of these lines become parallel.

13.3 Incidences between Large Sets of Points and Lines

We now return to the study of incidences between m point and n lines in $\mathbb{F}_q^2$. In Section 13.1, we saw that the Szemerédi–Trotter bound $O(m^{2/3}n^{2/3} + m + n)$ is false when taking all points and all lines of $\mathbb{F}_q^2$. This can be significantly generalized: The Szemerédi–Trotter bound is false when $m \cdot n$ is asymptotically larger than q^3. On the other hand, the Szemerédi–Trotter bound holds and is tight when $m \cdot n = \Theta(q^{3/2})$. These are consequences of the following result.

Theorem 13.3 *Let $\mathcal{P}$ be a set of m points and let $\mathcal{L}$ be a set of n lines, both in $\mathbb{F}_q^2$. Then*

$$I(\mathcal{P}, \mathcal{L}) = O\left(\frac{mn}{q} + \sqrt{mnq}\right).$$

When $mn = \Omega(q^3)$, every point-line configuration in $\mathbb{F}_q^2$ satisfies that $I(\mathcal{P}, \mathcal{L}) = \Theta(mn/q)$.

Theorem 13.3 has several rather different proofs. Here we present a proof of Vinh (2011) that is based on spectral graph theory. Other proofs rely on the Fourier transform or on basic additive combinatorics. We first briefly introduce a few concepts from spectral graph theory. A nice introduction to this topic can be found in West (2001, Section 8.6). Section A.2 in this current book provides a brief recap of basic graph theory notation.

The *adjacency matrix* M of a graph $G = (V, E)$ is a $|V| \times |V|$ matrix that is defined as follows. We set $N = |V|$ and write $V = \{v_1, \ldots, v_N\}$. Then the cell M_{ij} contains the number of edges in E between v_i and v_j. When there are no parallel edges in G, the matrix of M consists of ones and zeros.

We next consider the matrix $M^2 = M \cdot M$. The cell M_{ij}^2 contains the number of paths of length two in G between v_i and v_j. When there are no loops in G, the number of paths of length two from a vertex to itself is the degree of that vertex, so $M_{ii}^2 = \deg v_i$. For this property to still hold when there are loops in G, we define that a loop increases the degree of the corresponding vertex only by 1. That is, the degree of a vertex v is the number of distinct edges that are adjacent to v.

Spectral graph theory studies eigenvalues of matrices that are associated with graphs, such as adjacency matrices. Since we focus on adjacency matrices, we define the *eigenvalues of a graph* G to be the eigenvalues of the adjacency matrix of G. A graph G is *k-regular* if every vertex of G is of degree k. We require the following result about eigenvalues of regular graphs (for example, see Corollary 9.2.5 of Alon and Spencer, 2004). Let B, C be subsets of vertices that are not necessarily disjoint. Then $e(B, C)$ is the number of edges in E that have one endpoint in B and the other endpoint in C.

Lemma 13.4 *Let $G = (V, E)$ be a k-regular graph with eigenvalues $|\lambda_1| \geq |\lambda_2| \geq \cdots \geq |\lambda_{|V|}|$. Consider $B, C \subset V$ that are not necessarily disjoint. Then*

$$\left| e(B, C) - \frac{k}{n} |B||C| \right| \leq \lambda_2 \sqrt{|B||C|}.$$

The largest eigenvalue of a k-regular graph is k. Intuitively, Lemma 13.4 states that a regular graph with only one large eigenvalue behaves like a random graph. To see this, we create a graph G by starting with a set V of n vertices and adding edges at random, as follows. We allow the graph to have loops, but not parallel edges. For a constant k, we add each of the $|V|^2$ potential edges with probability k/n. The resulting graph is unlikely to be k-regular. However, G is similar to a k-regular graph in the sense that the expected degree of every vertex is k. For any vertex subsets $B, C \subset V$, there are $|B||C|$ potential edges with one endpoint in B and the other in C. Thus, the expected number of such edges is $\frac{k}{n} |B||C|$.

Proof of Theorem 13.3 We consider the projective plane $\mathbb{PF}_q^2$, as defined in Section 13.2. Recall that the number of points in $\mathbb{PF}_q^2$ is $q^2 + q + 1$. We construct the graph $G_q = (\mathbb{PF}_q^2, E)$ as follows. The graph contains a vertex for every point of $\mathbb{PF}_q^2$. There is an edge between the vertices $u = [u_1 : u_2 : u_3]$ and $v = [v_1 : v_2 : v_3]$ if and only if

$$u_1 v_1 + u_2 v_2 + u_3 v_3 \equiv 0. \tag{13.6}$$

This condition is equivalent to the point $[v_1 : v_2 : v_3]$ forming an incidence with the projective line that is defined by $u_1 x + u_2 y + u_3 z \equiv 0$. Since this is a homogeneous equation, it indeed defines a valid line $\mathbb{PF}_q^2$.

We fix $[u_1 : u_2 : u_3] \in \mathbb{PF}_q^2$ and consider the number of solutions to $u_1 x + u_2 y + u_3 z \equiv 0$. Since $[0 : 0 : 0] \notin \mathbb{PF}_q^2$, at least one of u_1, u_2, and u_3 is nonzero. Without loss of generality, we assume that $u_1 \not\equiv 0$. When $y \equiv z \equiv 0$ we get that $x \equiv 0$, and then $[x : y : z] \notin \mathbb{PF}_q^2$. There are $q^2 - 1$ choices for y and z where at most one of these variables is zero, and each choice uniquely determines x. Thus, there are $q^2 - 1$ solutions to $u_1 x + u_2 y + u_3 z \equiv 0$ in $\mathbb{F}_q^3 \setminus \{0_3\}$. Each point of $\mathbb{PF}_q^2$ corresponds to $q - 1$ points of $\mathbb{F}_q^3 \setminus \{0_3\}$. We conclude that the number of solutions to the equation is $(q^2 - 1)/(q - 1) = q + 1$. This implies that G_q is a $(q + 1)$-regular graph (recall that a loop increases the degree of the corresponding vertex by one).

For distinct points $[u_1 : u_2 : u_3], [u'_1 : u'_2 : u'_3] \in \mathbb{PF}_q^2$, we consider the system

$$u_1 x + u_2 y + u_3 z \equiv 0,$$

$$u'_1 x + u'_2 y + u'_3 z \equiv 0.$$

This system has a unique solution in $\mathbb{PF}_q^2$ (that is, it has $q - 1$ equivalent solutions). Thus every two vertices of G_q have exactly one common neighbor. In other words, every two vertices have one path of length two between them.

Let M be the adjacency matrix of G_q. We recall that M_{ij}^2 is the number of paths of length two between the vertices v_i and v_j. By the preceding paragraph, every cell of M^2 that is not on the main diagonal contains the value 1. Since G_q is $(q + 1)$-regular, there are $q + 1$ paths of length two from every vertex to itself. We get that

$$M_{ij}^2 = \begin{cases} 1, & i \neq j, \\ q + 1, & i = j. \end{cases} \tag{13.7}$$

The all 1s matrix of size $N \times N$ has the eigenvalue N with multiplicity one and the eigenvalue 0 with multiplicity $N - 1$. Indeed, the eigenvalue N has the eigenvector $(1, 1, \ldots, 1)$, and the eigenvalue 0 has the eigenvectors $(1, -1, 0, \ldots, 0), (1, 0, -1, 0, \ldots, 0)$, and so on. Increasing every element on the main diagonal of a matrix by c shifts all of the eigenvalues of that matrix by c. Thus, M^2 has the eigenvalue $(q^2 + q + 1) + q = (q + 1)^2$ with multiplicity 1 and the eigenvalue q with multiplicity $q^2 + q$.

We recall a basic fact from linear algebra: If the eigenvalues of M are $\lambda_1, \ldots, \lambda_N$ then the eigenvalues of M^2 are $\lambda_1^2, \ldots, \lambda_N^2$. In our case, we get that one eigenvalue of M is $q + 1$ and the absolute values of the other eigenvalues are $\sqrt{q}$. This fits the above claim that the largest eigenvalue of a k-regular graph is k.

We take each point $a = (a_x, a_y) \in \mathcal{P}$ to the point $[a_x : a_y : 1] \in \mathbb{PF}_q^2$ and note that no two points of $\mathcal{P}$ are equivalent after this change. We take a line $\ell \in \mathcal{L}$ that is defined by $b_y y + b_x x + b_1 \equiv 0$ to the point $[b_x : b_y : b_1] \in \mathbb{PF}_q^2$. Two distinct lines of $\mathcal{L}$ cannot lead to equivalent points. Let B be the set of vertices of G_q that correspond to points of $\mathcal{P}$. Let C be the set of vertices of G_q that correspond to lines of $\mathcal{L}$.

A point a is incident to a line ℓ if and only if $a_x b_x + a_y b_y + b_1 \equiv 0$. That is, $a \in \ell$ if and only if G_q contains an edge between the vertex of a and the vertex of ℓ. This implies that $I(\mathcal{P}, \mathcal{L}) = e(\mathcal{P}, \mathcal{L})$. By Lemma 13.4 with B, C, and $|\lambda_2| = \sqrt{q}$, we get that

$$\left| I(\mathcal{P}, \mathcal{L}) - \frac{q+1}{q^2+q+1} mn \right| = \left| e(\mathcal{P}, \mathcal{L}) - \frac{q+1}{q^2+q+1} mn \right| \leq \sqrt{qmn}. \quad (13.8)$$

Rearranging Equation (13.8) leads to $I(\mathcal{P}, \mathcal{L}) = O(mn/q + \sqrt{qmn})$. If $mn = \Omega(q^3)$ then $\frac{q+1}{q^2+q+1} mn = \Omega(\sqrt{qmn})$. In this case, Equation (13.8) becomes $I(\mathcal{P}, \mathcal{L}) = \Theta(mn/q)$. $\qquad \square$

Theorem 13.3 provides a tight bound when $mn = \Omega(q^3)$, and implies the Szemerédi–Trotter bound when both m and n are $\Theta(q^{3/2})$. However, the case of smaller m and n remains wide open. In Section 13.5, we study the current best bound in this case.

13.4 Planes in $\mathbb{F}_q^3$ and the Sum-Product Problem

Point-plane incidences: Rudnev (2018) proved a point-plane incidence bound in a three-dimensional space over any field. This bound is currently a main tool for studying incidence problems and related problems over finite fields. In this section we study Rudnev's result and use it to derive a sum-product bound over finite fields. In Section 13.5, we use Rudnev's result to obtain the current best point-line incidence bound in $\mathbb{F}_q^2$ when m and n are not very large or very small.

When considering point-plane incidences in $\mathbb{F}_q^3$, we have the same issue as in $\mathbb{R}^3$: By placing the points on a line ℓ and taking planes that contain ℓ, we get that every plane is incident to every point. To obtain a nontrivial incidence problem, we need to have additional restrictions. In this chapter, we assume that the no line in $\mathbb{F}_q^3$ contains k points. As usual, we first derive a weak incidence bound that relies on a combinatorial argument.

Lemma 13.5 *Let $\mathcal{P}$ be a set of m points and let H be a set of n planes, both in $\mathbb{F}_q^3$. If no line in $\mathbb{F}_q^3$ contains k points of $\mathcal{P}$, then*

$$I(\mathcal{P}, H) = O(n\sqrt{km} + m).$$

Proof Since no line contains k points of $\mathcal{P}$, the incidence graph of $\mathcal{P} \times H$ contains no $K_{k,2}$. We perform a *point-plane duality* by adapting the point-line duality from Section 13.1. In this case, the dual of a point $p = (p_x, p_y, p_z) \in \mathbb{F}_q^3$ is the line $p^* = \mathbf{V}(z - p_x x - p_y y + p_z)$. The dual of a plane $\Pi = \mathbf{V}(z - q_x x - q_y y + q_z)$ is the point $\Pi^* = (q_x, q_y, q_z) \in \mathbb{F}_q^3$. Let $\mathcal{P}^*$ be the set of planes that are dual to points of $\mathcal{P}$. Let H^* be the set of points that are dual to planes of H. Since duality preserves incidences, we have that $I(\mathcal{P}, H) = I(H^*, \mathcal{P}^*)$. Also, the incidence graph of $H^* \times \mathcal{P}^*$ contains no $K_{2,k}$.

We note that the proof of Lemma 3.4 also holds for points and planes in $\mathbb{F}_q^3$. By revising this proof to include the dependency on t, we obtain the bound

$$I(H^*, \mathcal{P}^*) = O\left(nm^{(s-1)/s}t^{1/s} + m\right).$$

(See also Exercise 3.9.) In Lemma 3.4, the roles of m and n in the bound are switched. This is because the dual space contains n points and m planes. To complete the proof, we set $s = 2$ and $t = k$.   □

Rudnev (2018) derived the following bound for the point-plane incidence problem. Rudnev's result holds over any field, but we only consider the case of $\mathbb{F}_q$.

Theorem 13.6 *Let* $q = p^r$ *for a prime* p *and a positive integer* r. *Let* $\mathcal{P}$ *be a set of* m *points and let* H *be a set of* n *planes, both in* $\mathbb{F}_q^3$, *such that* $n \geq m$. *Assume that* $m = O(p^2)$ *and that no line of* $\mathbb{F}_q^3$ *contains* k *points of* $\mathcal{P}$. *Then*

$$I(\mathcal{P}, H) = O(n\sqrt{m} + kn).$$

The proof of Theorem 13.6 is rather involved and requires multiple tools that do not appear in this book. For example, this proof requires working in the closure of the field $\mathbb{F}_q$ and an involved result of Kollár (2015). For that reason, we do not prove Theorem 13.6. For the proof, see Rudnev (2018). A simplified proof appears in de Zeeuw (2016).

At first, Theorem 13.6 might not seem impressive. This theorem improves upon Lemma 13.5 only by having a better dependency on k, while also adding two new restrictions. However, this better dependency on k led to improved bounds for many problems. We now study our first application of Theorem 13.6.

The sum-product problem: In Section 1.8, we study the sum-product problem over $\mathbb{R}$. Briefly, this problem conjectures that every finite $A \subset \mathbb{R}$ satisfies that $\max\{|A+A|, |AA|\}$ is large with respect to $|A|$. We now consider this problem for sets $A \subset \mathbb{F}_q$. As with point-line incidences, the sum-product problem exhibits unusual behavior when A is very large. In particular, when $A = \mathbb{F}_q$ we have $|A| = |A + A| = |AA| = q$. The problem becomes more interesting when $|A|$ is much smaller than q. The following result is by Roche-Newton et al. (2016).

Theorem 13.7 *Let* $q = p^r$ *and let* $A \subset \mathbb{F}_q$ *satisfy* $|A| \leq p^{5/8}$. *Then*

$$\max\{|A + A|, |AA|\} = \Omega(n^{6/5}).$$

The current best bound $\max\{|A + A|, |AA|\} = \Omega(n^{5/4})$ was obtained in Mohammadi and Stevens (2021) when $|A| = O(p^{1/2})$. As with point-line incidences, this bound is significantly weaker than current best bound over $\mathbb{R}$.

To move from the sum-product problem to an incidence problem, we rely on the concept of energy. This concept was briefly introduced in Section 12.2, but that section is not a prerequisite for the following. Given a finite set $A \subset \mathbb{F}_q$, the *additive energy* of A is

$$E(A) = \left| \left\{ (a_1, a_2, a_3, a_4) \in A^4 : a_1 + a_2 \equiv a_3 + a_4 \right\} \right|.$$

We have that $E(A) \geq |A|^2$, since there are $|A|^2$ quadruples with $a_1 = a_3$ and $a_2 = a_4$. There are $|A|^3$ choices of values for a_1, a_2, and a_3. After fixing these three variables, there is at most one valid choice for a_4. This implies that $E(A) \leq |A|^3$.

For $x \in A + A$, we set

$$r_A(x) = \left| \left\{ (a_1, a_2) \in A^2 : a_1 + a_2 \equiv x \right\} \right|.$$

In other words, $r_A(x)$ is the number of representations of x as a sum of two elements of A. Since every pair of A^2 contributes to exactly one $r_A(x)$, we have that $\sum_{x \in A+A} r_A(x) = |A|^2$. The number of quadruples $(a_1, a_2, a_3, a_4) \in A^4$ that satisfy $a_1 + a_2 \equiv a_3 + a_4 \equiv x$ is $r_A(x)^2$. Thus, $E(A) = \sum_x r_A(x)^2$. The Cauchy–Schwarz inequality implies that

$$E(A) = \sum_{x \in A+A} r_A(x)^2 \geq \frac{\left(\sum_x r_A(x) \right)^2}{|A + A|} = \frac{|A|^4}{|A + A|}. \tag{13.9}$$

One consequence of Equation (13.9) is that a small sum set implies a large additive energy. It may seem as if there is a reverse correlation between $E(A)$ and $|A + A|$. A few examples that fit such a correlation:

- If A is an arithmetic progression then $|A + A| = \Theta(|A|)$ and $E(A) = \Theta(|A|^3)$.
- If A is a random set that is much smaller than q, then with very high probability $|A + A| = \Theta(|A|^2)$ and $E(A) = \Theta(|A|^2)$.
- Let $0 < \alpha < 1$. Set $A = H + R$ with H being an arithmetic progression of size $\Theta(|A|^\alpha)$ and R being a random set of size $\Theta(|A|^{1-\alpha})$. Assume that A is much smaller than q. Then with very high probability, $|A + A| = \Theta(|A|^{2-\alpha})$ and $E(A) = \Theta(|A|^{2+\alpha})$.

Unfortunately, a large energy does not imply a small sum set. For example, consider $A = P \cup R$, where P is an arithmetic progression of size $|A|/2$ and R is a random set of size $|A|/2$. We assume that $|A|$ is much smaller than q. The elements of P lead to $E(A) = \Theta(|A|^3)$. Because of the elements of R, with very high probability $|A + A| = \Theta(|A|^2)$.

A large energy does imply that there exists a large subset $A' \subset A$ such that $A' + A'$ is small. In the above example, this subset is P. This property is called the *Balog–Szemerédi–Gowers theorem*, and is outside of the scope of this book. Our goal above is only to obtain an initial intuition about additive energy.

There are many variants of the concept of energy. For example, in the ESGK framework from Section 7.2, we may consider $|Q|$ as a distance energy. Instead of having two pairs of numbers with the same sum, we have two pairs of points that span the same distance. We now use another energy variant to prove a sum-product bound in $\mathbb{F}_q$.

Proof of Theorem 13.7 If $0 \in A$, then we remove this element from A. This does not change the asymptotic size of A and decreases the sizes of the sum set and product set. Let $A^{-1} = \{1/a : a \in A\}$. We have that

$$E(A)$$

$$= |A|^{-2} \left| \left\{ (a_1, \ldots, a_6) \in A^6 \ : \ a_1 + a_2 a_3 / a_3 \equiv a_4 + a_5 a_6 / a_6 \right\} \right|$$

$$\leq |A|^{-2} \left| \left\{ (a_1, b_1, c_1, a_2, b_2, c_2) \in (A \times AA \times A^{-1})^2 \ : \ a_1 + b_1 c_1 \equiv a_2 + b_2 c_2 \right\} \right|.$$
$$(13.10)$$

With the above in mind, we define the energy variant

$$E'(A) = \left| \left\{ (a_1, b_1, c_1, a_2, b_2, c_2) \in (A \times AA \times A^{-1})^2 \ : \ a_1 + b_1 c_1 \equiv a_2 + b_2 c_2 \right\} \right|.$$

We next adapt Elekes's sum-product argument from Theorem 1.14. We consider the point set

$$\mathcal{P} = \left\{ (a_1, b_2, c_1) \in A \times AA \times A^{-1} \right\},$$

and the set of planes

$$H = \left\{ \mathbf{V}(x + b_1 z - c_2 y - a_2) \ : \ (a_2, b_1, c_2) \in A \times AA \times A^{-1} \right\}.$$

A 6-tuple $(a_1, b_1, c_1, a_2, b_2, c_2)$ contributes to $E'(A)$ if and only if the point (a_1, b_2, c_1) is incident to the plane $\mathbf{V}(x + b_1 z - c_2 y - a_2)$. Thus, $E'(A) = I(\mathcal{P}, H)$. We wish to bound $I(\mathcal{P}, H)$ by using Theorem 13.6, and first check the conditions of this theorem hold.

We note that $m = |\mathcal{P}| = |H| = |A|^2 |AA|$. Since $\mathcal{P}$ is a Cartesian product of size $|A| \times |AA| \times |A|$, every line in $\mathbb{F}_q^3$ contains at most $|AA|$ points of $\mathcal{P}$. We may assume that $|AA| = O(|A|^{6/5})$, since otherwise we are done. Combining this with $|A| = O(p^{5/8})$ implies that

$$m = |A|^2 |AA| = O(A^{16/5}) = O(p^2).$$

We can thus apply Theorem 13.6 on $\mathcal{P}$ and H with $k = |AA|$, to obtain that

$$E'(A) = I(\mathcal{P}, H) = O\left(m^{3/2} + mk\right) = O\left(|A|^3|AA|^{3/2} + |A|^2|AA|^2\right)$$

$$= O\left(|A|^3|AA|^{3/2}\right).$$

Combining the above with Equation (13.10) leads to

$$E(A) \leq |A|^{-2}E'(A) = O\left(|A||AA|^{3/2}\right) = O\left(|A|^{14/5}\right).$$

Combining this with Equation (13.9) implies that $|A + A| = \Omega(|A|^{6/5})$. □

With some more work, we can improve the value of k in the proof of Theorem 13.7. However, a better bound on k does not lead to a better sum-product bound.

13.5 Incidences between Medium Sets of Points and Lines

In this section, we study point-line incidences in $\mathbb{F}_q^2$ when the numbers of points and lines are neither very large nor very small. This case is considered to be the main one. The current best bound is by Stevens and De Zeeuw (2017).

Theorem 13.8 *For $q = p^r$, let $\mathcal{P}$ be a set of m points and $\mathcal{L}$ be a set of n lines, both in $\mathbb{F}_q^2$. If $m^{7/8} < n < m^{8/7}$ and $n = O(p^{4/3})$, then*

$$I(\mathcal{P}, \mathcal{L}) = O\left(m^{11/15}n^{11/15}\right).$$

The bound $O(m^{11/15}n^{11/15})$ is somewhat close to the elementary upper bound from Lemma 13.1. Specifically, $11/15$ is one fifth of the way from the elementary $3/4$ to the conjectured $2/3$. Also, when n is asymptotically smaller than $m^{7/8}$ or asymptotically larger than $m^{8/7}$, the elementary bounds (13.2) and (13.3) are stronger. This can be seen as part of a more general phenomena: Incidence problems and related problems are significantly more difficult over finite fields.

Incidences with Cartesian products: The first step towards proving Theorem 13.8 is to consider the special case where the point set is a Cartesian product. In this case, we can use an energy argument that is similar to the proof of Theorem 13.7.

Theorem 13.9 *For $q = p^r$, let $\mathcal{L}$ be a set of n lines in $\mathbb{F}_q^2$. Consider $A, B \subset \mathbb{F}_q$ such that $a = |A|$, $b = |B|$, $a \leq b$, $ab^2 = O(n^3)$, and $an = O(p^2)$. Then*

$$I(A \times B, \mathcal{L}) = O\left(a^{3/4}b^{1/2}n^{3/4} + n\right).$$

To put this result in context, we consider the case of $a = b = \sqrt{m}$. In this case, Theorem 13.9 gives the bound $O(\mathcal{P}, \mathcal{L}) = O(m^{5/8}n^{3/4} + n)$. This bound is smaller than the bound of Lemma 13.1 by a factor of $m^{1/8}$ (in the special case where the point set is a lattice). This is a significant improvement for an incidence bound.

Proof of Theorem 13.9 We begin with two steps of pruning the set of lines $\mathcal{L}$. In the first step, we remove all of the vertical and horizontal lines from $\mathcal{L}$ (lines of the form $\mathbf{V}(x - c)$ or $\mathbf{V}(y - c)$). We note that every point of $A \times B$ is incident to at most one vertical line and to at most one horizontal line. Thus, the number of incidences decreases by at most $2ab$. The assumption $ab^2 = O(n^3)$ is equivalent to $a^{1/4}b^{1/2} = O(n^{3/4})$, which implies that $ab = O(a^{3/4}b^{1/2}n^{3/4})$. We conclude that removing the vertical and horizontal lines does not affect our incidence bound.

Every nonhorizontal line contains at most a points of $A \times B$, so we now have that $I(A \times B, \mathcal{L}) \leq an$. If $an = O(a^{3/4}b^{1/2}n^{3/4})$ then we are done. We may thus assume that $an = \Omega(a^{3/4}b^{1/2}n^{3/4})$. Rearranging this bound leads to $b = O(\sqrt{an})$.

In our second pruning step, we check whether $\mathcal{L}$ contains more than $\sqrt{an}$ lines that are all parallel or concurrent.[2] If such a set of lines exists, then we remove all of these lines from $\mathcal{L}$ and repeat this process. Let n_j be the number of lines that were removed during the jth iteration of this process. Excluding the point of concurrency, every point of $A \times B$ is incident to at most one line that was removed in the jth iteration. Thus, the number of incidences that are removed in the jth iteration is smaller than $n_j + ab$. Since we begin the second pruning step with at most n lines, the number of iterations is smaller than $n/\sqrt{an} = \sqrt{n/a}$. We conclude that, during the second pruning step, the number of removed incidences is smaller than

$$\sum_j (n_j + ab) = n + \sqrt{n/a} \cdot ab = n + b\sqrt{na}.$$

Since $b = O(\sqrt{an})$, the number of removed incidences is $O(n + b^{1/2}a^{3/4}n^{3/4})$. We next consider the dual set

$$\mathcal{L}^* = \{(s, t) \ : \ \mathbf{V}(y - sx + t) \in \mathcal{L}\}.$$

Since $\mathcal{L}$ does not contain vertical lines, every line of $\mathcal{L}$ has a dual point in $\mathcal{L}^*$. For $\beta \in B$, we set

$$r_\beta = |\{(\alpha, s, t) \in A \times \mathcal{L}^* \ : \ \beta \equiv \alpha s - t\}|.$$

---

[2] Recall that a set of lines is *concurrent* if there exists a point that is incident to all the lines. Such a point is called the *point of concurrency*.

Intuitively, r_β is the number of incidences that occur with points on the line $\mathbf{V}(y - \beta)$. We thus have that $I(A \times B, \mathcal{L}) = \sum_{\beta \in B} r_\beta$.

We define the energy variant

$$E = \left| \left\{ (\alpha, s, t, \alpha', s', t') \in (A \times \mathcal{L}^*)^2 \; : \; \alpha s - t \equiv \alpha' s' - t' \right\} \right|.$$

To complete the proof, we double count E. The number of 6-tuples $(A \times \mathcal{L}^*)^2$ that satisfy $\alpha s - t \equiv \alpha' s' - t' \equiv \beta$ is r_β^2. Thus, $E = \sum_{\beta \in B} r_\beta^2$. The Cauchy–Schwarz inequality implies that

$$E = \sum_{\beta \in B} r_\beta^2 \geq \frac{\left(\sum_{\beta \in B} r_\beta \right)^2}{b} = \frac{I(A \times B, \mathcal{L})^2}{b}. \tag{13.11}$$

To obtain an upper bound for E, we reduce the problem to a point-plane incidence problem in $\mathbb{F}_q^3$, as follows. Let the coordinates of $\mathbb{F}_q^3$ be x, y, z. We consider the point set

$$Q = \{ (\alpha, s', t') \in A \times \mathcal{L}^* \},$$

and the set of planes

$$H = \{ \mathbf{V}(xs - t - \alpha' y + z) \; : \; (s, t) \in \mathcal{L}^* \text{ and } \alpha' \in A \}.$$

We note that $E = I(Q, H)$. We wish to apply Theorem 13.6 with Q and H, and first check that the conditions of this theorem are satisfied. We have that $|Q| = a|\mathcal{L}| = |H|$ and that $|Q| = a|\mathcal{L}| = O(p^2)$. It remains to obtain an upper bound for the number of points of Q that can be on a common line. If a set of points from $\mathcal{L}^*$ are on the line $\mathbf{V}(y - cx + d) \subset \mathbb{F}_q^2$, then the corresponding lines of $\mathcal{L}$ intersect at the point $(c, d) \in \mathbb{F}_q^2$. By the second pruning step of $\mathcal{L}$, every line in $\mathbb{F}_q^2$ contains at most $\sqrt{an}$ points of $\mathcal{L}^*$. Thus, a line in $\mathbb{F}_q^3$ that is not parallel to the x-axis contains at most $\sqrt{an}$ points of Q. A line that is parallel to the x-axis contains at most a points of Q. Since $a \leq b = O(\sqrt{an})$, every line in $\mathbb{F}_q^3$ contains $O(\sqrt{an})$ points of Q.

By the above, we may apply Theorem 13.6 on Q and H, with $k = O(\sqrt{an})$. This leads to

$$E = I(Q, H) = O\left(|H|\sqrt{|Q|} + k|H| \right) = O\left(a^{3/2} n^{3/2} \right).$$

Combining this with Equation (13.11) implies that $I(A \times B, \mathcal{L}) = O(a^{3/4} b^{1/2} n^{3/4})$. □

Theorem 13.9 provides a point-line incidence bound only when the point set is a Cartesian product. However, we can rely on it to obtain a general point-line incidence bound. We now show that, when there are many incidences, there exists a Cartesian product that contains many of the points.

Finding Cartesian products: Let u and v be distinct points of $\mathbb{F}_q^2$. Two disjoint sets $\mathcal{L}_u, \mathcal{L}_v$ of lines in $\mathbb{F}_q^2$ form a (u, v)-*Cartesian product* if they satisfy the following: Every line of $\mathcal{L}_u$ is incident to u, every line of $\mathcal{L}_v$ is incident to v, and the line that is incident to both u and v is in neither set.

A (u, v)-Cartesian product may not seem like a Cartesian product. We can turn it into a Cartesian product, as follows. We move from the affine plane $\mathbb{F}_q^2$ to the projective plane $\mathbb{PF}_q^2$, as described in Section 13.2. We then perform a projective transformation that takes u to $[1 : 0 : 0]$ and v to $[0 : 1 : 0]$. This transformation takes the lines of $\mathcal{L}_u$ to horizontal lines and the lines of $\mathcal{L}_v$ to vertical lines. Then, the intersection points of the line pairs from $\mathcal{L}_u \times \mathcal{L}_v$ are a Cartesian product. For more details, see the proof of Theorem 13.8.

We prove Theorem 13.8 by induction. As discussed in Chapter 8, using $O(\cdot)$-notation in proofs by induction can be problematic. We thus avoid using $O(\cdot)$-notation starting now.

Lemma 13.10   *The following holds for any real constants $0 < c_1 < c_2$. Let $\mathcal{P}$ be a set of m points and let $\mathcal{L}$ be a set of n lines, both in $\mathbb{F}_q^2$. Let*

$$r \geq \max\left\{4n/c_1 m, 4/c_1, \left(2^5 n^2/c_1^3 m\right)^{1/3}\right\}$$

satisfy that every point of $\mathcal{P}$ is incident to at least $c_1 r$ lines of $\mathcal{L}$ and to at most $c_2 r$ such lines. Then there exist distinct points $u, v \in \mathbb{F}_q^2$, a subset $\mathcal{P}' \subset \mathcal{P}$, and a (u, v)-Cartesian product $\mathcal{L}_u, \mathcal{L}_v \subset \mathcal{L}$ that satisfy the following. Every point of $\mathcal{P}'$ is incident to a line of $\mathcal{L}_u$ and to a line of $\mathcal{L}_v$. Also, $|\mathcal{L}_u| \leq c_2 r$, $|\mathcal{L}_v| \leq c_2 r$, and $|\mathcal{P}'| \geq m\frac{c_1^4 r^4}{2^7 n^2}$.

Proof We set $x = I(\mathcal{P}, \mathcal{L})$, and let $\mathcal{L}_+$ be the set of lines of $\mathcal{L}$ that are incident to at least $x/2n$ points of $\mathcal{P}$. Since there are at most n lines in $\mathcal{L}\backslash\mathcal{L}_+$, we get that

$$I(\mathcal{P}, \mathcal{L}_+) = I(\mathcal{P}, \mathcal{L}) - I(\mathcal{P}, \mathcal{L}\backslash\mathcal{L}_+) \geq x - n \cdot \frac{x}{2n} = x/2. \tag{13.12}$$

If every point of $\mathcal{P}$ is incident to fewer than $x/2m$ lines of $\mathcal{L}_+$, then

$$I(\mathcal{P}, \mathcal{L}_+) < m \cdot \frac{x}{2m} = x/2.$$

Since this contradicts Equation (13.12), there exists a point $u \in \mathcal{P}$ that is incident to at least $x/2m$ lines of $\mathcal{L}_+$. Since every point of $\mathcal{P}$ is incident to at least $c_1 r$ lines of $\mathcal{L}$, we have that $x \geq mc_1 r$. Thus, u is incident to at least $c_1 r/2$ lines of $\mathcal{L}_+$.

We set

$$\hat{\mathcal{P}} = \{v \in \mathcal{P}\backslash\{u\} \ : \ \text{there exists } \ell \in \mathcal{L} \text{ such that } u, v \in \ell\},$$

and $\hat{m} = |\hat{\mathcal{P}}|$. Every line of $\mathcal{L}_+$ is incident to at least $x/2n \geq mc_1r/2n$ points of $\mathcal{P}$. Thus, every line of $\mathcal{L}_+$ is incident to at least $mc_1r/2n - 1$ points of $\mathcal{P}\backslash\{u\}$. Combining this with the assumption $r \geq 4n/c_1m$ leads to

$$\hat{m} \geq \frac{c_1r}{2}\left(\frac{mc_1r}{2n} - 1\right) \geq \frac{mc_1^2r^2}{8n}. \tag{13.13}$$

To find a valid v, we repeat the above analysis with $\hat{\mathcal{P}}$ instead of $\mathcal{P}$. That is, we set $\hat{x} = I(\hat{\mathcal{P}}, \mathcal{L})$, and let $\hat{\mathcal{L}}_+$ be the set of lines of $\mathcal{L}$ that are incident to at least $\hat{x}/2n$ points of $\hat{\mathcal{P}}$. We have that

$$I(\hat{\mathcal{P}}, \hat{\mathcal{L}}_+) = I(\hat{\mathcal{P}}, \mathcal{L}) - I(\hat{\mathcal{P}}, \mathcal{L}\backslash\hat{\mathcal{L}}_+) \geq \hat{x} - n \cdot \frac{\hat{x}}{2n} = \hat{x}/2. \tag{13.14}$$

If every point of $\hat{\mathcal{P}}$ is incident to fewer than $\hat{x}/2\hat{m}$ lines of $\hat{\mathcal{L}}_+$, then

$$I(\hat{\mathcal{P}}, \hat{\mathcal{L}}_+) < m' \cdot \frac{\hat{x}}{2\hat{m}} = \hat{x}/2.$$

Since this contradicts Equation (13.14), there exists a point $v \in \hat{\mathcal{P}}$ that is incident to at least $\hat{x}/2\hat{m}$ lines of $\hat{\mathcal{L}}_+$. Since every point of $\hat{\mathcal{P}}$ is incident to at least c_1r lines of $\mathcal{L}$, we have that $\hat{x} \geq \hat{m}c_1r$. Thus v is incident to at least $c_1r/2$ lines of $\hat{\mathcal{L}}_+$.

Let ℓ_{uv} be the line that is incident to both u and v. We set

$$\mathcal{P}' = \{w \in \hat{\mathcal{P}}\backslash\ell_{uv} : \text{ there exists } \hat{\ell} \in \hat{\mathcal{L}} \text{ such that } w, v \in \hat{\ell}\}.$$

Every line of $\hat{\mathcal{L}}_+$ is incident to at least $\hat{x}/2n \geq \hat{m}c_1r/2n$ points of $\hat{\mathcal{P}}$. Thus every line of $\hat{\mathcal{L}}_+$ is incident to at least $\hat{m}c_1r/2n - 1$ points of $\hat{\mathcal{P}}\backslash\{v\}$. Combining this with Inequality (13.13) and with the assumption $r \geq \max\left\{4/c_1, (2^5n^2/c_1^3m)^{1/3}\right\}$ leads to

$$|\mathcal{P}'| \geq \left(\frac{c_1r}{2} - 1\right)\left(\frac{\hat{m}c_1r}{2n} - 1\right) \geq \left(\frac{c_1r}{2} - 1\right)\left(\frac{mc_1^3r^3}{16n^2} - 1\right) \geq \frac{mc_1^4r^4}{2^7n^2}.$$

Let $\mathcal{L}_u$ be the lines of $\mathcal{L}\backslash\{\ell_{uv}\}$ that are incident to u. Let $\mathcal{L}_v$ be the lines of $\mathcal{L}\backslash\{\ell_{uv}\}$ that are incident to v. By an assumption of the lemma, we have that $|\mathcal{L}_u| \leq c_2r$ and $|\mathcal{L}_v| \leq c_2r$. By definition, every point of $\mathcal{P}'$ is incident to a line of $\mathcal{L}_v$. Since every point of $\hat{\mathcal{P}}$ is incident to a line of $\mathcal{L}_u$, so is every point of $\mathcal{P}' \subset \hat{\mathcal{P}}$.   □

We are now ready to prove the general point-line incidence bound in $\mathbb{F}_q^2$.

Proof of Theorem 13.8 Our goal is to prove that there exists a constant c that satisfies

$$I(\mathcal{P}, \mathcal{L}) < cm^{11/15}n^{11/15}. \tag{13.15}$$

Below we set c to be sufficiently large. However, the value of c does not depend on m and n.

By Equation (13.2), we have the bound $I(\mathcal{P}, \mathcal{L}) = O(m\sqrt{n} + n)$. Combining this bound with the assumption $m > n^{7/8}$ implies that $I(\mathcal{P}, \mathcal{L}) \leq \beta m \sqrt{n}$, for some $\beta \in \mathbb{R}$. We note that $\beta m \sqrt{n} < c m^{11/15} n^{11/15}$ when $c > \beta m^{4/15}/n^{7/30}$. Since $m < n^{8/7}$, we have that $\beta m^{4/15}/n^{7/30} < \beta n^{1/14}$. Thus, Inequality (13.15) holds when $\sqrt{c} > \beta$ and $\sqrt{c} > n^{1/14}$. In other words, Inequality (13.15) holds when c is sufficiently large and $n < c^7$. For the remainder of the proof, we assume that $n \geq c^7$.

We now prove that Inequality (13.15) holds whenever $n^{4/11} \leq m \leq n^{8/7}$. We consider a fixed $n \geq c^7$ and prove the bound by induction on m. For the induction basis, we consider the case where $n^{4/11} \leq m \leq n^{7/8}$. In this case, when c is sufficiently large, the bound (13.2) implies Inequality (13.15).

For the induction step, we consider the case of $m > n^{7/8}$. The induction hypothesis is that Inequality (13.15) holds for every smaller m that is at least $n^{4/11}$. We assume for contradiction that Inequality (13.15) is false in this case. Then there exist a set $\mathcal{P}$ of m points and a set $\mathcal{L}$ of n lines, such that $I(\mathcal{P}, \mathcal{L}) \geq c m^{11/15} n^{11/15}$. As long as $I(\mathcal{P}, \mathcal{L}) > 2 c m^{11/15} n^{11/15}$, we move points of $\mathcal{P}$ so that they are no longer incident to lines of $\mathcal{L}$. We may thus assume that

$$c m^{11/15} n^{11/15} \leq I(\mathcal{P}, \mathcal{L}) \leq 2 c m^{11/15} n^{11/15}.$$

Points that form approximately $I(\mathcal{P}, \mathcal{L})/m$ incidences: We set $r = I(\mathcal{P}, \mathcal{L})/m$ and consider the sets

$$\mathcal{P}_{\text{poor}} = \{p \in P \ : \ p \text{ is incident to fewer than } r/4 \text{ lines of } \mathcal{L}\},$$

$$\mathcal{P}_{\text{rich}} = \{p \in P \ : \ p \text{ is incident to more than } 8r \text{ lines of } \mathcal{L}\},$$

$$\mathcal{P}_{\text{mid}} = \mathcal{P} \backslash \left(\mathcal{P}_{\text{poor}} \cup \mathcal{P}_{\text{rich}} \right).$$

By definition,

$$I(\mathcal{P}_{\text{poor}}, \mathcal{L}) < m \cdot r/4 = I(\mathcal{P}, \mathcal{L})/4.$$

Claim 13.11 *If $\mathcal{P}' \subset \mathcal{P}$ satisfies $|\mathcal{P}'| \leq 8m$, then $I(\mathcal{P}', \mathcal{L}) < I(\mathcal{P}, \mathcal{L})/4$.*

Proof If $|\mathcal{P}'| \geq n^{4/11}$, then the induction hypothesis implies that

$$I(\mathcal{P}', \mathcal{L}) < c|\mathcal{P}'|^{11/15} n^{11/15} \leq c(m/8)^{11/15} n^{11/15} < I(\mathcal{P}, \mathcal{L})/4.$$

If $|\mathcal{P}'| \leq n^{4/11}$, then Equation (13.2) and a sufficiently large c imply that

$$I(\mathcal{P}', \mathcal{L}) = O(n) < I(\mathcal{P}, \mathcal{L})/4. \qquad \square$$

We observe that $|\mathcal{P}_{\text{rich}}| < I(\mathcal{P}, \mathcal{L})/8r = m/8$. Claim 13.11 implies that $I(\mathcal{P}_{\text{rich}}, \mathcal{L}) < I(\mathcal{P}, \mathcal{L})/4$. If $|\mathcal{P}_{\text{mid}}| \leq m/8$, then Claim 13.11 also implies that $I(\mathcal{P}_{\text{rich}}, \mathcal{L}) < I(\mathcal{P}, \mathcal{L})/4$. This leads to

$$I(\mathcal{P}, \mathcal{L}) = I(\mathcal{P}_{\text{poor}}, \mathcal{L}) + I(\mathcal{P}_{\text{mid}}, \mathcal{L}) + I(\mathcal{P}_{\text{rich}}, \mathcal{L}) < 3I(\mathcal{P}, \mathcal{L})/4.$$

We may thus assume that $|\mathcal{P}_{\text{mid}}| > m/8$.

Finding Cartesian products: We set $\mathcal{P}_1 = \mathcal{P}_{\text{mid}}$, $m_1 = |\mathcal{P}_1|$, $c_1 = 1/4$, and $c_2 = 8$. By definition, every point of $\mathcal{P}_1$ is incident to at least $c_1 r$ lines of $\mathcal{L}$ and to at most $c_2 r$ such lines. To apply Lemma 13.10, it remains to show that

$$r \geq \max \left\{ 4n/c_1 m_1, 4/c_1, \left(2^5 n^2/c_1^3 m_1 \right)^{1/3} \right\}. \tag{13.16}$$

We recall that $r = I(\mathcal{P}, \mathcal{L})/m \geq cn^{11/15} m^{-4/15}$. Since $m_1 > m/8$ and $m > n^{4/11}$, we get that

$$4n/c_1 m_1 < 32n/c_1 m < 32n^{11/15}/c_1 m^{4/15}.$$

When c is sufficiently large, we indeed have that $r \geq 4n/c_1 m_1$. It is not difficult to verify that the other two cases of Inequality (13.16) also hold. One of these cases also requires the assumption $m < n^{8/7}$.

By the above, we can apply Lemma 13.10 with $\mathcal{P}_1, \mathcal{L}, c_1, c_2$, and r. We obtain points $u_1, v_1 \in \mathcal{P}_1$, a set $\mathcal{P}_1' \subset \mathcal{P}_1$, and a (u_1, v_1)-Cartesian product $\mathcal{L}_{u_1}, \mathcal{L}_{v_1} \subset \mathcal{L}$, as stated in that lemma. We then set $\mathcal{P}_2 = \mathcal{P}_1 \backslash \mathcal{P}_1'$.

We repeat the above process, as follows. At the jth iteration, we apply Lemma 13.10 with $\mathcal{P}_j$ and $\mathcal{L}$. We obtain points $u_j, v_j \in \mathcal{P}_{\text{mid}}$, a set $\mathcal{P}_j' \subset \mathcal{P}_{\text{mid}}$, and a (u_j, v_j)-Cartesian product $\mathcal{L}_{u_j}, \mathcal{L}_{v_j} \subset \mathcal{L}$. We then set $\mathcal{P}_{j+1} = \mathcal{P}_j \backslash \mathcal{P}_j'$. The process stops when $|\mathcal{P}_{j+1}| \leq m/8$. The values of c_1, c_2, and r remain unchanged throughout this process. As long as $|\mathcal{P}_{j+1}| > m/8$, the conditions of Inequality (13.16) remain valid, so we may keep applying Lemma 13.10.

Let s be the number of times that we applied Lemma 13.10 in the above process. By that lemma, for every $1 \leq j \leq s$ we have that

$$|\mathcal{P}_j'| \geq |\mathcal{P}_j| \frac{c_1^4 r^4}{2^7 n^2} > \frac{m}{8} \cdot \frac{c_1^4 r^4}{2^7 n^2} = \frac{mr^4}{2^{18} n^2}.$$

Since the sets $\mathcal{P}_j'$ are pairwise disjoint, we get that

$$s \leq \frac{m}{mr^4/2^{18} n^2} = \frac{2^{18} n^2}{r^4}. \tag{13.17}$$

Incidences with a Cartesian product: We fix a value of j between 1 and s and consider $I(\mathcal{P}_j, \mathcal{L})$. By Lemma 13.10, every point of $\mathcal{P}_j'$ is incident to a line of $\mathcal{L}_{u,j}$ and to a line of $\mathcal{L}_{v,j}$. We move from the affine plane $\mathbb{F}_q^2$ to the projective plane $\mathbb{PF}_q^2$, as explained in Section 13.2. We then perform a projective transformation τ_j that takes u_j to $[1 : 0 : 0]$ and v_j to $[0 : 1 : 0]$. Asking τ_j to take one point to another leads to a system of three linear equations, as shown in Equation (13.5). By counting the number of variables and the number of linear equations, we get that such a transformation τ_j always exists.

As explained in Section 13.2, the transformation τ_j takes every line of $\mathcal{L}_{u,j}$ to a horizontal line and every line of $\mathcal{L}_{v,j}$ to a vertical line. This implies that τ_j takes $\mathcal{P}_j$ to points of a Cartesian product of size $(c_2 r) \times (c_2 r) = 8r \times 8r$. We denote this Cartesian product as G_j. We then return to the affine plane $\mathbb{F}_q^2$ by taking the intersection of $\mathbb{PF}_q^3$ with the plane $\mathbf{V}(z - 1)$. Let $\mathcal{L}_j$ be the set of lines of $\mathcal{L}$ after the above transformations. We recall that no point of $\mathcal{P}_j$ is on the line that is incident to both u_j and v_j. This implies that τ_j does not take any point of $\mathcal{P}_j$ to the line at infinity. Thus, $I(\mathcal{P}_j, \mathcal{L}) \leq I(G_j, \mathcal{L}_j)$.

We now verify that the conditions of Theorem 13.9 hold for G_j and $\mathcal{L}_j$. The condition on the size of the Cartesian product becomes $(8r)^3 < n^3$, or equivalently $r < n/8$. We have that $r = I(\mathcal{P}, \mathcal{L})/m \leq 2cn^{11/15}m^{-4/15} < 2cn^{11/15}$. Thus, $r < n/8$ when $2c < n^{4/15}/8$. This is indeed the case, since $n \geq c^7$ and c is sufficiently large. Since $r = I(\mathcal{P}, \mathcal{L})/m \leq 2cn^{11/15}m^{-4/15}$, $m > n^{7/8}$, and $n = O(p^{4/3})$, we have that

$$8rn \leq 16cn^{26/15}m^{-4/15} < 16cn^{26/15}n^{-7/30} = 16cn^{3/2} = O(p^2).$$

Let α be the constant that is hidden by the $O(\cdot)$-notation in the bound of Theorem 13.9. By applying Theorem 13.9, we obtain that

$$I(\mathcal{P}_j, \mathcal{L}) \leq I(G_j, \mathcal{L}_j) \leq \alpha \left((8r)^{3/4}(8r)^{1/2}n^{3/4} + n \right). \tag{13.18}$$

Wrapping up: Combining Equations (13.17) and (13.18) leads to

$$\sum_{j=1}^{s} I(\mathcal{P}_j, \mathcal{L}) \leq \frac{n^2 2^{18}}{r^4} \cdot \alpha \left((8r)^{5/4}n^{3/4} + n \right) = \alpha \left(2^{22}\frac{n^{11/4}}{r^{11/4}} + \frac{n^3 2^{18}}{r^4} \right)$$

$$= \alpha \left(2^{22}\frac{n^{11/4}m^{11/4}}{I(\mathcal{P}, \mathcal{L})^{11/4}} + \frac{n^3 m^4 2^{18}}{I(\mathcal{P}, \mathcal{L})^4} \right).$$

Since $I(\mathcal{P}, \mathcal{L}) \geq cm^{11/15}n^{11/15}$, when c is sufficiently large we get that

$$\sum_{j=1}^{s} I(\mathcal{P}_j, \mathcal{L}) < I(\mathcal{P}, \mathcal{L})/4.$$

By definition, $|\mathcal{P}_{s+1}| \leq m/8$. Claim 13.11 implies that $I(\mathcal{P}_{s+1}, \mathcal{L}) < I(\mathcal{P}, \mathcal{L})/4$. We conclude that

$$I(\mathcal{P}, \mathcal{L}) = I(\mathcal{P}_{\text{poor}}, \mathcal{L}) + I(\mathcal{P}_{\text{rich}}, \mathcal{L}) + \sum_{j=1}^{s+1} I(\mathcal{P}_j, \mathcal{L}) < I(\mathcal{P}, \mathcal{L}).$$

This contradiction completes the induction step. □

13.6 Exercises

Exercise 13.1   When moving from $\mathbb{F}_q^2$ to $\mathbb{PF}_q^2$, one new point is added to each line. The coordinates of the new point depend on the slope of the line, but not on the y-intercept. We define a circle in $\mathbb{F}_q^2$ as $V((x-a)^2 + (y-b)^2 - r^2)$, where $a, b, r \in \mathbb{F}_q$. When moving from $\mathbb{F}_q^2$ to $\mathbb{PF}_q^2$, what new points are added to a circle? What properties of a circle determine the coordinates of these new points?

Exercise 13.2   Let $k \geq 3$, let $r \geq 2$, and let H be a set of n planes in $\mathbb{F}_q$. Let $\mathcal{P}$ be a set of points, such that no k points are collinear and every point is incident to at least r planes of H. Prove that

$$|\mathcal{P}| = O\left(\frac{n^2 k}{r^2} + \frac{n}{r}\right).$$

Exercise 13.3   Let $k \geq 3$, let $r = \Omega(\sqrt{nk})$, and let H be a set of n planes in $\mathbb{F}_q$. Let $\mathcal{P}$ be a set of points such that no k points are collinear and every point is incident to at least r planes of H. Prove that

$$|\mathcal{P}| = O\left(\frac{n^2}{r^2} + \frac{nk}{r}\right).$$

(Hint: How is this related to Exercise 13.2?)

Exercise 13.4   We proved Theorem 13.7 by applying Theorem 13.6. Provide an alternative proof by applying Theorem 13.9. You may not use Theorem 13.6.

Exercise 13.5   Let $A \subset \mathbb{F}_q$ satisfy $|A| = O(p^{2/3})$. Recall that

$$A + AA = \{a + bc \ : \ a, b, c \in A\}.$$

(a) Prove that $|A + AA| = \Omega(|A|^{3/2})$ by using Theorem 13.9.
(b) Prove the same result by using Theorem 13.6 instead of Theorem 13.9. (Hint: Use the energy E' from the proof of Theorem 13.7. Use the Cauchy–Schwarz inequality to find a lower bound for this energy.)

Exercise 13.6 Let $A \subset \mathbb{F}_q$ satisfy $|A| = O(p^{2/3})$. Recall that

$$A(A + A) = \{a(b + c) : a, b, c \in A\}.$$

(a) Prove that $|A(A + A)| = \Omega(|A|^{3/2})$ by using Theorem 13.9.
(b) When trying to imitate the proof of Exercise 13.5(b), we get to an energy that is defined by $a(b + c) \equiv a'(b' + c')$. Explain why we cannot turn this into a point-plane incidence problem in $\mathbb{F}_q^3$ and then apply Theorem 13.6, as before.
(c) Fix the issue from part (b) by considering the equation $k + ac \equiv k + a'c'$, where $k, k' \in AA$. (Hint: The set AA does not affect the number of points.)

13.7 Open Problems

Let $\mathcal{P}$ be a set of m points and let $\mathcal{L}$ be a set of n lines, both in $\mathbb{F}_q^2$. We know that the Szemerédi–Trotter bound $I(\mathcal{P}, \mathcal{L}) = O(|\mathcal{P}|^{2/3}|\mathcal{L}|^{2/3} + |\mathcal{P}| + |\mathcal{L}|)$ does not hold when $m \cdot n$ is asymptotically larger than q^3. We know that this bound does hold when $m \cdot n = \Theta(q^3)$, and also when $m = n = O(\log \log \log p)$ and q is prime. There is a huge gap between these two extreme cases. The current best bound in this gap, stated in Theorem 13.8, might be far from being tight.

Open Problem 13.12 *Let $\mathcal{P}$ be a set of m points and let $\mathcal{L}$ be a set of n lines, both in $\mathbb{F}_q^3$. Let $m \cdot n$ be asymptotically smaller than q^3. Find a tight upper bound for $I(\mathcal{P}, \mathcal{L})$.*

Since point-line incidences have many applications, progress towards Open Problem 13.12 is likely to lead to progress on other problems. For example, a new point-line incidence bound may lead to a new sum-product bound in $\mathbb{F}_q$.

14

Algebraic Families, Dimension Counting, and Ruled Surfaces

Algebraic geometry seems to have acquired the reputation of being esoteric, exclusive, and very abstract, with adherents who are secretly plotting to take over all the rest of mathematics. In one respect this last point is accurate.

David Mumford (1999, Appendix).

In this chapter, we discuss advanced tools and techniques that rely on additional concepts from algebraic geometry. These tools could be helpful for people who do research work in incidence theory and related topics. A reader who is new to this field might prefer to skip this chapter.

We sometimes wish to consider families of varieties, such as the set of circles in $\mathbb{R}^2$ or the set of planes in $\mathbb{R}^3$ that not are incident to the origin. In this chapter, we rigorously define such families. We also generalize the idea of point-line duality to every family of varieties. We then see how these notions could be used to prove various results. In particular, we derive a new incidence bound and prove various properties of surfaces in $\mathbb{R}^3$ and $\mathbb{C}^3$.

14.1 Families of Varieties

Consider a set of polynomials $\Gamma \subset \mathbb{R}[x_1, \ldots, x_d]$, where the coefficients of each polynomial are themselves polynomials in $\mathbb{R}[s_1, \ldots, s_k]$. We refer to $s_1, \ldots, s_k$ as the *parameters* of Γ. For example, we consider

$$\Gamma_1 = \left\{ (x - a)^2 + (y - b)^2 - r^2 \right\} \subset \mathbb{R}[x, y],$$

with parameters $a, b, r \in \mathbb{R}$. Every assignment of values to the parameters a, b, r turns $\mathbf{V}((x - a)^2 + (y - b)^2 - r^2)$ into a circle or a point in $\mathbb{R}^2$.

For a set of polynomials $\Gamma = \{f_1, \ldots, f_\ell\}$, we set $\mathbf{V}(\Gamma) = \mathbf{V}(f_1, \ldots, f_\ell)$. For example, we consider

$$\Gamma_2 = \left\{ (x - p_x)^2 + (y - p_y)^2 + (z - p_z)^2 - r^2, a_x x + a_y y + a_z z + b \right\} \subset \mathbb{R}[x, y, z],$$

204

with parameters $p_x, p_y, p_z, r, a_x, a_y, a_z, b \in \mathbb{R}$. Assigning values to p_x, p_y, p_z, r turns $\mathbf{V}((x - p_x)^2 + (y - p_y)^2 + (z - p_z)^2 - r^2)$ into a sphere or a point in $\mathbb{R}^3$. Assigning values to a_x, a_y, a_z, b turns $\mathbf{V}(a_x x + a_y y + a_z z + b)$ into a plane in $\mathbb{R}^3$. Thus, after fixing the values of all the parameters, $\mathbf{V}(\Gamma_2)$ is a circle, a point, or an empty set in $\mathbb{R}^3$. When $a_x = a_y = a_z = b = 0$ and $r \neq 0$, the set $\mathbf{V}(\Gamma_2)$ is a sphere in $\mathbb{R}^3$.

The space $\mathbb{R}^d$ that contains the varieties $\mathbf{V}(\Gamma)$ is called the *primal space*. The space $\mathbb{R}^k$ whose coordinates are the parameters of Γ is the *dual space*. In the preceding example, the primal space $\mathbb{R}^3$ contains the circles and points that are formed by $\mathbf{V}(\Gamma_2)$. The dual space $\mathbb{R}^8$ has the coordinates p_x, p_y, p_z, r, a_x, a_y, a_z, b.

Consider a set of polynomials $\Gamma \subset \mathbb{R}[x_1, \ldots, x_d]$ with parameters $s_1, \ldots, s_k \in \mathbb{R}$. For a point $s \in \mathbb{R}^k$, we let $\Gamma(s)$ be the set of polynomials of Γ after assigning the values of s to the parameters $s_1, \ldots, s_k$. The *algebraic family of varieties* of Γ is

$$\left\{ \mathbf{V}(\Gamma(s)) \ : \ s \in \mathbb{R}^k \right\}.$$

For brevity, we usually remove the word "algebraic" and refer to a *family of varieties*.

Here, we give couple of examples of families of varieties:

(a) The family of points and circles in $\mathbb{R}^2$ is defined by Γ_1. Every circle originates from two points of $\mathbb{R}^3$: one where $r > 0$ and another where $r < 0$.

(b) The family of points, circles, and spheres in $\mathbb{R}^3$ is defined by Γ_2. Every element of this family corresponds to infinitely many points of $\mathbb{R}^8$.

In Section 14.4 we further generalize the definition of a family.

Duality: We consider a family F of varieties in $\mathbb{R}^d$, which is defined by a set of polynomials Γ with parameters $s_1, \ldots, s_k$. The *dual* of a point $p \in \mathbb{R}^d$ is

$$p^* = \left\{ s \in \mathbb{R}^k \ : \ p \in \mathbf{V}(\Gamma(s)) \right\}.$$

In other words, p^* is the set of points that parameterize varieties that contain p. We claim that p^* is a variety. Indeed, p^* is defined by the polynomials of Γ, but with $s_1, \ldots, s_k$ as variables and with the coordinates of p assigned to $x_1, \ldots, x_d$.

Let $U \subset \mathbb{R}^d$ be a variety from F. A point s in the dual space $\mathbb{R}^k$ is *dual* to U if $U = \mathbf{V}(\Gamma(s))$. The variety U may have more than one dual point. In example (a) above, every circle in $\mathbb{R}^2$ has two dual points. In example (b), every circle in $\mathbb{R}^3$ has infinitely many dual points. We denote by U^* an arbitrary point that is dual to U.

We consider a variety U from F and a point $p \in \mathbb{R}^d$. Then p is incident to U if and only if the point U^* is incident to the variety p^*. Indeed, in both cases we get the same set of equations from Γ. This is a generalization of the point-line duality from Section 1.10.

The *primal complexity* of F is the maximum complexity of a variety of F. The *dual complexity* of F is the maximum complexity of a variety that is dual to a point $p \in \mathbb{R}^d$. Equivalently, the dual complexity is the minimum D such that Γ is defined by at most D polynomials of degree at most D in $s_1, \ldots, s_k$. While the primal complexity of F is determined by the degrees in $x_1, \ldots, x_k$, the dual complexity of F is determined by the degrees in $s_1, \ldots, s_k$. For brevity, we denote the *complexity* of F as the maximum of the primal and dual complexities of F.

As an example, consider the family F that is defined by

$$\Gamma = \left\{ a_1 x^2 + a_2 xy + a_3 y^2 + a_4 x + a_5 y + a_6 \right\}.$$

This family consists of all curves of degree at most two in $\mathbb{R}^2$, some finite sets, and the entire plane $\mathbb{R}^2$. With this Γ, the primal complexity of F is two and the dual complexity is one. Thus, the dual varieties are hyperplanes in $\mathbb{R}^6$.

14.2 An Incidence Bound for Large Parameters

We now explore our first application for families of varieties. Throughout this book, we have studied many incidence bounds in $\mathbb{R}^d$, when the incidence graph contains no $K_{s,t}$. For example, see Theorems 11.3, 8.6, and 11.1. We also saw many applications for such bounds. In these applications, we usually have that $s = 2$ or $s = 3$. For example, see Theorems 10.5, 12.2, and 12.3.

When s is large, our incidence bounds become almost trivial. For example, for a set $\mathcal{P}$ of m points and a set $\mathcal{V}$ of n varieties, Theorem 11.3 gives the bound

$$I(\mathcal{P}, \mathcal{V}) = O\left(m^{\frac{(d-1)s}{ds-1} + \varepsilon} n^{\frac{d(s-1)}{ds-1}} + m + n \right).$$

As s grows, this bound becomes closer to the bound $O(m^{1-1/s}n + m)$. This is the nongeometric bound from Lemma 8.1.

When s is large, we can obtain stronger incidence bounds by considering the number of parameters that define the family. In other words, by considering the dimension of the dual space. Our first example is the following result, which is by Braß and Knauer (2003) (see also Apfelbaum and Sharir, 2007).

Theorem 14.1 *Let $\mathcal{P}$ be a set of m points and let H be a set of n hyperplanes, both in $\mathbb{R}^d$. If the incidence graph of $\mathcal{P} \times H$ contains no $K_{s,s}$, then*

$$I(\mathcal{P}, H) = O_{s,d}\left(m^{d/(d+1)} n^{d/(d+1)} + m + n \right).$$

For point-hyperplane incidences, Theorem 14.1 is stronger than Theorem 11.3 when $s > d$. Significantly more general results can be found in Do (2019) and Fox et al. (2017). We now derive a special case of these results, with a different proof.

Theorem 14.2 *Let $\mathcal{P}$ be a set of m points and let $\mathcal{V}$ be a set of n varieties, both in $\mathbb{R}^d$. The varieties of $\mathcal{V}$ belong to a family F with k parameters and complexity D. The incidence graph of $\mathcal{P} \times \mathcal{V}$ contains no $K_{s,t}$. Then, for every $\varepsilon > 0$, we have that*

$$I(\mathcal{P}, \mathcal{V}) = O_{s,t,d,D,k,\varepsilon}\left(m^{\frac{(k-1)d}{kd-1}+\varepsilon} n^{\frac{k(d-1)}{kd-1}} + m^{1+\varepsilon} + n\right).$$

Proof We rely on $\varepsilon, \varepsilon^*, \varepsilon' > 0$, such that ε' is sufficiently small with respect to ε^* and ε^* is sufficiently small with respect to ε. More precise relations are described below. Since s may be large, the bound of Lemma 8.1 might not be strong enough for our purposes. Thus, our first goal is to derive an alternative initial bound.

Deriving a weak bound: Let $\mathcal{P}' \subset \mathcal{P}$ be a set of M points. Let $\mathcal{V}' \subset \mathcal{V}$ be a set of N varieties. We apply Theorem 11.3 with ε', to obtain that

$$I(\mathcal{P}', \mathcal{V}') = O_{s,t,D,\varepsilon',d}\left(M^{\frac{(d-1)s}{ds-1}+\varepsilon'} N^{\frac{d(s-1)}{ds-1}} + M + N\right). \tag{14.1}$$

We first consider the case where

$$N^d \le M^{1+\frac{\varepsilon' d(ds-1)}{d-1}}.$$

Taking both sides to the power of $(s-1)/(ds-1)$ leads to

$$N^{\frac{d(s-1)}{ds-1}} \le M^{\frac{s-1}{ds-1}+\frac{\varepsilon' d(s-1)}{d-1}}.$$

Multiplying both sides by $M^{\frac{(d-1)s}{ds-1}+\varepsilon'}$ gives that

$$M^{\frac{(d-1)s}{ds-1}+\varepsilon'} N^{\frac{d(s-1)}{ds-1}} \le M^{\frac{s-1}{ds-1}+\frac{\varepsilon' d(s-1)}{d-1}+\frac{(d-1)s}{ds-1}+\varepsilon'} < M^{1+ds\varepsilon'}.$$

By combining this with Equation (14.1) and setting $1 + ds\varepsilon' \le \varepsilon^*$, we get that

$$I(\mathcal{P}', \mathcal{V}') = O_{s,t,D,\varepsilon,d}(M^{1+\varepsilon^*} + N).$$

We now consider the case where

$$N^d > M^{1+\frac{\varepsilon' d(ds-1)}{d-1}}.$$

By switching sides and raising both sides to the power $(d-1)/d(ds-1)$, we obtain that

$$M^{\frac{d-1}{d(ds-1)}+\varepsilon'} < N^{\frac{d-1}{ds-1}}.$$

Combining this with Equation (14.1) leads to

$$I(\mathcal{P}', \mathcal{V}') = O_{s,t,D,\varepsilon',d}\left(M^{\frac{(d-1)s}{ds-1}+\varepsilon'}N^{\frac{d(s-1)}{ds-1}} + M + N\right)$$

$$= O_{s,t,D,\varepsilon',d}\left(M^{\frac{d-1}{d}}M^{\frac{d-1}{d(ds-1)}+\varepsilon'}N^{\frac{d(s-1)}{ds-1}} + M + N\right)$$

$$= O_{s,t,D,\varepsilon',d}\left(M^{\frac{d-1}{d}}N^{\frac{d-1}{ds-1}}N^{\frac{d(s-1)}{ds-1}} + M + N\right)$$

$$= O_{s,t,D,\varepsilon',d}\left(M^{\frac{d-1}{d}}N + M\right).$$

To recap, in both of the above cases, we obtain that

$$I(\mathcal{P}', \mathcal{V}') = O_{s,t,D,\varepsilon,d}\left(M^{\frac{d-1}{d}}N + M^{1+\varepsilon^*}\right). \tag{14.2}$$

The dual space: We rely on the duality that is described in Section 14.1. We define the point set

$$\mathcal{V}^* = \{U^* \,:\, U \in \mathcal{V}\} \subset \mathbb{R}^k.$$

We also define the set of varieties

$$\mathcal{P}^* = \{p^* \,:\, p \in \mathcal{P}\}.$$

Every variety of $\mathcal{P}^*$ is of complexity at most D. Note that we now have n points and m varieties.

As described in Section 14.1, we have that $I(\mathcal{P}, \mathcal{V}) = I(\mathcal{P}^*, \mathcal{V}^*)$. Thus, it suffices to derive the bound

$$I(\mathcal{P}^*, \mathcal{V}^*) = O_{s,t,d,k,D,\varepsilon}\left(m^{\frac{(k-1)d}{kd-1}+\varepsilon}n^{\frac{k(d-1)}{kd-1}} + m^{1+\varepsilon} + n\right).$$

When the incidence graph of $\mathcal{V}^* \times \mathcal{P}^*$ contains no $K_{d,t}$, Theorem 11.3 implies a stronger bound than the above. Unfortunately, the incidence graph of $\mathcal{V}^* \times \mathcal{P}^*$ may contain $K_{d,t}$.

Theorem 11.3 is an immediate corollary of Theorem 11.6. By inspecting the proof of Theorem 11.6, we note that the assumption about no $K_{s,t}$ is used in two places:

- to derive the weak bound $n = O(m^s)$ in Equation (11.8),
- when we study the incidences that contribute to I_1. That is, incidences between a point that is contained in a component of the partition and a variety that contains this component.

The second case, of incidences that contribute to I_1, can be handled as before. Handling this case requires no $K_{s,t}$ in the incidence graph, but works also when s and t are large. The incidence graph of $\mathcal{V}^* \times \mathcal{P}^*$ contains no $K_{t,s}$. It remains to address the first case.

Instead of Equation (11.8), we use Equation (14.2). It may seem as if the variables were switched in those two bounds. However, while in Theorem 11.3 there are m points and n varieties, we currently have n points and m varieties. Thus, the only difference between Equations (11.8) and (14.2) is the additional ε^* in the exponent. Repeating the remainder of the proof of Theorem 11.3 leads to the asserted bound. Checking this is left as an exercise for the reader (Exercise 14.2). □

14.3 Complexification and Constructible Sets

For our more advanced use of families of varieties, we require a few additional concepts and tools from algebraic geometry. In this section we briefly present these concepts and tools.

Complexification: We recall that a variety in $\mathbb{C}^d$ is defined just as in $\mathbb{R}^d$: For polynomials $f_1, \ldots, f_k \in \mathbb{C}[x_1, \ldots, x_d]$, we have that

$$\mathbf{V}(f_1, \ldots, f_k) = \left\{ (a_1, \ldots, a_d) \in \mathbb{C}^d \ : \ f_j(a_1, \ldots, a_d) = 0 \text{ for all } 1 \leq j \leq k \right\}.$$

The *dimension* of a variety in $\mathbb{C}^d$ can also be defined as in $\mathbb{R}^d$ (see Section 4.2). The situation is different with respect to the degree of a variety. Unlike the case of $\mathbb{R}^d$, there is a standard and intuitive definition for the degree of a variety in $\mathbb{C}^d$. This definition has many equivalent formulations. For example, the *degree* of a d'-dimensional variety $U \subset \mathbb{C}^d$ is the number of intersection points between U and a generic $(d - d')$-dimensional flat. For the current chapter, we do not need a good understanding of this degree definition. Thus, we do not discuss this definition in detail and do not provide examples. For more information, see for example Harris (2013, Chapter 18).

Given a point $p \in \mathbb{R}^d$, we can also consider p as a point in $\mathbb{C}^d$ with no imaginary terms. Similarly, given a variety $U \subset \mathbb{R}^d$, we can also consider U as a point set in $\mathbb{C}^d$. The *complexification* of U, denoted $U_\mathbb{C}$, is the smallest variety $U \subset \mathbb{C}^d$ that contains U. Every variety in $\mathbb{C}^d$ that contains U also contains $U_\mathbb{C}$. Such a complexification always exists, and U is precisely the set of points of $U_\mathbb{C}$ that consist of real coordinates. For a set $S \subset \mathbb{C}^d$, we let $\Re(S)$ denote the set of points of S that consist of real coordinates. We can now rephrase the above as: Every variety $U \subset \mathbb{R}^d$ satisfies that $U = \Re(U_\mathbb{C})$.

Lemma 14.3
(a) Let $U \subset \mathbb{R}^d$ be a variety of dimension d' and complexity D. Then $U_\mathbb{C} \subset \mathbb{C}^d$ is a variety of dimension d' and degree $O_{D,d}(1)$.
(b) Let $W \subset \mathbb{C}^d$ be a variety of dimension d' and degree D. Then $\Re(W) \subset \mathbb{R}^d$ is a variety of dimension at most d' and complexity $O_{D,d}(1)$.

It is not true that every variety $W \subset \mathbb{C}^d$ satisfies that $W = (\mathfrak{R}(W))_{\mathbb{C}}$. For example, let z_1, z_2 be the coordinates of the plane $\mathbb{C}^2$ and consider the complex line $\ell = \mathbf{V}(z_1 - iz_2)$. Since $\mathfrak{R}(\ell) = \{(0, 0)\}$, we get that $(\mathfrak{R}(\ell))_{\mathbb{C}} = \{(0, 0)\}$. For more information about the interaction between real and complex varieties, see Whitney (1957).

Constructible sets: We recall from Section 4.5 that the *Zariski closure* of a set $S \subset \mathbb{R}^d$, denoted $\overline{S}$, is the smallest variety in $\mathbb{R}^d$ that contains S. If a variety $U \subset \mathbb{R}^d$ contains S then U also contains $\overline{S}$. For example, the Zariski closure of the segment $\{(k, k) : 0 \leq k \leq 1\} \subset \mathbb{R}^2$ is the line $\mathbf{V}(x - y)$.

A set $X \subset \mathbb{R}^d$ is *constructible* if there exist nonempty varieties $X_1, \ldots, X_\ell \subset \mathbb{R}^d$ such that $\dim X_1 > \dim X_2 > \cdots > \dim X_\ell$ and

$$X = \Big(((X_1 \backslash X_2) \cup X_3) \backslash X_4 \ldots \Big). \tag{14.3}$$

For example, removing two great circles from a sphere leads to a constructible set. On the other hand, a hemisphere is not a constructible set. See Figure 14.1.

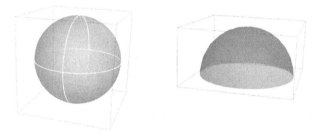

Figure 14.1 (Left) The set $\mathbf{V}(x^2 + y^2 + z^2 - 1) \backslash \mathbf{V}(xz)$ is constructible. (Right) The hemisphere $\{(x, y, z) \in \mathbb{R}^3 : x^2 + y^2 + z^2 = 1 \text{ and } z \geq 0\}$ is not constructible.

The dimension of a constructible set X as defined in Equation (14.3) is

$$\dim(X) = \dim(\overline{X}) = \dim(X_1).$$

Let D_j denote the complexity of X_j. The *complexity* of X is $\min\{D_1 + D_2 + \cdots + D_\ell\}$, where the minimum is taken over all representations of X of the form in Equation (14.3).

The notions of Zariski closure and constructible sets are defined in the same way in $\mathbb{C}^d$. The only difference is that, in $\mathbb{C}^d$, the complexity of X is $\min\{\deg X_1 + \deg X_2 + \cdots + \deg X_\ell\}$. Projections of constructible sets in $\mathbb{C}^d$ are well behaved.

Lemma 14.4 *Let $X \subset \mathbb{C}^d$ be a constructible set of dimension d' and complexity D. Let $\pi \colon \mathbb{C}^d \to \mathbb{C}^e$ be a projection on e of the d coordinates* *of* $\mathbb{C}^d$. *Then* $\pi(X)$ *is a constructible set of dimension at most* d' *and complexity* $O_{D,d}(1)$.

For more information about Lemma 14.4, see, for example, Harris (2013, Chapter 3). This lemma does not hold in $\mathbb{R}^d$. For example, the projection of the circle $\mathbf{V}(x^2 + y^2 - 1) \subset \mathbb{R}^2$ on the x-axis is the interval $[-1, 1]$, which is not constructible. On the other hand, the projection of the complex circle $\mathbf{V}(x^2 + y^2 - 1) \subset \mathbb{C}^2$ on the x-axis is $\mathbb{C}$.

The inverse of a projection is also well behaved in $\mathbb{C}^d$.

Lemma 14.5 *Let* $U \subset \mathbb{C}^d$ *be a variety. For a projection* $\pi \colon \mathbb{C}^d \to \mathbb{C}^e$, *let* $W = \pi(U)$. *Then, for a generic point* $p \in W$, *we have that* $\pi^{-1}(p) \cap U$ *is a constructible set of dimension* $\dim U - \dim W$.

For more information about Lemma 14.5, see Harris (2013, Corollary 11.13) and the paragraph following it. For the claim that the set is constructible, see Harris (2013, Chapter 3).

Let $X \subset \mathbb{R}^d$ be a constructible set as described in Equation (14.3). The *complexification* of X is

$$X_{\mathbb{C}} = \Big((((X_1)_{\mathbb{C}} \backslash (X_2)_{\mathbb{C}}) \cup (X_3)_{\mathbb{C}}) \backslash (X_4)_{\mathbb{C}} \dots \Big).$$

Let $Y \subset \mathbb{C}^d$ be a constructible set of the form $(((Y_1 \backslash Y_2) \cup Y_3) \backslash Y_4 \dots)$. It is not difficult to check that

$$\mathfrak{R}(Y) = \Big((((\mathfrak{R}(Y_1) \backslash \mathfrak{R}(Y_2)) \cup \mathfrak{R}(Y_3)) \backslash \mathfrak{R}(Y_4) \dots \Big).$$

The following result is a variant of Lemma 14.3 for constructible sets. It is immediate from Lemma 14.3 and the above definitions.

Corollary 14.6
(a) Let $U \subset \mathbb{R}^d$ *be a constructible set of dimension* d' *and complexity* D. *Then* $U_{\mathbb{C}} \subset \mathbb{C}^d$ *is a constructible set of dimension* d' *and complexity* $O_{D,d}(1)$.
(b) Let $W \subset \mathbb{C}^d$ *be a constructible set of dimension* d' *and complexity* D. *Then* $\mathfrak{R}(W) \subset \mathbb{R}^d$ *is a constructible set of dimension at most* d' *and complexity* $O_{D,d}(1)$.

14.4 Families with Sets of Parameters

We now further generalize the concept of a family of varieties from Section 14.1. As in Section 14.1, we consider a set of polynomials $\Gamma \subset \mathbb{R}[x_1, \dots, x_d]$. The coefficients in the polynomials of Γ are polynomials in $\mathbb{R}[s_1, \dots, s_k]$. Let $S \subset \mathbb{R}^k$ be a constructible set in the dual space. For a point $s \in S$, we let $\Gamma(s)$

be the set of polynomials of Γ after assigning the coordinates of $s \in \mathbb{R}^k$ to the parameters $s_1, \ldots, s_k$. The *family of varieties* of Γ and S is

$$\{V(\Gamma(s)) \; : \; s \in S\}.$$

We consider a couple of families of varieties with a set of parameters as examples. First, we consider $\Gamma_1 = \{(x - a)^2 + (y - b)^2 - r^2\}$, the parameters a, b, r, and the constructible set $S_1 = \mathbb{R}^3 \backslash V(r)$. This defines the family of circles in $\mathbb{R}^2$. Unlike example (a) in Section 14.1, the current family does not include sets that consist of a single point.

Next, we consider

$$\Gamma_2 = \{x + a_y y + a_z z + a_0, x + b_y y + b_z z + b_0\} \subset \mathbb{R}^3,$$

the parameters $a_y, a_z, a_0, b_y, b_z, b_0$, and the constructible set

$$S_2 = \mathbb{R}^6 \backslash (V(a_0) \cup V(a_y - b_y, a_z - b_z)).$$

This defines the family of lines in $\mathbb{R}^3$ that are not incident to the origin. Removing $V(a_0)$ eliminates the lines that are incident to the origin. Removing $V(a_y - b_y, a_z - b_z)$ eliminates planes and the empty set.

Duality: When we have a set of parameters S, we need to slightly revise the definition of a dual. We fix a family of varieties F that is defined by a set of polynomials Γ and a constructible set $S \subset \mathbb{R}^k$. The *dual* of a point p in the primal space $\mathbb{R}^d$ is

$$p^* = \{s \in S \; : \; p \in V(\Gamma(s))\}.$$

In other words, p^* is the set of points of S that parameterize varieties that contain p. We claim that p^* is a constructible set. Indeed, let $W \subset \mathbb{R}^k$ be the variety defined by the polynomials of Γ, but with $s_1, \ldots, s_k$ as variables and with the coordinates of p assigned to $x_1, \ldots, x_d$. Then $p^* = W \cap S$.

Let $U \subset \mathbb{R}^d$ be a variety from the family F. A point $s \in S$ in the dual space $\mathbb{R}^k$ is *dual* to U if $U = V(\Gamma(s))$. As before, the variety U may have more than one dual point. We denote by U^* an arbitrary point from S that is dual to U.

In addition to the primal and dual complexities of F, we now also have the complexity of the constructible set S. For simplicity, we define the *complexity* D of F to be the maximum of the primal complexity, the dual complexity, and the complexity of S. We note that, for every $p \in \mathbb{R}^d$, the complexity of a dual set p^* is $O_D(1)$.

Families of complex varieties: The above still holds for families of varieties in $\mathbb{C}^d$. In this case, $\Gamma \subset \mathbb{C}[x_1, \ldots, x_d]$, the parameters are in $\mathbb{C}$, the dual space is $\mathbb{C}^k$, and the set of parameters S is constructible with respect to $\mathbb{C}^k$. The dimension and complexity of a family are defined as in the real case. It is not difficult to verify that duality works as before.

Application: Unions of lines in $\mathbb{R}^3$ and $\mathbb{C}^3$: We now study a first application of families with sets of parameters. This application demonstrates a useful technique that is becoming more common in recent years. An additional application is presented in the following section.

We recall a definition from Section 10.3: A line $\ell \subset \mathbb{R}^3$ *is parallel to the xy-plane* if ℓ is contained in a plane that is parallel to the xy-plane. In other words, ℓ is parallel to the xy-plane if all the points of ℓ have the same z-coordinate. We rely on the same definition in $\mathbb{C}^3$.

The following holds both in $\mathbb{R}^3$ and in $\mathbb{C}^3$. Every line that is not parallel to the xy-plane intersects the plane $\mathbf{V}(z)$ and the plane $\mathbf{V}(z-1)$. Thus, every such line is uniquely defined by two points $(a, b, 0)$, $(c, d, 1)$. In particular, this line can be parameterized by $x = a + (c - a)z$ and $y = b + (d - b)z$.[1]

The preceding paragraph provides a way to define the family of lines that are not parallel to the xy-plane, both in $\mathbb{R}^3$ and in $\mathbb{C}^3$. In both cases, the parameters are a, b, c, d and the defining polynomials are

$$\Gamma = \{x - a - (c - a)z, y - b - (d - b)z\}. \tag{14.4}$$

Let F be a one-dimensional family of lines in $\mathbb{R}^3$ or in $\mathbb{C}^3$. That is, we use the line parameterization from Equation (14.4), but with a one-dimensional set of parameters S. Here, either $S \subset \mathbb{R}^4$ or $S \subset \mathbb{C}^4$. Since we have a one-dimensional family of one-dimensional objects, we expect their union to be two-dimensional. The following lemma makes this intuition rigorous. The lemma also provides an important distinction between families in complex spaces and families in real spaces.

Lemma 14.7

(a) Let F be a one-dimensional family of lines in $\mathbb{C}^3$, of complexity D. Then the union of the lines of F is a two-dimensional constructible set of complexity $O_D(1)$.

(b) Let F be a one-dimensional family of lines in $\mathbb{R}^3$, of complexity D. Then the union of the lines of F is contained *in a two-dimensional variety of degree $O_D(1)$.*

---

[1] The most common way of parameterizing lines in $\mathbb{R}^3$ or $\mathbb{C}^3$ is probably by using *Plücker coordinates*. This parameterization has many nice properties and also includes lines that are parallel to the xy-plane. Rudnev and Selig (2016) provide a nice introduction to Plücker coordinates in the context of distinct distances and incidences. For our purposes, we do not need this more involved tool. All the proofs in this chapter hold also when using Plücker coordinates.

Proof (a) By definition, F does not contain lines that are parallel to the xy-plane. Let Γ be as in Equation (14.4) and let $S \subset \mathbb{C}^4$ be the one-dimensional set of parameters of F.

Let

$$W = \left\{ (p, s) \in \mathbb{C}^3 \times S \ : \ p \in \mathbf{V}(\Gamma(s)) \right\} \subset \mathbb{C}^7.$$

Here, $\mathbf{V}(\Gamma(s))$ is a line in $\mathbb{C}^3$. We note that W is a constructible set of complexity $O_D(1)$. Let $\pi \colon \mathbb{C}^7 \to \mathbb{C}^4$ be a projection on the last four coordinates. We note that $S = \pi(W)$. Also, for every $p \in S$, the set $\pi^{-1}(p)$ is a line in $\mathbb{C}^7$. Since $\dim S = 1$ and a line is one-dimensional, Lemma 14.5 implies that $\dim W = 2$.

Let $U \subset \mathbb{C}^3$ be the projection of W on the first three coordinates of $\mathbb{C}^7$. Lemma 14.4 implies that U is a constructible set of complexity $O_D(1)$ and of dimension at most two. We note that U is the union of all the lines of F. By Lemma 4.10, if $\dim U \le 1$ then $\overline{(U)}$ contains $O_D(1)$ lines. Since F consists of infinitely many lines, we conclude that $\dim U = 2$.

(b) Let Γ be as in Equation (14.4) and let $S \subset \mathbb{R}^4$ be the one-dimensional set of parameters of F. The complexification $S_\mathbb{C} \subset \mathbb{C}^4$ is a one-dimensional constructible set of complexity $O_D(1)$. The sets Γ and $S_\mathbb{C}$ define a family $F_\mathbb{C}$ of lines in $\mathbb{C}^3$. By definition, $F_\mathbb{C}$ contains the complexifications of all lines of F. The family $F_\mathbb{C}$ includes many additional lines, for which at least one of the parameters a, b, c, d is not real.

By part (a), the union the lines of $F_\mathbb{C}$ is a two-dimensional constructible set U of complexity $O_D(1)$. We set $U_\mathbb{R} = \Re(U)$. Corollary 14.6 implies that $U_\mathbb{R}$ is a constructible set of complexity $O_D(1)$. Since $U_\mathbb{R}$ contains the real parts of the lines of $F_\mathbb{C}$, it contains all the lines of F. We complete the theorem by taking the Zariski closure of $U_\mathbb{R}$. $\square$

Lemma 14.7(b) states that the union of a one-dimensional family of lines in $\mathbb{R}^3$ is contained in a constructible set. It might seem as if the proof of the lemma states that this union *is* a constructible set, but this is not always true. The issue is that many of the complex lines of $F_\mathbb{C}$ may contain a single real point. Such lines are not complexifications of lines from F, but still contribute points to the constructible set that we obtain in $\mathbb{R}^3$. When removing those points, the remaining set may no longer be constructible.

Claim 14.8 *There exists a one-dimensional family F of lines in $\mathbb{R}^3$, such that the union of the lines is not a constructible set.*

Proof We consider the plane $\Pi = \mathbf{V}(y) \subset \mathbb{R}^3$ (that is, the xz-plane). Let C be the circle of radius 1 in $\mathbb{R}^3$ that is centered at the origin and contained in Π. Let F be the family of lines that are contained in Π and tangent to C.

See Figure 14.2. More precisely, F is the family of such lines that are not parallel to the xy-plane.

Figure 14.2 Lines that are tangent to a circle.

By inspecting Equation (14.4), we note that every line of F is defined by $b = d = 0$. By using elementary Euclidean geometry in Π, we obtain that $a^2 - (c - a)^2 = 1$. Thus, F is indeed defined by a one-dimensional constructible set of parameters in $\mathbb{R}^4$. We also note that

$$\bigcup_{\ell \in F} \ell = \left\{ (p, 0, q) \; : \; p^2 + q^2 \geq 1 \right\} \setminus \{(0, 0, -1), (0, 0, 1)\}.$$

In other words, the union of the lines is the plane Π after removing an open disk of radius one and two points. The points $(0, 0, -1)$ and $(0, 0, 1)$ are not in the union since they are contained in tangent lines that are parallel to the xy-plane. We conclude that $\bigcup_{\ell \in F} \ell$ is not a constructible set. □

14.5 Properties of Ruled Surfaces

Let U be an irreducible two-dimensional variety in $\mathbb{R}^3$ or in $\mathbb{C}^3$. The variety U is *ruled* if, for every point $p \in U$, there exists a line that is contained in U and incident to p. For example, the conical surface and the circular cylinder are both ruled. See Figure 14.3.

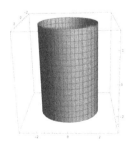 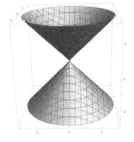

Figure 14.3 (Left) The circular cylinder $\mathbf{V}(x^2 + y^2 - 1)$, and (Right) The conical surface $\mathbf{V}(x^2 + y^2 - z^2)$. Both are ruled.

The variety U is *doubly ruled* if, for every point $p \in U$, there exist *two lines* that are contained in U and incident to p. Figure 14.4 depicts two doubly ruled surfaces. Similarly, a surface is *infinitely ruled* if, for every point $p \in U$, there exist *infinitely many lines* that are contained in U and incident to p. A plane is infinitely ruled. By definition, a variety that is doubly ruled is also ruled. Similarly, a variety that is infinitely ruled is also doubly ruled.

Figure 14.4 (Left) The hyperboloid $\mathbf{V}(x^2+y^2-z^2-1)$, and (Right) The hyperbolic paraboloid $\mathbf{V}(z - xy)$. Both are doubly ruled.

Ruled surfaces are quite useful in incidence geometry. For example, recall that Guth and Katz proved that every n points in $\mathbb{R}^2$ span $\Omega(n/\log n)$ distinct distances (Theorem 7.1). In this book, we prove the slightly weaker bound of $\Omega(n^{1-\varepsilon})$ distinct distances (Theorem 9.1). The proof of the stronger bound of Guth and Katz relies on ruled surfaces.

Ruled surfaces have many interesting and useful properties. Unfortunately, in $\mathbb{R}^3$ we are only familiar with proofs of these properties that require significant additional background. For this reason, we prove these properties over $\mathbb{C}^3$. Some of the following proofs also hold in $\mathbb{R}^3$, but not all. Deriving properties of ruled surfaces allows us to use families with parameters in several different ways.

It is not difficult to characterize the infinitely ruled surfaces.

Lemma 14.9 *Let U be an irreducible two-dimensional variety in $\mathbb{C}^3$ that is infinitely ruled. Then U is a plane.*

Proof Claim 4.7 states that every hypersurface in $\mathbb{R}^d$ has a regular point. The proof of that claim also holds in $\mathbb{C}^3$. Let p be a regular point of U. Then every line that is contained in U and incident to p is also contained in $T_p U$. We set $W = U \cap T_p U$.

Assume for contradiction that $\dim W < 2$. Since U is infinitely ruled, W contains infinitely many lines. By Lemma 4.10, which also holds in $\mathbb{C}^d$, a one-dimensional variety contains a finite number of lines. This contradiction implies that $\dim W = 2$. Since U is irreducible, this implies that $U = T_p U$, which is a plane. □

We now establish several basic properties of surfaces that are not planes.

Lemma 14.10 *Let U be an irreducible two-dimensional variety in $\mathbb{C}^3$ that is not a plane. Let F be the set of lines that are contained in U and are not parallel to the xy-plane.*

(a) The set F is an algebraic family of lines.
(b) Let p be a point of U. Then the set of lines of F that are incident to p is a family of dimension at most one.
(c) If p is a regular point of U, then finitely many lines of F are incident to p.
(d) The family of lines F is of dimension at most one.

Proof (a) Let $f \in \mathbb{C}[x, y, z]$ be a polynomial that satisfies $U = \mathbf{V}(f)$. We consider a line ℓ that is parameterized by $(a, b, c, d) \in \mathbb{C}^4$ and set $g(z) = f(a + (c - a)z, b + (d - b)z, z)$. By Equation (14.4), we have that $\ell \subset U$ if and only if g is identically zero. We write $g(z) = \sum_{j=0}^r \alpha_j z^j$, where $r = \deg f$. Then $\ell \subset U$ if and only if $\alpha_1 = \cdots = \alpha_r = 0$. By the definition of $g(z)$, we can think of every α_j as a polynomial in $\mathbb{C}[a, b, c, d]$. We set $S = \mathbf{V}(\alpha_0, \ldots, \alpha_r)$ and note that S is the set of parameterizations of lines that are contained in U and are not parallel to the xy-plane. Thus, F is a family with the parameter set S.

(b) We write $p = (p_x, p_y, p_z)$. A line that is parameterizied by a, b, c, d is incident to p if and only if $p_x = a + (c - a)p_z$ and $p_y = b + (d - b)p_z$. Thus, the family of lines that are contained in U and incident to p is parameterized by

$$S_p = S \cap \mathbf{V}(p_x - a - (c - a)p_z, p_y - b - (d - b)p_z).$$

Assume for contradiction that $\dim S_p \geq 2$. We arbitrarily take q_z to be a number that is different from the z-coordinate of p. We set $\Pi = \mathbf{V}(z - q_z)$ and note that this is a plane that is parallel to the xy-plane and does not contain p. We also set

$$T = \left\{ (a, b, c, d, a + (c - a)q_z, b + (d - b)q_z, q_z) \in S_p \times \mathbb{C}^3 \right\}.$$

Intuitively, an element of T is a line ℓ from S_p together with the intersection point $\ell \cap \Pi$. It is not difficult to check that T is a variety in $\mathbb{C}^7$. Since the lines of S_p are not parallel to the xy-plane, every such line intersects $\mathbf{V}(z - q_z)$ at a single point. Thus, the projection of T on the first four coordinates is a bijection between T and S_p. Lemma 14.5 implies that $\dim T \geq 2$.

Let Π_U be the projection of T on the last three coordinates. Since two lines have at most one point in common, distinct lines from S_p cannot have the same intersection point with Π. This implies that the projection from T to Π_U is a bijection. Thus, Lemma 14.5 implies that $\dim \Pi_U = 2$. By Lemma 14.4, the set Π_U is constructible. We also note that $\Pi_U = \Pi \cap U$. Since U is irreducible and is not a plane, $\Pi \cap U$ is a variety of dimension at most one. This contradiction implies that $\dim S_p \leq 1$.

(c) Every line that is contained in U and incident to p is also contained in $T_p U$. Since U is not a plane, $U \cap T_p U$ is a variety of dimension at most one. By Lemma 4.10, which also holds in $\mathbb{C}^d$, the variety $U \cap T_p U$ contains finitely many lines. We conclude that p is incident to a finite number of lines that are contained in U.

(d) We consider the set

$$W = \left\{ (a, b, c, d, p_x, p_y, p_z) \in \mathbb{C}^7 \; : \; (a, b, c, d) \in S, \right.$$

$$\left. p_x = a + (c - a)p_z, \text{ and } p_y = b + (d - b)p_z \right\}.$$

Intuitively, every element of W consists of a line $\ell \in F$ and a point on ℓ.

Let $\pi_4 \colon \mathbb{C}^7 \to \mathbb{C}^4$ be the projection on the first four coordinates. We note that $\pi_4(W) = S$. Also, for every $s \in S$, the inverse projection $\pi_4^{-1}(s) \cap W$ is a line. Combining this with Lemma 14.5 implies that $\dim W = \dim S + 1$.

Let W_3 be the projection of W on the last three coordinates of $\mathbb{C}^3$. By Lemma 14.4, the set W_3 is constructible. By observing that $W_3 \subseteq U$, we get that $\dim W_3 \leq 2$. Theorem 4.11 states that a generic point of a variety in $\mathbb{R}^d$ is regular. This also holds for varieties in $\mathbb{C}^d$. Thus, part (c) of the current lemma states that a generic point of U is incident to finitely many lines of F. Then, Lemma 14.5 implies that $\dim W \leq 2$.

In the preceding paragraphs, we obtained that $\dim W = \dim S + 1$ and that $\dim W \leq 2$. Combining these two bounds leads to $\dim S \leq 1$. $\square$

The following lemma is a step towards characterizing the doubly ruled surfaces.

Lemma 14.11   *Let U be an irreducible two-dimensional variety in $\mathbb{C}^3$ that is doubly ruled. Then there exist three lines $\ell_1, \ell_2, \ell_3 \subset U$ such that infinitely many lines that are contained in U intersect $\ell_1, \ell_2,$ and ℓ_3. We may also assume that $\ell_1 \cap \ell_2 \cap \ell_3 = \emptyset$.*

Proof If U is a plane, then the lemma is obtained by taking ℓ_1, ℓ_2, ℓ_3 to be three parallel lines in this plane. For the rest of the proof, we assume that U is not a plane. Throughout the proof, we only consider lines that are not parallel to the xy-plane.

As stated in a previous proof, it is not difficult to check that the proof of Claim 4.7 also holds in $\mathbb{C}^3$. That is, U contains a regular point p. Since U is doubly ruled, there exist distinct lines $\ell_{p,1}$ and $\ell_{p,2}$ that are contained in U and incident to p. Let F_1 be the set of lines that are contained in U and intersect $\ell_{p,1}$. Let F_2 be the set of lines that are contained in U and intersect $\ell_{p,2}$.

Three pairwise intersecting lines in $\mathbb{R}^3$, with three distinct intersection points, are contained in a common plane. An example is depicted in Figure 14.5. This claim also holds in $\mathbb{C}^3$. Indeed, the three intersection points define a unique plane Π. A line that is not contained in Π intersects this plane in at most one point, so all three lines are contained in Π.

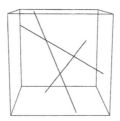

Figure 14.5 Three pairwise intersecting lines lie in a common plane.

Let Π be the plane that contains both $\ell_{p,1}$ and $\ell_{p,2}$. By the preceding paragraph, every line from $F_1 \cap F_2$ that is not incident to p is contained in Π. Since U is not a plane, $U \cap \Pi$ is a variety of dimension at most one. Such a variety contains finitely many lines that are not incident to p. Also, Lemma 14.10(c) implies that finitely many lines are incident to p and contained in U. We conclude that $F_1 \cap F_2$ consists of finitely many lines.

Studying F_1: We now prove that F_1 is a family of lines. Lemma 14.10(a) states that the lines that are contained in U form a one-dimensional family F. Let S be the constructible set of parameters of F. Let $f_1, f_2 \in \mathbb{C}[x, y, z]$ be two linear polynomials that define $\ell_{p,1}$. We set

$$T = \{(a, b, c, d, q_x, q_y, q_z) \in S \times \mathbb{C}^3 \ : \ f_1(q_x, q_y, q_z) = f_2(q_x, q_y, q_z) = 0,$$
$$q_x = a + (a - c)q_z, \quad q_y = b + (d - b)q_z\}.$$

Intuitively, every element of T is a line parameterization $(a, b, c, d) \in \mathbb{C}^4$ and a point $q = (q_x, q_y, q_z)$ that satisfy the following. The point q is on $\ell_{1,p}$ and on the line $\ell = \mathbf{V}(\Gamma(a, b, c, d))$. In particular, ℓ and $\ell_{1,p}$ intersect.

Let S_1 be the projection of T on the first four coordinates. Then S_1 is the set of parameters that define lines of S that intersect $\ell_{1,p}$. By Lemma 14.4, S_1 is a constructible set. We note that $S_1 \subset S$, so $\dim S_1 \leq \dim S = 1$. Since U is doubly ruled, every point of $\ell_{1,p}$ intersects at least one line of F. This means that S_1 is infinite, so $\dim S_1 = 1$.

We note that S_1 is the union of one-dimensional irreducible components, possibly with finitely many points removed. We arbitrarily keep a single one-dimensional component and discard the others. We refer to the resulting subset of S_1 as $S_{1,\mathrm{irr}}$, which is short for "irreducible." Finitely many points are removed from the one-dimensional component of $S_{1,\mathrm{irr}}$, as in S_1. By definition, $S_{1,\mathrm{irr}}$ is a one-dimensional constructible set. This set defines a one-dimensional family of lines $F_{1,\mathrm{irr}}$, which is a subset of F_1.

The three lines: By Lemma 14.7(a), the union of the lines of $F_{1,\mathrm{irr}}$ is a two-dimensional constructible set $U_1 \subset U$. This implies that $U = U_1 \backslash A$, where A is a constructible set of dimension at most one. The variety $\overline{A}$ contains a finite number of lines, since $\dim \overline{A} \leq 1$. We also recall that $F_1 \cap F_2$ consists of finitely many lines. Since F_2 consists of infinitely many lines, there exist distinct ℓ_1, ℓ_2, ℓ_3 in $F_2 \backslash F_1$ that are not contained in $\overline{A}$.

To make sure that $\ell_1 \cap \ell_2 \cap \ell_3 = \emptyset$, we choose these three lines as follows. We first fix two distinct lines ℓ_1, ℓ_2 that do not have the same intersection point with $\ell_{1,p}$. If $\ell_1 \cap \ell_2 = \emptyset$, then we arbitrarily choose ℓ_3 from the lines of $F_2 \backslash F_1$ that are not contained in $\overline{A}$. It remains to consider the case where ℓ_1 and ℓ_2 intersect at a point q. By definition, we have that $q \notin \ell_{1,p}$. Let Π_q be the plane that contains $\ell_{1,p}$ and q. A line that intersects $\ell_{1,p}$ and contains q is also contained in Π_q. Since U is not a plane, $\dim(U \cap \Pi_q) \leq 1$, so finitely many lines from F_1 contain q. Thus, we may choose $\ell_3 \in F_2 \backslash F_1$ that is not incident to q and is not contained in $\overline{A}$.

For every $j \in \{1, 2, 3\}$, let $F_{1,\mathrm{irr}}^j$ be the set of lines of $F_{1,\mathrm{irr}}$ that intersect ℓ_j. By repeating the above projection argument that involves T, we get that $F_{1,\mathrm{irr}}^j$ is a family of lines. Let $S_{1,\mathrm{irr}}^j$ be the constructible set of parameters that defines $F_{1,\mathrm{irr}}^j$. Since ℓ_j is not contained in $\overline{A}$, the intersection $\ell_j \cap U_1$ is infinite. In every point of $\ell_j \cap U_1$, the line ℓ_j intersects a distinct line of $F_{1,\mathrm{irr}}$. This implies that $S_{1,\mathrm{irr}}^j$ is infinite, which in turn implies that $\dim S_{1,\mathrm{irr}}^j \geq 1$. Since $S_{1,\mathrm{irr}}^j \subset S_{1,\mathrm{irr}}$, we conclude that $\dim S_{1,\mathrm{irr}}^j = 1$.

We recall that $S_{1,\mathrm{irr}}$ consists of a single one-dimensional component, possibly with a finite set of points removed. Thus, for every $j \in \{1, 2, 3\}$, there exists a finite set B_j that satisfies $S_{1,\mathrm{irr}}^j = S_{1,\mathrm{irr}} \backslash B_j$. We set $S_{\mathrm{pruned}} = S_{1,i} \backslash (B_1 \cup B_2 \cup B_3)$. Since $S_{\mathrm{pruned}} \subset S_{1,\mathrm{irr}}^1 \cap S_{1,\mathrm{irr}}^2 \cap S_{1,\mathrm{irr}}^3$, every point in S_{pruned} corresponds to a line of F_1 that intersects ℓ_1, ℓ_2, and ℓ_3. Since S_{pruned} consists of a one-dimensional components, possibly with a finite set of points removed, S_{pruned} is infinite. We conclude that there exist infinitely many lines that are contained in U and intersect ℓ_1, ℓ_2, and ℓ_3.   □

We are now ready to characterize the doubly ruled surfaces in $\mathbb{C}^3$. We first recall a few definitions. As in the real case, two lines in $\mathbb{C}^3$ are *skew* if no plane contains both lines. Equivalently, two lines are skew if they are neither parallel nor intersecting. Up to rotations and translations, a *hyperbolic paraboloid* in $\mathbb{C}^3$ is a variety of the form $\mathbf{V}(z - y^2/a^2 + x^2/b^2)$, for nonzero $a, b \in \mathbb{C}$. Up to rotations and translations, a *hyperboloid of one sheet* in $\mathbb{C}^3$ is a variety of the form $\mathbf{V}(x^2/a^2 + y^2/b^2 - z^2/c^2 - 1)$, for nonzero $a, b, c \in \mathbb{C}$. Hyperbolic paraboloids and hyperboloids of one sheet are irreducible two-dimensional varieties that are doubly ruled. See Figure 14.4.

Theorem 14.12 *Let U be an irreducible two-dimensional variety in $\mathbb{C}^3$ that is doubly ruled. Then U is either a plane, a hyperbolic paraboloid, or a hyperboloid of one sheet.*

Proof By Lemma 14.11, there exist distinct lines $\ell_1, \ell_2, \ell_3 \subset U$ such that infinitely many lines are contained in U and intersect ℓ_1, ℓ_2, and ℓ_3. Moreover, $\ell_1 \cap \ell_2 \cap \ell_3 = \emptyset$. We first assume that ℓ_1, ℓ_2, and ℓ_3 are contained in a common plane $\Pi \subset \mathbb{C}^3$. If a line intersects a plane in more than one point, then the line is contained in the plane. Thus, the infinitely many lines that intersect ℓ_1, ℓ_2, and ℓ_3 are also contained in $U \cap \Pi$. This implies that $U = \Pi$.

We next consider the case where two of the lines are contained in a plane $\Pi \subset \mathbb{C}^3$, but not the third line. If the third line does not intersect Π, then no line intersects ℓ_1, ℓ_2, and ℓ_3, which is impossible. If the third line intersects Π at a point, then infinitely many lines are contained in $U \cap \Pi$. This implies that $U = \Pi$, which is impossible since one of ℓ_1, ℓ_2, ℓ_3 is not contained in Π. We conclude that this case cannot happen.

It remains to consider the case where ℓ_1, ℓ_2, and ℓ_3 are pairwise skew. We recall a definition from Section 5.2: A *regulus* is the union of all lines in $\mathbb{R}^3$ that intersect three *pairwise-skew* lines ℓ_1, ℓ_2, ℓ_3. We extend this definition to $\mathbb{C}^3$. Lemma 5.6 states that every regulus in $\mathbb{R}^3$ is contained in an irreducible variety of dimension two and degree two. The proof of the lemma also holds in $\mathbb{C}^3$. Thus U is contained in a two-dimensional variety of degree two. Since U is an irreducible two-dimensional variety, we conclude that $\deg U = 2$.

The irreducible degree two varieties in $\mathbb{C}^3$ are ellipsoids, elliptic cones, elliptic cylinders, elliptic paraboloids, hyperbolic cylinders, hyperbolic paraboloids, hyperboloids of one and two sheets, and parabolic cylinders. For example, see Hilbert and Cohn-Vossen (2021, Section 1.3). It is not difficult to verify that, out of those varieties, the only ones that are doubly ruled are hyperbolic paraboloids and hyperboloids of one sheet. $\square$

For more properties of ruled surfaces, see Exercises 14.4–14.6. For a more advanced discussion, see Kollár (2015).

14.6 Exercises

Some of the following exercises that involve ruled surfaces rely on the following lemma. You may use this lemma without proving it. The proof of part (b) of the lemma is identical to proofs from Section 14.5.

Lemma 14.13   *Let U be an irreducible two-dimensional variety in $\mathbb{C}^3$. Let X be a two-dimensional constructible set that is contained in U.*
(a) If every point of X is incident to a line that is contained in U, then U is ruled.
(b) If every point of X is incident to at least two lines that are contained in U, then U is doubly ruled.

Exercise 14.1   In Section 14.1, we define a family that contains all the circles in $\mathbb{R}^3$, but also points, spheres, and the empty set.
(a) Show that the set of circles in $\mathbb{R}^3$ is a family with a set of parameters. This family may also contain the empty set, but no points or spheres.
(b) Show that the set of circles in $\mathbb{R}^3$ that are incident to the origin is a family with a set of parameters. This family should not include any elements except for circles that are incident to the origin.

Exercise 14.2   Complete the proof of Theorem 14.2. That is, assume that ε^* is sufficiently small with respect to ε. Then repeat the proof of Theorem 11.3 with an extra ε^* in the exponent of the weak bound.

Exercise 14.3   Let F be a family of lines that are not parallel to the xy-plane in $\mathbb{C}^3$. Assume that F is one-dimensional and that no point in $\mathbb{C}^3$ is incident to infinitely many lines of F. Let

$$X = \left\{ (a, b, 0) \in \mathbb{C}^3 \ : \ (a, b, 0) \text{ is incident to at least one line of } F \right\}.$$

Prove that X is a one-dimensional contructible set. In your proof, you may not apply Lemma 14.7. Instead, adapt the projection trick that was used in the proofs of Lemmas 14.7 and 14.10.

Exercise 14.4   Let U be an irreducible two-dimensional variety in $\mathbb{C}^3$ that is ruled but not doubly ruled. Prove that at most one point of U is incident to infinitely many lines that are contained in U.

Exercise 14.5   Let U be an irreducible two-dimensional variety in $\mathbb{C}^3$ that is ruled but not doubly ruled. A line $\ell \subset U$ is *exceptional* if ℓ contains infinitely many points that are incident to another line that is in U. Note that a line that contains one point that is incident to infinitely many lines that are in U may not be exceptional. Prove that U contains at most two exceptional lines.

Exercise 14.6 Let U be an irreducible two-dimensional variety in $\mathbb{C}^3$ that is not ruled. Prove that U contains finitely many lines.

Exercise 14.7 Let U be an irreducible two-dimensional variety in $\mathbb{C}^3$ that is not doubly ruled. Let F be the family of lines that are contained in U and let S be the set of parameters of F. By Lemma 14.10, we have that $\dim S \leq 1$. Prove that $\overline{S}$ contains at most one one-dimensional component.

Exercise 14.8 Provide an example of the following claim: Let U be an irreducible two-dimensional variety in $\mathbb{R}^3$ and let p be a regular point of U. Let Π be a plane that contains p, such that $\Pi \cap U$ is a curve γ. Then p may be a singular point of γ. (Hint: How is this exercise related to the topics of the current chapter?)

14.7 Open Problems

The current chapter is about a technique rather than about problems, so it does not exactly lead to main open problems. This chapter does include one interesting open problem: incidences with large parameters s and t. The most interesting case of this problem might be the following one.

Open Problem 14.14 *Let $\mathcal{P}$ be a set of m points and let H be a set of n planes, both in $\mathbb{R}^3$. Assume that the incidence graph of $\mathcal{P} \times H$ contains no $K_{s,s}$ for some $s \in \mathbb{N}$. How large can $I(\mathcal{P}, H)$ be?*

The current best upper bound for Open Problem 14.14, which is provided by Theorem 14.1, is $O(m^{3/4}n^{3/4} + m + n)$. The current best lower bound, by Braß and Knauer (2003), is $\Omega(m^{7/10}n^{7/10} + m + n)$. Both bounds remain unchanged from before the introduction of the new polynomial methods.

We conclude this chapter with a question by the author of this book, which might be known and might not be of interest to many researchers. In Section 14.5, we prove properties of ruled surfaces in $\mathbb{C}^3$. While these properties are known to also hold over $\mathbb{R}^3$, the proofs from Section 14.5 no longer work in $\mathbb{R}^3$.

Open Problem 14.15 *Are there proofs of the ruled surfaces properties in $\mathbb{R}^3$ that are simple enough to be presented in an elementary book such as this one?*

One main issue prevents the proofs from Section 14.5 to hold in $\mathbb{R}^3$: Throughout that section, we rely on the fact that a projection of a complex variety is a constructible set. This property does not hold for real varieties. One unfortunate consequence of this is presented in Claim 14.8.

Some proofs from Section 14.5 do extend to $\mathbb{R}^3$ without any changes. For example, this is the case for the proof of Lemma 14.9.

Appendix

Preliminaries

In this appendix, we briefly cover a few basic concepts that are used throughout the book. Section A.1 is a basic introduction to asymptotic notation. Section A.2 is a brief reminder of graph theory notation. Section A.3 consists of the Cauchy–Schwarz inequality and Hölder's inequality.

A.1 Asymptotic Notation

Asymptotic notation originated from the study of algorithms and has been gradually spreading to other fields. Beyond combinatorics, it is now regularly used in some parts of number theory, harmonic analysis, and more. It is ubiquitous in this book.

Where did asymptotic notation come from? To measure the running time of an algorithm, we count the number of operations that the algorithm performs. An operation could be adding two numbers, reading a value from the memory, printing a letter on the screen, and so on. Different operations take rather different amounts of time and also depend on the hardware that is being used. However, counting the number of operations is the most standard approach.

The running time of an algorithm is measured with respect to the size of the input. Assume that when receiving an input of size n, algorithm A makes n^8 operations. Algorithm B produces the same result by making $10^{10} \cdot n^2$ operations. How do we decide which algorithm is more efficient? Table A.1 contains the number of steps for several values of n.

When measuring the efficiency of algorithms, we consider the case where the input is very large. When the input is small, most algorithms would run quickly, so efficiency is less important. Moreover, it is becoming more and more common for algorithms to deal with huge inputs. It is not difficult to check that

Table A.1 *The number of operations made by algorithm A and algorithm B.*

n	n^8	$10^{10} \cdot n^2$
1	10^{10}	1
10	10^8	10^{12}
100	10^{16}	10^{14}
10^7	10^{56}	10^{24}

algorithm A performs more operations than algorithm B when $n \geq 47$. For that reason, we consider algorithm B to be more efficient. This is not accurate and the above ignores multiple technicalities. However, this is good enough for the discussion below.

In the context of Table A.1, performing 10^{24} operations is considered reasonable. On the other hand, 10^{56} operations is orders of magnitude beyond the capabilities the fastest supercomputers that currently exist. In addition, it is not uncommon to have inputs that are significantly larger than 10^7.

The above example demonstrates one of the main principles of running time analysis: The most important detail is the dependency in n. Algorithm B was more efficient because n^2 grows slower than n^8. Having a factor of 10^{10} is much less significant. Asymptotic notation presents only the dependency in n. Intuitively, this notation allows us to focus on the main part of a bound.

The $O(\cdot)$-notation　　We consider functions $f, g \colon \mathbb{N} \to \mathbb{R}$. Intuitively, the statement $f(n) = O(g(n))$ says that $f(n)$ does not grow faster than $g(n)$. Formally, $f(n) = O(g(n))$ means that there exist constants c and n_0 that satisfy the following: For every $n \geq n_0$, we have that $f(n) \leq c \cdot g(n)$. That is, when n is sufficiently large, $f(n)$ is at most $g(n)$ times some constant. Note that c does not depend on n.

We consider a few examples:

- The statement $10n^2 + 1{,}000 = O(n^2)$ is true, since we can take $c = 100$ and $n_0 = 20$. Indeed, let $f(n) = 10n^2 + 1{,}000$ and $g(n) = n^2$. Then, every $n \geq 20$ satisfies that $f(n) \leq 100 \cdot g(n)$. More intuitively, $f(n)$ does not grow faster than $g(n)$, since both have a quadratic dependency in n.
- The statement $n^2 = O(10^{10}n \log n)$ is false. Indeed, let $f(n) = n^2$ and $g(n) = 10^{10}n \log n$. For every constant c, every sufficiently large n satisfies $f(n) > c \cdot g(n)$. More intuitively, $f(n)$ grows faster than $g(n)$ since n^2 grows faster than $n \log n$.
- The statement $10^{10} = O(1)$ is true, since we can take $c = 10^{10}$ and $n_0 = 1$. Indeed, let $f(n) = 10^{10}$ and $g(n) = 1$. Then every $n \geq 1$ satisfies that $f(n) \leq 10^{10} \cdot g(n)$. Every quantity that does not depend on n is $O(1)$.

If the statement $f(n) = O(g(n))$ is true, then there are infinitely many choices of c and n_0 that satisfy the above condition. There is no need to find the minimum valid values for c and n_0. After getting used to working with $O(\cdot)$-notation, one does not think about c and n_0 at all. It is easier to compare the fastest growing factors of the two expressions.

Consider the statement $x = O(n^2 + n)$. This statement implies that there exist c, n_0 such that $x \leq c(n^2 + n)$ for every $n \geq n_0$. Since $n^2 + n \leq 2n^2$, we also have that $x \leq 2cn^2$ for every $n \geq n_0$. That is, it suffices to write that $x = O(n^2)$. In general, when an $O(\cdot)$-notation contains a constant number of terms, we may remove all but the fastest growing term. This is not true when the number of terms depends on n. For example, consider the expression $O(n + n + \cdots + n)$ where the number of terms is $4 \log n$. Since $n + n + \cdots + n = 4n \log n$, the above expression should be simplified to $O(n \log n)$.

We sometimes obtain a bound of the form $x = O(f(n) + g(n))$, without knowing which of these two terms is larger. In such a case, it is helpful to consider each option separately. That is, we first assume that $g(n) = O(f(n))$, in which case $x = O(f(n))$. We then assume that $f(n) = O(g(n))$, in which case $x = O(g(n))$. If we obtain the same conclusion in both cases, then this conclusion always holds. See also Exercise A.1.

Additional notation. It is not unusual to have $f, g : \mathbb{N} \to \mathbb{R}$ that satisfy both $f(n) = O(g(n))$ and $g(n) = O(f(n))$. For example, both $n^2 + 100n = O(10n^2)$ and $10n^2 = O(n^2 + 100n)$ are true, since both expressions are dominated by n^2. In such cases, we write $f(n) = \Theta(g(n))$. Intuitively, the statement $f(n) = \Theta(g(n))$ says that $f(n)$ and $g(n)$ grow at the same rate.

The statement $f(n) = \Omega(g(n))$ is equivalent to $g(n) = O(f(n))$. Intuitively, this means that $f(n)$ grows at least as fast as $g(n)$. Formally, $f(n) = \Omega(g(n))$ means that there exist constants c, n_0, such that for any $n \geq n_0$, we have $f(n) \geq c \cdot g(n)$. We note that $f(n) = \Theta(g(n))$ holds if and only if $f(n) = O(g(n))$ and $f(n) = \Omega(g(n))$.

An expression of the form $f(n) = O_{s,t}(g(n))$ means that $f(n) = O(g(n))$, where the hidden constant c may depend on the variables s and t. For example, the statement $10s^{100}n^2 + 5t^2n = O_{s,t}(n^2)$ is true. Indeed, in this case we may take $n_0 = 1$ and $c = 10s^{100} + 5t^2$. We usually write $f(n) = O_{s,t}(g(n))$ when n is arbitrarily large but s and t are constants, so we care mainly about the dependency in n. Constant variables appear inside of the $O(\cdot)$-notation when they affect the growth rate of n. For example, we have that $s^5 n^s + s^s = O_s(n^s)$.

A.2 Graph Theory

In this section, we briefly recall the graph theory notation that is used in this book. For a nice introduction to graph theory, see for example West (2001). We denote a graph as $G = (V, E)$, where V is the set of vertices and E is the set of edges. In particular, V is a set of vertex names and E is a set of pairs from V^2. See Figure A.1(a). In this book we only use *undirected* graphs. That is, the edges in our graphs do not have a direction.

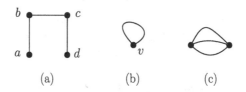

(a)   (b)   (c)

Figure A.1 (a) The graph $G = (V, E)$, where $V = \{a, b, c, d\}$ and $E = \{(a, b), (b, c), (c, d)\}$. (b) A loop is an edge of the form (v, v). (c) Three parallel edges.

A *loop* is an edge with both endpoints at the same vertex. See Figure A.1(b). Edges are *parallel* if they have the same endpoints. See Figure A.1(c). Unless stated otherwise, the graphs in this book do not contain loops or parallel edges.

The *degree* of a vertex v is the number of edges that have v as an endpoint. For example, in Figure A.1(a) the degree of a is 1 and the degree of b is 2. A *complete graph* K_s is a graph with s vertices and an edge between every two vertices. See Figure A.2(a). All of the vertex degrees in K_s are $s - 1$.

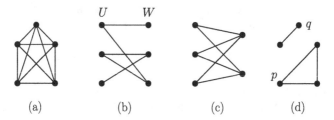

(a)   (b)   (c)   (d)

Figure A.2 (a) The complete graph K_5. (b) In a bipartite graph, every edge has one endpoint in U and the other in W. (c) The complete bipartite graph $K_{3,2}$. (d) The graph is not connected, since we cannot start at p and travel along edges until we reach q.

A *bipartite graph* has two disjoint sets of vertices and every edge has one endpoint in each set. We denote such a graph as $G = (U \cup W, E)$, where U and W are the sets of vertices. In other words, in a bipartite graph we have that $E \subseteq U \times W$. See Figure A.2(b). The *complete bipartite graph* $K_{s,t}$ is a bipartite graph with vertex sets of sizes s and t, and all of the edges that have one endpoint in each set. See Figure A.2(c). In $K_{s,t}$, the vertices of one set are of degree t and the vertices of the other set are of degree s.

A graph $G = (V, E)$ is *connected* if, for every two vertices $v_j, v_k \in V$, we can start at v_j and travel along edges until we reach v_k. The graph in Figure A.2(d) is not connected, since we cannot start at p and travel along edges until we reach q. The graphs in parts (a), (b), and (c) of Figure A.2 are all connected.

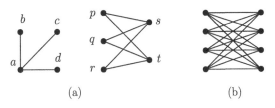

(a)   (b)

Figure A.3 (a) The graph on the left is a subgraph of the graph on the right. For example, we can set $f(a) = s, f(b) = p, f(c) = q$, and $f(d) = r$. (b) Bipartite graphs do not contain K_3.

Consider two graphs $G = (V, E)$ and $H = (U, F)$. We write $U = \{u_1, \ldots, u_m\}$. We say that H is a *subgraph* of G if there exists an injection $f: U \to V$ such that $(u_j, u_k) \in F$ implies that $(f(u_j), f(u_k)) \in E$. For an example, see Figure A.3(a). For brevity, we sometimes write that G does not contain H. For example, bipartite graphs do not contain K_3. Indeed, at least two vertices of K_3 must be on the same side of the bipartite graph. By definition, these two vertices do not have an edge between them. See Figure A.3(b). For more details about graphs that do not contain K_3, see Theorem A.2.

A.3 Inequalities

We rely on the following inequality throughout this book.

Theorem A.1 (The Cauchy–Schwarz inequality) *Consider a positive integer n and two sequences of real numbers $a_1, a_2, \ldots, a_n$ and $b_1, b_2, \ldots, b_n$. Then*

$$\sum_{j=1}^{n} |a_j b_j| \leq \sqrt{\left(\sum_{j=1}^{n} a_j^2\right)\left(\sum_{j=1}^{n} b_j^2\right)}.$$

The following result is by Mantel (1907). It demonstrates how the Cauchy–Schwarz inequality is used to obtain combinatorial results.

Theorem A.2 *Consider a graph $G = (V, E)$ that does not contain K_3 as a subgraph. Then $|E| \leq |V|^2/4$.*

Proof We set $|V| = n$ and $V = \{v_1, \ldots, v_n\}$. For $1 \le j \le n$, let d_j be the degree of v_j. Since every edge of E increases the degree of two vertices by 1, we have that $\sum_{j=1}^{n} d_j = 2|E|$.

We consider an edge $(v_j, v_k) \in E$. Since G does not contain K_3 as a subgraph, for every $v_m \in V \setminus \{v_j, v_k\}$ at least one of the edges (v_j, v_m) and (v_k, v_m) is not in E. This implies that $d_j + d_k \le n$, which in turn leads to

$$\sum_{(v_j, v_k) \in E} (d_j + d_k) \le n|E|.$$

In the left side of the above equation, a term d_j appears in exactly d_j elements of the sum. We thus have that

$$n|E| \ge \sum_{v_j \in V} d_j^2. \tag{A.1}$$

By applying the Cauchy–Schwarz inequality with $a_j = d_j$ and $b_j = 1$, we obtain that

$$\sum_{j=1}^{n} d_j \le \sqrt{\left(\sum_{j=1}^{n} d_j^2\right)\left(\sum_{j=1}^{n} 1\right)} = \sqrt{\left(\sum_{j=1}^{n} d_j^2\right) \cdot n}.$$

Rearranging and recalling that $\sum_{j=1}^{n} d_j = 2|E|$ leads to

$$\sum_{j=1}^{n} d_j^2 \ge 4|E|^2/n.$$

Combining this with Equation (A.1) gives

$$n|E| \ge \sum_{v_j \in V} d_j^2 \ge 4|E|^2/n.$$

We conclude that $|E| \le n^2/4$. □

We sometimes rely on *Hölder's inequality*, which generalizes the Cauchy–Schwarz inequality.

Theorem A.3 (Hölder's Inequality) *Consider a positive integer n and two sequences of real numbers $a_1, a_2, \ldots, a_n$ and $b_1, b_2, \ldots, b_n$. Let $1 < p, q$ satisfy $1/p + 1/q = 1$. Then*

$$\sum_{i=1}^{n} |a_i b_i| \le \left(\sum_{i=1}^{n} a_i^p\right)^{1/p} \left(\sum_{i=1}^{n} b_i^q\right)^{1/q}.$$

A.4 Exercises

Exercise A.1   Prove that the bound $f(n)^2 = O(g(n)^2 + h(n)^2)$ implies the bound $f(n) = O(g(n) + h(n))$.

Exercise A.2
(a) Prove that $k^n = O(n!)$ holds for every $k \in \mathbb{R}$.
(b) Prove that $n! = O(n^n)$.

Exercise A.3   Find a function $f(n)$ that satisfies both of the following properties:

- $f(n) = \Omega(n^k)$ for all $k > 0$,
- $f(n) = O(k^n)$ for all $k > 1$.

Exercise A.4   Consider a bipartite graph $G = (U \cup W, E)$. Prove that the sum of the degrees of the vertices of U is equal to the sum of the degrees of the vertices of W.

Exercise A.5   Let $k \geq 2$ and let $n = 2k$. We consider all sets of k distinct integers from $\{1, 2, 3, \ldots, n\}$. For example, if $k = 2$ then $n = 4$ and there are six sets of k integers from $\{1, 2, 3, 4, 5, 6\}$: $\{1, 2\}$, $\{1, 3\}$, $\{1, 4\}$, $\{2, 3\}$, $\{2, 4\}$, and $\{3, 4\}$. We build a graph $G = (V, E)$, such that V contains a vertex for every set. There is an edge between two vertices if the corresponding sets have at least one number in common. For example, there is an edge between $\{1, 2\}$ and $\{1, 4\}$, but there is no edge between $\{1, 2\}$ and $\{3, 4\}$.

Assume that k is very large. Is G connected? Prove your answer.

Exercise A.6   We saw that bipartite graphs do not contain K_3. Find a graph H such that no bipartite graph contains H, and K_3 is not a subgraph of H.

Exercise A.7   Consider $a_1, \ldots, a_n \in \mathbb{R}$ that satisfy $a_1 + \cdots + a_n = 1$. Prove that $a_1^2 + \cdots + a_n^2 \geq 1/n$. Explain why a better lower bound is not possible.

References

Ajtai, Miklós, Chvátal, Vašek, Newborn, Monroe M., and Szemerédi, Endre. 1982. Crossing-free subgraphs. *Annals of Discrete Mathematics*, **12**, 9–12.

Alon, Noga. 1999. Combinatorial nullstellensatz. *Combinatorics, Probability and Computing*, **8**(1–2), 7–29.

Alon, Noga, and Spencer, Joel H. 2004. *The Probabilistic Method*. John Wiley & Sons.

Apfelbaum, Roel, and Sharir, Micha. 2007. Large complete bipartite subgraphs in incidence graphs of points and hyperplanes. *SIAM Journal on Discrete Mathematics*, **21**(3), 707–725.

Atiyah, Michael. 2005. *Michael Atiyah Collected Works*. Vol. 6. Oxford University Press.

Bardwell-Evans, Sam, and Sheffer, Adam. 2019. A reduction for the distinct distances problem in $\mathbb{R}^d$. *Journal of Combinatorial Theory, Series A*, **166**, 171–225.

Barone, Sal, and Basu, Saugata. 2012. Refined bounds on the number of connected components of sign conditions on a variety. *Discrete & Computational Geometry*, **47**(3), 577–597.

Basu, Saugata, and Raz, Orit E. 2018. An o-minimal Szemerédi–Trotter theorem. *The Quarterly Journal of Mathematics*, **69**(1), 223–239.

Basu, Saugata, and Sombra, Martin. 2016. Polynomial partitioning on varieties of codimension two and point-hypersurface incidences in four dimensions. *Discrete & Computational Geometry*, **55**(1), 158–184.

Bateman, Michael, and Katz, Nets. 2012. New bounds on cap sets. *Journal of the American Mathematical Society*, **25**(2), 585–613.

Beck, József. 1983. On the lattice property of the plane and some problems of Dirac, Motzkin and Erdős in combinatorial geometry. *Combinatorica*, **3**(3–4), 281–297.

Behrend, Felix A. 1946. On sets of integers which contain no three terms in arithmetical progression. *Proceedings of the National Academy of Sciences of the United States of America*, **32**(12), 331.

Berndt, Bruce C., and Rankin, Robert Alexander. 1995. *Ramanujan. Letters and commentary*. American Mathematical Society.

Bochnak, Jacek, Coste, Michel, and Roy, Marie-Françoise. 2013. *Real algebraic geometry*. Vol. 36 of A Series of Modern Surveys in Mathematics. Springer Science & Business Media.

Bombieri, Enrico, and Pila, Jonathan. 1989. The number of integral points on arcs and ovals. *Duke Mathematical Journal*, **59**(2), 337–357.

Bóna, Miklós. 2006. *A Walk Through Combinatorics: An Introduction to Enumeration and Graph Theory*. World Scientific.

Bourgain, Jean, and Demeter, Ciprian. 2015. New bounds for the discrete Fourier restriction to the sphere in 4D and 5D. *International Mathematics Research Notices*, **2015**(11), 3150–3184.

Boyer, Carl B. 1949. The invention of analytic geometry. *Scientific American*, **180**(1), 40–45.

Braß, Peter, and Knauer, Christian. 2003. On counting point-hyperplane incidences. *Computational Geometry*, **25**(1–2), 13–20.

Breuillard, Emmanuel, Green, Ben, and Tao, Terence. 2011. Approximate subgroups of linear groups. *Geometric and Functional Analysis*, **21**(4), 774.

Bruner, Ariel, and Sharir, Micha. 2018. Distinct distances between a collinear set and an arbitrary set of points. *Discrete Mathematics*, **341**(1), 261–265.

Carbery, Anthony, and Iliopoulou, Marina. 2020. Joints formed by lines and a k-plane, and a discrete estimate of Kakeya type. *Discrete Analysis*, 18361.

Charalambides, Marcos. 2013. A note on distinct distance subsets. *Journal of Geometry*, **104**(3), 439–442.

Chazelle, Bernard, Edelsbrunner, Herbert, Guibas, Leonidas J., et al. 1992. Counting and cutting cycles of lines and rods in space. *Computational Geometry*, **1**(6), 305–323.

Chernikov, Artem, Galvin, David, and Starchenko, Sergei. 2020. Cutting lemma and Zarankiewicz's problem in distal structures. *Selecta Mathematica*, **26**(2), 1–27.

Clarkson, Kenneth L., Edelsbrunner, Herbert, Guibas, Leonidas J., Sharir, Micha, and Welzl, Emo. 1990. Combinatorial complexity bounds for arrangements of curves and spheres. *Discrete & Computational Geometry*, **5**(2), 99–160.

Cox, David, Little, John, and O'Shea, Donal. 2013. *Ideals, Varieties, and Algorithms: An Introduction to Computational Algebraic Geometry and Commutative Algebra*. Springer Science & Business Media.

Croot, Ernie, Lev, Vsevolod F., and Pach, Péter Pál. 2017. Progression-free sets in are exponentially small. *Annals of Mathematics*, **185**, 331–337.

Demeter, Ciprian. 2014. Incidence theory and restriction estimates. *arXiv preprint arXiv:1401.1873*.

Demeter, Ciprian. 2020. *Fourier Restriction, Decoupling and Applications*. Vol. 184 of Cambridge Studies in Advanced Mathematics. Cambridge University Press.

Dhar, Manik, and Dvir, Zeev. 2020. Proof of the Kakeya set conjecture over rings of integers modulo square-free N. *arXiv preprint arXiv:2011.11225*.

Di Benedetto, Daniel, Solymosi, Jozsef, and White, Ethan. 2020. Combinatorics of intervals in the plane I: trapezoids. *arXiv preprint arXiv:2005.09003*.

Do, Thao. 2019. Representation complexities of semialgebraic graphs. *SIAM Journal on Discrete Mathematics*, **33**(4), 1864–1877.

Do, Thao, and Sheffer, Adam. 2021. A general incidence bound in $\mathbb{R}^d$. *European Journal of Combinatorics*, **95**, 103330.

Dvir, Zeev. 2009. On the size of Kakeya sets in finite fields. *Journal of the American Mathematical Society*, **22**(4), 1093–1097.

Dvir, Zeev. 2012. Incidence theorems and their applications. *Foundations and Trends® in Theoretical Computer Science*, **6**(4), 257–393.

Dvir, Zeev, and Gopi, Sivakanth. 2015. On the number of rich lines in truly high dimensional sets. Pages 584–598 of: *31st International Symposium on Computational Geometry, SoCG 2015*. Schloss Dagstuhl-Leibniz-Zentrum fur Informatik GmbH, Dagstuhl Publishing.

Dvir, Zeev, Kopparty, Swastik, Saraf, Shubhangi, and Sudan, Madhu. 2013. Extensions to the method of multiplicities, with applications to Kakeya sets and mergers. *SIAM Journal on Computing*, **42**(6), 2305–2328.

Edel, Yves. 2004. Extensions of generalized product caps. *Designs, Codes and Cryptography*, **31**(1), 5–14.

Elekes, György. 1997. On the number of sums and products. *Acta Arithmetica*, **81**, 365–367.

Elekes, György. 2001. Sums versus products in number theory, algebra and Erdős geometry. *Paul Erdős and his Mathematics II*, **11**, 241–290.

Elekes, György, and Szabó, Endre. 2012. How to find groups? (and how to use them in Erdős geometry?). *Combinatorica*, **32**(5), 537–571.

Ellenberg, Jordan S., and Gijswijt, Dion. 2017. On large subsets of with no three-term arithmetic progression. *Annals of Mathematics*, **185**, 339–343.

Erdős, Paul. 1946. On sets of distances of *n* points. *The American Mathematical Monthly*, **53**(5), 248–250.

Erdős, Paul. 1985. Problems and results in combinatorial geometry. *Annals of the New York Academy of Sciences*, **440**(1), 1–11.

Erdős, Paul. 1986. On some metric and combinatorial geometric problems. *Discrete Mathematics*, **60**, 147–153.

Erdős, Paul. 1993. On some of my favourite theorems. Pages 97–132 of: Combinatorics, *Paul Erdős is Eighty*, vol. 2. János Bolyai Mathematical Society.

Erdős, Paul, and Purdy, George. 1971. Some extremal problems in geometry. *Journal of Combinatorial Theory, Series A*, **10**, 246–252.

Erdős, Paul, and Szemerédi, Endre. 1983. On sums and products of integers. Pages 213–218 of: *Studies in Pure Mathematics*. Springer.

Farber, Miriam, Ray, Saurabh, and Smorodinsky, Shakhar. 2014. On totally positive matrices and geometric incidences. *Journal of Combinatorial Theory, Series A*, **128**, 149–161.

Fox, Jacob, Pach, János, Sheffer, Adam, Suk, Andrew, and Zahl, Joshua. 2017. A semi-algebraic version of Zarankiewicz's problem. *Journal of the European Mathematical Society*, **19**(6), 1785–1810.

Fox, Jacob, Pach, János, and Suk, Andrew. 2018. More distinct distances under local conditions. *Combinatorica*, **38**(2), 501–509.

Gibson, Christopher G. 1998. *Elementary Geometry of Algebraic Curves: An Undergraduate Introduction*. Cambridge University Press.

Giusti, Marc. 1984. Some effectivity problems in polynomial ideal theory. Pages 159–171 of: *International Symposium on Symbolic and Algebraic Manipulation*. Springer.

Green, Ben, and Tao, Terence. 2013. On sets defining few ordinary lines. *Discrete & Computational Geometry*, **50**(2), 409–468.

Grosswald, Emil. 1985. *Representations of Integers as Sums of Squares*. Springer.

Grosu, Codruţ. 2014. $\mathbb{F}_p$ is locally like $\mathbb{C}$. *Journal of the London Mathematical Society*, **89**(3), 724–744.

Guth, Larry. 2015a. Distinct distance estimates and low degree polynomial partitioning. *Discrete & Computational Geometry*, **53**(2), 428–444.

Guth, Larry. 2015b. Polynomial partitioning for a set of varieties. *Mathematical Proceedings of the Cambridge Philosophical Society*, **159**(3), 459–469.

Guth, Larry. 2016. *Polynomial Methods in Combinatorics*. Vol. 64 of University Lecture Series. American Mathematical Society.

Guth, Larry, and Katz, Nets Hawk. 2010. Algebraic methods in discrete analogs of the Kakeya problem. *Advances in Mathematics*, **225**(5), 2828–2839.

Guth, Larry, and Katz, Nets Hawk. 2015. On the Erdős distinct distances problem in the plane. *Annals of Mathematics*, **181**(1), 155–190.

Hablicsek, Márton, and Scherr, Zachary. 2016. On the number of rich lines in high dimensional real vector spaces. *Discrete & Computational Geometry*, **55**(4), 955–962.

Halmos, Paul R. 1970. How to write mathematics. *L'Enseignement Mathématique*, **16**(2), 123–152.

Harris, Joe. 2013. *Algebraic Geometry: A First Course*. Vol. 133 of Graduate Texts in Mathematics. Springer Science & Business Media.

Hickman, Jonathan, and Wright, James. 2018. The Fourier restriction and Kakeya problems over rings of integers modulo N. *Discrete Analysis* 2018: 11, 1–54.

Hilbert, David, and Cohn-Vossen, Stephan. 2021. *Geometry and the Imagination*. American Mathematical Society.

Iosevich, Alex. 2008. Fourier analysis and geometric combinatorics. Pages 321–335 of: *Topics In Mathematical Analysis*. World Scientific.

Jackson, John. 1821. *Rational Amusement for Winter Evenings, Or, A Collection of Above 200 Curious and Interesting Puzzles and Paradoxes Relating to Arithmetic, Geometry, Geography, &c: With Their Solutions, and Four Plates, Designed Chiefly for Young Persons*. Longman, Hurst, Rees, Orme, and Brown.

Kaplan, Haim, Sharir, Micha, and Shustin, Eugenii. 2010. On lines and joints. *Discrete & Computational Geometry*, **44**(4), 838–843.

Kaplan, Haim, Matoušek, Jiří, Safernová, Zuzana, and Sharir, Micha. 2012. Unit distances in three dimensions. *Combinatorics, Probability and Computing*, **21**(4), 597–610.

Kelley, Walter G., and Peterson, Allan C. 2010. *The Theory of Differential Equations: Classical and Qualitative*. Springer Science & Business Media.

Kollár, János. 2015. Szemerédi–Trotter-type theorems in dimension 3. *Advances in Mathematics*, **271**, 30–61.

Landau, Edmund. 1909. Über die Einteilung der positiven ganzen Zahlen in vier Klassen nach der Mindestzahl der zu ihrer additiven Zusammensetzung erforderlichen Quadrate. *Archiv der Mathematik und Physik*, **13**, 305–312.

Larman, David G, Rogers, C. Ambrose, and Seidel, Johan J. 1977. On two-distance sets in Euclidean space. *Bulletin of the London Mathematical Society*, **9**(3), 261–267.

Lee, John M. 2013. *Introduction to Smooth Manifolds*. Vol. 218 of Graduate Texts in Mathematics. Springer-Verlag.

Leighton, Frank Thomson. 1983. *Complexity Issues in VLSI: Optimal Layouts For the Shuffle-Exchange Graph and Other Networks*. MIT Press.

Lo, Chi-Yuan, Matoušek, Jiří, and Steiger, William. 1994. Algorithms for ham-sandwich cuts. *Discrete & Computational Geometry*, **11**(4), 433–452.

Mantel, Willem. 1907. Problem 28. *Wiskundige Opgaven*, **10**(60–61), 320.

Matoušek, Jiří. 2011. The number of unit distances is almost linear for most norms. *Advances in Mathematics*, **226**(3), 2618–2628.

Matoušek, Jiří. 2013. *Lectures on Discrete Geometry*. Vol. 212 of Graduate Texts in Mathematics. Springer Science & Business Media.

Meshulam, Roy. 1995. On subsets of finite abelian groups with no 3-term arithmetic progressions. *Journal of Combinatorial Theory, Series A*, **71**(1), 168–172.

Milnor, J. 1964. On the Betti numbers of real varieties. *Proceedings of the American Mathematical Society*, **15**(2): 275–280.

Mohammadi, Ali, and Stevens, Sophie. 2021. Attaining the exponent 5/4 for the sum-product problem in finite fields. *arXiv preprint arXiv:2103.08252*.

Moser, Leo. 1952. On the different distances determined by n points. *The American Mathematical Monthly*, **59**(2), 85–91.

Moshkovitz, Dana. 2010. An alternative proof of the Schwartz-Zippel lemma. Page 96 of: *Electronic Colloquium on Computational Complexity*, vol. 17.

Mumford, David. 1999. *The Red Book of Varieties and Schemes: Includes the Michigan Lectures (1974) on Curves and Their Jacobians*. Vol. 1358 of Lecture Notes in Mathematics. Springer Science & Business Media.

Naslund, Eric, and Sawin, Will. 2017. Upper bounds for sunflower-free sets. *Forum of Mathematics, Sigma*, **5**, e15. doi: 10.1017/fms.2017.12.

Pach, János, and Sharir, Micha. 1992. Repeated angles in the plane and related problems. *Journal of Combinatorial Theory, Series A*, **59**(1), 12–22.

Pach, János, and Sharir, Micha. 1998. On the number of incidences between points and curves. *Combinatorics, Probability and Computing*, **7**(1), 121–127.

Pach, János, and Sharir, Micha. 2004. Geometric incidences. *Contemporary Mathematics*, **342**, 185–224.

Pach, János, and Tardos, Gábor. 2002. Isosceles triangles determined by a planar point set. *Graphs and Combinatorics*, **18**(4), 769–779.

Pach, János, and de Zeeuw, Frank. 2017. Distinct distances on algebraic curves in the plane. *Combinatorics, Probability and Computing*, **26**(1), 99–117.

Pascal, Blaise. 1640. Essay pour les coniques. *Œuvres (Brunschvigg et Boutroux, eds.)*, **1**.

Plücker, Julius. 1847. Note sur le theorème de Pascal. *Journal für die Reine und Angewandte Mathematik*, **34**, 337–340.

Quilodrán, René. 2010. The Joints Problem in $\mathbb{R}^n$. *SIAM Journal on Discrete Mathematics*, **23**(4), 2211–2213.

Ramanujan, Srinivasan. 1916. Some formulae in the analytic theory of numbers. *Messenger of Mathematics*, **45**, 81–84.

Raz, Orit E., and Sharir, Micha. 2017. The number of unit-area triangles in the plane: Theme and variation. *Combinatorica*, **37**(6), 1221–1240.

Raz, Orit E., Roche-Newton, Oliver, and Sharir, Micha. 2015. Sets with few distinct distances do not have heavy lines. *Discrete Mathematics*, **338**(8), 1484–1492.

Raz, Orit E., Sharir, Micha, and de Zeeuw, Frank. 2016a. Polynomials vanishing on Cartesian products: The Elekes–Szabó theorem revisited. *Duke Mathematical Journal*, **165**(18), 3517–3566.

Raz, Orit E., Sharir, Micha, and Solymosi, József. 2016b. Polynomials vanishing on grids: The Elekes-Rónyai problem revisited. *American Journal of Mathematics*, **138**(4), 1029–1065.

Rédei, László. 1970. *Luckenhafte Polynome uber Endlichen Korpern*. Birkhauser.

Reid, Constance. 1970. *Hilbert*. Springer-Verlag.

Roche-Newton, Oliver, Rudnev, Misha, and Shkredov, Ilya D. 2016. New sum-product type estimates over finite fields. *Advances in Mathematics*, **293**, 589–605.

Rota, Gian-Carlo. 2008. *Indiscrete Thoughts*. Springer Science & Business Media.

Roth, Klaus F. 1953. On certain sets of integers. *Journal of the London Mathematical Society*, **1**(1), 104–109.

Rudnev, Misha. 2018. On the number of incidences between points and planes in three dimensions. *Combinatorica*, **38**(1), 219–254.

Rudnev, Misha, and Selig, JM. 2016. On the use of the Klein quadric for geometric incidence problems in two dimensions. *SIAM Journal on Discrete Mathematics*, **30**(2), 934–954.

Saraf, Shubhangi, and Sudan, Madhu. 2008. An improved lower bound on the size of Kakeya sets over finite fields. *Analysis & PDE*, **1**(3), 375–379.

Schwartz, Jacob T. 1980. Fast probabilistic algorithms for verification of polynomial identities. *Journal of the ACM*, **27**(4), 701–717.

Sharir, M. 2009. On distinct distances and incidences: Elekes's transformation and the new algebraic developments. In *Annales Univ. Sci. Budapest*, **52**, 75–102.

Sharir, Micha, and Zahl, Joshua. 2017. Cutting algebraic curves into pseudo-segments and applications. *Journal of Combinatorial Theory, Series A*, **150**, 1–35.

Sharir, Micha, Sheffer, Adam, and Solymosi, József. 2013. Distinct distances on two lines. *Journal of Combinatorial Theory, Series A*, **120**(7), 1732–1736.

Sharir, Micha, Sheffer, Adam, and Solomon, Noam. 2016. Incidences with curves in $\mathbb{R}^d$. *The Electronic Journal of Combinatorics*, P4–16.

Sheffer, Adam. 2014. Distinct distances: open problems and current bounds. *arXiv:1406.1949*.

Sheffer, Adam. 2016. Lower bounds for incidences with hypersurfaces. *Discrete Analysis*, 2016:16.

Sheffer, Adam, and Zahl, Joshua. 2021. Distinct distances in the complex plane. *Transactions of the American Mathematical Society*, **367**, 6691–6725.

Sheffer, Adam, Zahl, Joshua, and de Zeeuw, Frank. 2016. Few distinct distances implies no heavy lines or circles. *Combinatorica*, **36**(3), 349–364.

Sheffer, Adam, Szabó, Endre, and Zahl, Joshua. 2018. Point-curve incidences in the complex plane. *Combinatorica*, **38**(2), 487–499.

Solymosi, József. 2006. Dense arrangements are locally very dense. I. *SIAM Journal on Discrete Mathematics*, **20**(3), 623–627.

Solymosi, József. 2009. Bounding multiplicative energy by the sumset. *Advances in Mathematics*, **222**(2), 402–408.

Solymosi, József, and Tao, Terence. 2012. An incidence theorem in higher dimensions. *Discrete & Computational Geometry*, **48**(2), 255–280.

Solymosi, József, and Vu, Van H. 2008. Near optimal bounds for the Erdős distinct distances problem in high dimensions. *Combinatorica*, **28**(1), 113–125.

Spencer, Joel, Szemerédi, Endre, and Trotter, William T. 1984. Unit distances in the Euclidean plane. Pages 294–304 of: *Graph Theory and Combinatorics*. Academic Press.

Stevens, Sophie, and de Zeeuw, Frank. 2017. An improved point-line incidence bound over arbitrary fields. *Bulletin of the London Mathematical Society*, **49**(5), 842–858.

Stillwell, John. 2008. *Naive Lie Theory*. Springer Science & Business Media.

Stone, Arthur H., and Tukey, John W. 1942. Generalized "sandwich" theorems. *Duke Mathematical Journal*, **9**(2), 356–359.

Sylvester, James Joseph. 1868. Mathematical question 2571. *Educational Times*.

Székely, László A. 1997. Crossing numbers and hard Erdős problems in discrete geometry. *Combinatorics, Probability and Computing*, **6**(3), 353–358.

Szemerédi, Endre, and Trotter, William T. 1983. Extremal problems in discrete geometry. *Combinatorica*, **3**(3), 381–392.

Tao, Terence. 2014. Algebraic combinatorial geometry: the polynomial method in arithmetic combinatorics, incidence combinatorics, and number theory. *EMS Surveys in Mathematical Sciences*, **1**(1), 1–46.

Tao, Terence. 2016. A symmetric formulation of the Croot-Lev-Pach-Ellenberg-Gijswijt capset bound. *blog post*. https://terrytao.wordpress.com/2016/05/18/a-symmetric-formulation-of-the-croot-lev-pach-ellenberg-gijswijt-capset-bound/.

Tao, Terence, and Vu, Van H. 2006. *Additive Combinatorics*. Vol. 105 of Cambridge Studies in Advanced Mathematics. Cambridge University Press.

Thom, R. 2015. Sur l'homologie des variétés algébriques réelles. In *Differential and Combinatorial Topology*, pages 255–265. Princeton University Press.

Tidor, Jonathan, Yu, Hung-Hsun Hans, and Zhao, Yufei. 2020. Joints of varieties. *arXiv preprint arXiv:2008.01610*.

Tóth, Csaba D. 2015. The Szemerédi–Trotter theorem in the complex plane. *Combinatorica*, **35**(1), 95–126.

Valtr, Pavel. 2005. Strictly convex norms allowing many unit distances and related touching questions. *Preprint*.

Vinh, Le Anh. 2011. The Szemerédi–Trotter type theorem and the sum-product estimate in finite fields. *European Journal of Combinatorics*, **32**(8), 1177–1181.

Walsh, Miguel N. 2020. The polynomial method over varieties. *Inventiones mathematicae*, **222**, 469–512.

Warren, Hugh E. 1968. Lower bounds for approximation by nonlinear manifolds. *Transactions of the American Mathematical Society*, **133**(1), 167–178.

West, Douglas Brent. 2001. *Introduction to Graph Theory*. Vol. 2. Prentice Hall.

Whitney, Hassler. 1957. Elementary structure of real algebraic varieties. *Annals of Mathematics*, **66**(3), 545–556.

Wolff, Thomas. 1999. Recent work connected with the Kakeya problem. In: *Prospects in Mathematics 1996*. American Mathematical Society.

Yu, Hung-Hsun Hans, and Zhao, Yufei. 2019. Joints tightened. *arXiv preprint arXiv:1911.08605*.

Zahl, Joshua. 2013. An improved bound on the number of point-surface incidences in three dimensions. *Contributions to Discrete Mathematics*, **8**(1) 100–121.

Zahl, Joshua. 2015. A Szemerédi–Trotter type theorem in $\mathbb{R}^4$. *Discrete & Computational Geometry*, **54**(3), 513–572.

Zahl, Joshua. 2016. A note on rich lines in truly high dimensional sets. *Forum of Mathematics, Sigma*, **4**, 1–13.

de Zeeuw, Frank. 2016. A short proof of Rudnev's point-plane incidence bound. *arXiv preprint arXiv:1612.02719*.

Zhang, Ruixiang. 2020. A proof of the Multijoints Conjecture and Carbery's generalization. *Journal of the European Mathematical Society*, **22**(8), 2405–2417.

Zippel, Richard. 1979. Probabilistic algorithms for sparse polynomials. Pages 216–226 of: *International Symposium on Symbolic and Algebraic Manipulation*. Springer.

Index

Printed in the United States
by Baker & Taylor Publisher Services